CEREAL TECHNOLOGY

First Edition

First printing, 1970
Second printing, 1999

by

Samuel A. Matz

December 1999

CEREAL TECHNOLOGY

First edition, Second printing

Copyright 1999 by SAMUEL A. MATZ

ISBN 0-942849-22-1

PAN-TECH INTERNATIONAL, INC.
P. O. Box 4548
McAllen, TX 78502

Printed in the United States of America

Preface

This book is based on the technology section of *The Chemistry and Technology of Cereals as Food and Feed,* which was published in 1959 and is now out of print. In combination with the 1969 volume *Cereal Science, Cereal Technology* covers the same subjects as the earlier book. Publication as two volumes instead of one was decided upon as a convenience to the user and because of cost considerations.

The text of every chapter was completely rewritten and many new tables and illustrations were added. In most cases, the chapter authors who contributed to the 1959 book submitted revised manuscripts. I modified, brought up-to-date, and expanded the chapters of Drs. Rumsey and Kester, deceased. Dr. Goodwin's duties as Scientific Director of the Southwest Research Institute made it impossible for him to contribute a new Corn Wet-milling chapter, so this subject, as revised, was combined with the Cereal Oil text to form a new chapter. The discussion of prepared mixes written by C. G. Harrel (also deceased) is somewhat outdated and is omitted from the present volume, although many of the topics are included in other chapters.

Illustrations, data, and other valuable information were contributed by many persons who are credited in the appropriate place. Special thanks are due to Mr. Kenneth Huber of Cargill, Inc., who reviewed the section on Corn Wet-milling, and to P. D. Baynes of Schlitz Brewing Company, who reviewed the Brewing manuscript and made many helpful suggestions. I also appreciate the assistance of Mr. L. H. Bradee of Schlitz, who granted permission for use of their extensive library on brewing science.

Samuel A. Matz

Chicago, Illinois
March 1970

Contents

CHAPTER   PAGE

1. MILLING. *Robert A. Larsen*   1

2. BAKING. *Samuel A. Matz*   43

3. FEED MANUFACTURE. *Bruce W. Smith*   91

4. MALTING. *Paul R. Witt, Jr.*   129

5. BREWING. *Donald W. Ohlmeyer and
Samuel A. Matz*   173

6. MANUFACTURE OF BREAKFAST CEREALS.
Samuel A. Matz   221

7. MACARONI PRODUCTS. *Charles M. Hoskins*   246

8. STARCH AND OIL PRODUCTION FROM CEREALS.
Samuel A. Matz   300

9. RICE PROCESSING. *Ernest B. Kester and
Samuel A. Matz*   338

INDEX.   382

CHAPTER 1

Robert A. Larsen | # Milling

INTRODUCTION

Flour and products made from flour have been used by man for centuries. No doubt the bland taste of food made from cereal grains has much to do with this fact. The first articles made from flour were unleavened breads probably much like the modern Indian chapatties (Nath *et al.* 1957). Later, the Egyptians introduced yeast leavened bread (Thorpe 1927). Chemical leavening did not appear until the nineteenth century (Anon. 1940).

The primary and compelling reason for the continued use of cereal grains as a food is not taste but the ease of growing cereals and their storage stability (Oxley 1948). Early man discovered the power of drying as a means of preserving cereal grains. Sun drying techniques were used to avoid spoilage.

Milling, in spite of its antiquity, is a major manufacturing industry in the modern world. Formerly, the production of flour from grain, like spinning and weaving, was done in the home. Millstones were highly valued household articles. Eventually, custom milling came into existence and this trend culminated in the large, complex flour mills that are typical today.

The original millstones were known as "saddle stones" and essentially were a crude mortar and pestle arrangement designed to crush wheat into smaller particles. Such stones have been found in ruins that date back to 4000 B.C. It was discovered that movement of one millstone on another in a semicircular fashion was a more efficient means of reducing grain to flour. The grain was fed through an opening in the center of the upper stone and crude flour flowed from the sides.

There followed the idea of rotating the top stone in a circular fashion. Such a mill was called a quern. Power to run such mills often came from animals and the lever principle allowed the use of larger and larger stones. By 100 B.C., the water wheel had been invented, and wind and water have been prime sources of energy for milling until comparatively recent times.

ROBERT A. LARSEN is Executive Vice President of IDS Venture Capital.

The invention of the steam engine by James Watt in 1769 marked a new era. Watt himself was responsible for the first application of steam power to flour milling. Watt's invention made it possible to build extremely large flour mills but in practice, inadequate transportation made it necessary to limit the size of the mill to fit the needs of the local area which the mill served. The coming of the railroad changed all this.

It soon became necessary to design mills in which the product was mechanically conveyed from place to place. About 1875 the screw conveyor and the bucket elevator were first used. This was followed by use of steel roller mills and, with the application of air currents to help remove the seed coating (bran) from flour, the equipment for modern mill design was in hand.

Not all cereal grains are pulverized to a flour. Rice and barley, for example, are milled in such a way that only the husk is removed (pearling), and oats are rolled into a flattened form. Our purpose is to discuss the milling of cereals into flour only. Rice, barley, and oat processing will be described elsewhere.

RECEPTION AND STORAGE OF WHEAT

Wheat selection is important to the flour miller. Without sound wheat—dry, unsprouted, and free from insect contamination—the miller is powerless to produce an adequate product.

Whether a wheat is to be used for bread production or for cakes, cookies, pastry, or biscuit making depends primarily upon the protein content of the wheat. While recent developments in attrition grinding of flour and the subsequent air classification into fractions of varying protein content has reduced the miller's dependence on wheat variety (Graham 1965), selection of wheat for protein content remains important.

Bread flours normally contain at least 11% protein. Cake and pastry flours fall between 8 and 11% protein. Cake flours are generally at the lower end of this protein scale and biscuit flours at the upper end. Since flour has less protein than the grain from which it is milled, wheat is bought at about 1% higher protein content than wanted in the finished flour.

It is up to the wheat buyer to provide the miller with the proper wheats for milling. He mainly looks for proper protein content. Improper drying in the field leads to a musty odor caused by mold growth. Sprouted wheat results from damp conditions during harvest. Generally, the wheat buyer knows which areas of the country are afflicted with sprouted or tough wheat in a given year. He avoids purchasing from these areas. An abnormally high maltose test (see chemical methods) also indicates sprouting conditions. Insect contamination can be detected by soft x-ray analysis of the wheat, and this method of control is now finding wide acceptance.

In most countries, wheat and other cereal grains are subject to government inspection and grading. Such a grading system helps the grain buyer since it not only takes into consideration the soundness of wheat for milling purposes, but also limits the amount of contamination of other cereal seeds, weed seeds, and foreign matter. Lower grade designations are made for wheats of high moisture content, usually above 14 to 15%, because such high moisture wheats do not keep well.

Wheat is stored at the mill in large bins. The amount of storage space at the mill varies greatly. Most mills store at least 6 weeks supply of wheat with some mills having over 42 weeks storage capacity.

On arrival at the mill, wheat will contain foreign impurities, the extent of the contamination being less if adequate government standards and inspection exist. Some wheat shipments contain sticks, stones, string, and even bits of cloth. These must be removed before wheat goes into storage. Dust and smaller pieces of foreign matter are left to the cleaning done just prior to milling.

Therefore, after the wheat is removed from the carrier and is weighed, it goes over sieves which remove all coarse impurities by sifting. One such type is an oscillating inclined sieve. Usually air is drawn through the sieve at several points to remove lighter impurities by aspiration.

Another separator which is widely used depends upon a revolving wire mesh drum. The wheat falling on the mesh (about $\frac{3}{8}$ by $\frac{3}{4}$ in. rectangular perforations) goes through the openings and is trapped in the center and carried away by baffles. The large rubble cannot pass the screen and is discarded.

Next, the wheat, if too wet, is dried before storage. Wheat drying is more important today because of the tendency to store more wheat at the mill for longer times and with consequently greater opportunities for heating. Heat production results from the growth of microorganisms in the grain. Heat of respiration accumulates in the confined storage space of the grain silo. If the moisture of the grain is kept low (below 14.5%), the organisms cannot thrive (Anderson and Alcock 1954).

Another reason for the modern trend towards moist grain is that most wheat is harvested by a combine machine which cuts, threshes, and sacks the wheat all at the same time. In this operation, the sun drying of wheat before threshing has been abandoned. The result often is wheat that is still damp.

Wheat is dried by the application of heat. Both hot air and radiators are used. In either case, the evaporated moisture is carried away by air currents. Wheat cannot be heated indiscriminately in order to drive off moisture. Excessive temperatures damage the properties of gluten for bread baking. Lockwood (1952) states that a drying section must be large enough to "allow time for the internal moisture to diffuse at a temperature of not over 115°F when the moisture content is over 17%, and not over 130°F when the moisture content is under 17%."

Since uniform heating is desired, hot water radiators through which the grain passes are preferred to hot air. Hot air tends to channel and cause hot pockets. Normally, moisture is removed to the last 0.5 to 1.0% above the desired dryness. The evaporation of this last amount of moisture is used to cool the wheat back to ambient temperature.

Thus, the three stages of drying are these. The wheat is preheated to the desired temperature. Next, water is vaporized by heat. Finally, the grain is cooled by water evaporation. Modern wheat driers, therefore, are composed of three sections, each handling one of these tasks.

After drying, the wheat is ready for storage. Most storage facilities are built of concrete because it is cheap, easy to maintain, and fire- and verminproof. In metal bins, condensation can occur on the wall during periods of declining temperature while grain near the wall may heat and drive moisture into the interior of the bin during hot spells. Wood is not practical for large bins.

Bins hold from 50 to 1,000 tons of grain, are up to 20 ft in diam and 60 to 100 ft deep. A big problem in bin design is to avoid stratifying the wheat by weight. The heavy wheat goes to the bottom, the light to the top. This density separation can cause serious complications in milling. Grading is avoided by using distributors or baffles to divide the grain as it enters from the top of the silo or is removed from the bottom. Since, upon discharge, the wheat does not "pull down" as a solid but funnels through the center of the pile, the baffles aid in removing wheat approximately in the sequence in which it was loaded.

THE CLEANING HOUSE

As grain is needed for milling, it is withdrawn from the storage areas and is brought into the mill. First, the wheat is sent through a cleaning operation prior to the actual milling. The section of the mill where this is done is called the cleaning house.

Since the gross impurities have already been removed at the bins, the techniques used in the cleaning house must be more refined than those used earlier. Whatever the machine used, it either removes loose foreign material or it removes dirt adhering to the surface of the grain. Normally, dirt removal from the surface of the grain is done last using friction, aspiration, or water washing techniques. Loose material is separated by sieves, magnets, and rapidly moving air streams.

Wheat is weighed as it enters the cleaning house and thereafter goes through a separator. This is the same type of separator as used in the silo operation except that it is set to remove fine impurities and dust. Small pieces of sticks, stones, sand, and dust are sifted away and light impurities such as wheat chaff are removed by air current.

The next operation involves magnetic separation to remove bits of adventitious metal which may have found their way into the wheat mix.

Although aspiration is provided in the first sifting machine, there is usually a second aspiration and possibly more as the wheat passes through the cleaning house. No aspirator is totally efficient, so in order to do an adequate job, several treatments are necessary.

Another cleaning machine is designed to remove cereal grains other than wheat and to remove foreign seeds, particularly weed seeds. This is done by catching the unwanted seeds in specially designed pockets in a revolving metal plate. These indentations are designed so that only the weed seeds or oats or barley fit into the pockets. The wheat, therefore, passes through the machine while the other seeds are lifted up and carried away. In such a disk separator, different types of plates can be positioned to catch various seed types and the machine can be set to discharge wheat at any place along the route. Normally, the revolving disks are set at right angles to the flow of wheat.

Scouring involves the removal of dirt on the surface of the wheat by friction. The scourers differ widely in design. Usually, the wheat is removed by paddles against an emery-coated surface, the severity of the treatment being controlled by the clearance between the paddles and the stationary emery surface. In some cases, a light washing is performed before the wheat enters the scouring machine in order to toughen the wheat so it does not break during the mechanical operation. The dirt and beeswing (the outer coating of the wheat bran) are removed by air aspiration.

The final cleaning step is a water wash. The water dissolves the dirt and permits stones and bits of metal to sink. In some cases, it appears to reduce microbiological contamination of the wheat, and in all cases the washer tends to add about 1% water to the original moisture content of the wheat.

There are numerous types of washers. Usually, the wheat is conveyed through a trough of water where dirt is floated away and the stones sink. This operation is particularly effective in removing dirt from the crease of wheat kernels, dirt that is missed by dry scouring. Following the washing, excess moisture on the surface of the wheat is removed by centrifugal forces.

TEMPERING (CONDITIONING)

Tempering, as the word is used in the United States, refers to the addition of water to the bran and endosperm. The bran becomes tough and rubbery while the endosperm becomes less vitreous. This improves milling efficiency.

In European circles, the concept of tempering is broadened to include not only the toughening of bran but also the changing of the physical and chemical characteristics of the endosperm during milling. This broader concept of tempering is called conditioning (Gehle 1952). In conditioning, a controlled heating process is employed in addition to the moistening procedure.

Wheats can be broadly classified into the classes of hard wheat and soft wheat according to their milling characteristics. The former takes about 20% more horsepower to grind to a given particulate size. When hard wheats and soft wheats are found in the same wheat mix—and this is quite common, particularly in European countries—the miller is confronted with a difficult problem. Shall he set the rolls close and overgrind the soft wheat, or shall he set them wide and undergrind the hard wheat. Obviously, neither alternative is desirable. Conditioning provides a partial answer.

If hard wheats are maintained in the presence of moisture under the proper conditions and for appropriate periods of time, the endosperm chunks become softer. They grind more like soft wheats because the water is allowed to penetrate the endosperm and it becomes more mellow. By this technique, the milling problems of hard and soft wheat mixtures are considerably simplified.

Tempering—as it is still practiced in the United States—involves adding water to grain to raise the moisture to 15 to 19% for hard wheats and 14.5 to 17% for soft wheats and allowing the wheat to lie in tempering bins (with little or no temperature control) for periods of 18 up to 72 hr. During this time, the water enters the bran and diffuses inward causing the bran to lose its friable characteristic and to become leathery in texture. Usually, tempering is done in successive steps since it is impractical to add more than a few percent of water to wheat at one time.

Conditioning, in contrast to tempering, always involves the use of heat since quick diffusion of water into the endosperm as well as the bran is the purpose. It has been established that water enters the endosperm through the germ end of the grain and also through breaks in the bran coat (Sugden 1956). If properly controlled, the bran is toughened, the contents separate more easily from the bran and the endosperm can be ground to a powder with less horsepower requirement. Since heat affects gluten quality, its application must be controlled. Normally, a temperature of 115°F cannot be exceeded without causing detectable changes in the baking quality of the flour.

Under some conditions, this temperature limit is deliberately exceeded in order to change the protein. Higher temperatures cause the gluten to become more resistant to extension and to break more easily during the extension (Kent-Jones and Amos 1957). Normally, gluten treatment during conditioning involves temperatures between 115°F and 180°F. The drier the wheat, the higher the temperature used and the longer the time required.

Wheat conditioners involve four sections. The first section heats the wheat to the proper temperature. The second section adds moisture and holds the wheat for the proper time. The third section cools the wheat to room temperature and the final section provides a holding bin where the moisture in the wheat is allowed to equilibrate before milling. Conditioners are sold as the first 3 units while the holding bin is separate, but all 4 parts must be considered as one in planning such an installation.

In addition to being able to mellow endosperm, conditioners save much time compared to tempering. All of the conditioning water can be added at once. The wheat goes through the preheat section, the water addition section, and the cooling section in 1.5 hr. or less. Millers differ as to the time the wheat is held in the holding bins. Times from 8 to 18 hrs are quoted with the longer times being used for the harder wheats.

The construction of conditioners is much like that of driers and indeed the same device serves a dual purpose in some instances. Radiators rather than hot air are the preferred method of heating. Cooling is done by allowing water evaporation from the surface of the wheat or by the application of cold air.

In one method of conditioning, the wheat is heated very rapidly by the direct injection of steam and is held at a temperature of 120°F for about 1 min. Following this, there is a rapid cooling by plunging the wheat into cold water and thence to a centrifugal machine to remove the surface water. This method of conditioning is extremely fast and has the advantage that the cooling process, since it is not done by cool air or evaporation, does not tend to dry the bran coat.

THE GRINDING OF WHEAT

The grinding of wheat is done between pairs of rolls. These rolls, since they are moving in opposite directions, moving at a different rate of speed one from the other, and set with an appreciable gap between them, do not grind the wheat primarily by crushing. Rather, the reduction of the wheat is by shearing forces which, because of the set of the rolls, are relatively gentle. The rapidly moving roll runs about 2.5 times faster than the slower one and at speeds from 250 to 450 rpm. Faster speeds are preferred in US mills.

The purpose of flour milling is to first separate the endosperm from bran and germ in as large chunks as possible and then reduce the size of the endosperm chunks to flour-sized particles through a series of milling steps.

Hence, the roller milling area is divided into two sections, the break section and the reduction section. In the first, the bran is broken open and the endosperm is milled away. This system quite often involves four or more sets of rolls each taking stock from the preceding one. After each break, the mixture of free bran, free endosperm, free germ, and bran containing adhering endosperm is sifted. The bran having endosperm still attached goes to the next break roll, and the process is repeated until as much endosperm has been separated from the bran as is possible.

The Break Milling System

The surface of break rolls is always fluted to obtain the necessary grinding effect. These saw-tooth flutes run spirally around the roll and the number of flutes per inch increase from the first to the fourth break. The first break runs 10 to 12 flutes per inch, the fourth about 28 flutes.

In Europe, the rolls are not set horizontally as in the United States but rather on a diagonal. Each design has an advantage. The diagonal rolls allow more machines in a given amount of floor space but the horizontal rolls permit more uniform feeding of grain to the rolls and thus allow faster roll speeds and more production per roll.

Courtesy of Buehler Corp.

FIG. 1. A ROLLER MILL FOR FLOUR

After each set of break rolls there is a sifting system (called scalping). The scalping system is a combination of a sieving operation (plansifters) and air aspiration (purifiers).

As the name implies, a plansifter has flat sieves piled in tiers, one above the other. The action of the sifter is rotary in a plane parallel with the floor. As the sifter moves in about a 3.5-in. circle, the small-sized particles spill through to the sieve below while the oversized particles travel across the sieve to a collecting trough and are removed. As many as 12 sieves can be piled one on top of the other and there are four separate compartments in one plansifter.

In plansifters, large pieces of bran with adhering endosperm are first removed and these are sent to the next break roll. On the

finer sieves, bran and germ are scalped off. Below these are flour
sieves, the finest being small indeed. The sieves run 196 mesh per
inch and have openings of about 0.06 mm.

The resulting flour and endosperm chunks (middlings) still contain minute size bran particles which are removed by sending the
product through a purifier where air currents carry the bran away.
A purifier is essentially a long oscillating sieve, inclined downwards
and becoming coarser from head to tail. Passing through the sieve
in the direction from floor to ceiling are air currents. The buoyancy
effect of the air causes the flour to stratify into endosperm chunks
of different size.

Thus, in addition to removal of bran, a purifier can be and is
used to make a coarse separation of middlings. The endosperm
chunks (middlings) fall through the sieve openings, collect, and
are sent to the appropriate reduction rolls. The overs are a composite of bran and bran plus endosperm. These go back to the
break rolls or to millfeed stock. Aspirated materials go to millfeed.

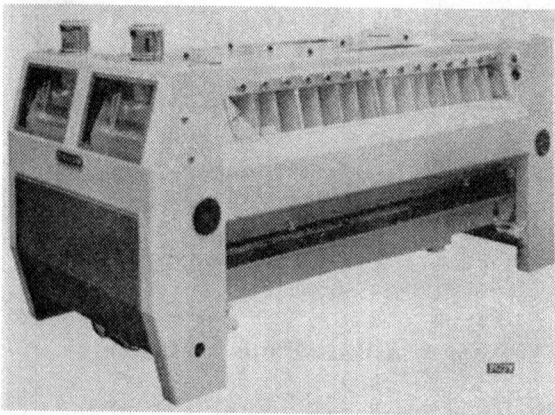

Courtesy of Buehler Corp.

FIG. 2. A PURIFIER FOR FLOUR

As many as 12 purifiers are quite normally found for 4 break
rolls in the scalping system. For example, the fine middlings from
the first, second, third, and fourth breaks normally go through a
double purification step while the coarser middlings need only one
purification treatment.

The Reduction Rolls

The reduction system comprises two parts, roll mills and sifting
machines. The major difference from the break system is that the

surface of the reduction mills is smooth rather than grooved. The purpose of reduction rolls is to reduce endosperm middlings to flour size and facilitate the removal of the last remaining particles of bran and germ.

The roll stands in a reduction system are divided into coarse rolls and fine rolls. Coarse and fine does not refer to the condition of the roll surface since the surfaces are smooth. Rather, it refers to the setting of the rolls, whether they are set wide to produce coarse grinding or close to produce fine grinding. The coarse rolls are used only to produce middlings of uniform size for later reduction to flour. Hence, these rolls are often called sizing rolls and the middlings sent to them are called "chunks." Middlings sized on the sizing rolls are then sent to the fine rolls to be ground to flour.

After each reduction, the resulting product is sent to sifters where a classification by particle size is done. Finished flour is removed and oversized material is sent back to the reduction rolls for further processing. Normally, for every reduction roll stand, there is at least one sifting device to take care of its output.

Plansifters are used behind the reduction rolls and their purpose is to divide the stock into coarse middlings, fine middlings, and flour. The coarse middlings are returned to the coarse (or sizing) rolls, and the fine middlings are returned to the fine roll, while the flour is removed from the milling system.

Purifiers are often used behind the coarse reduction rolls. The

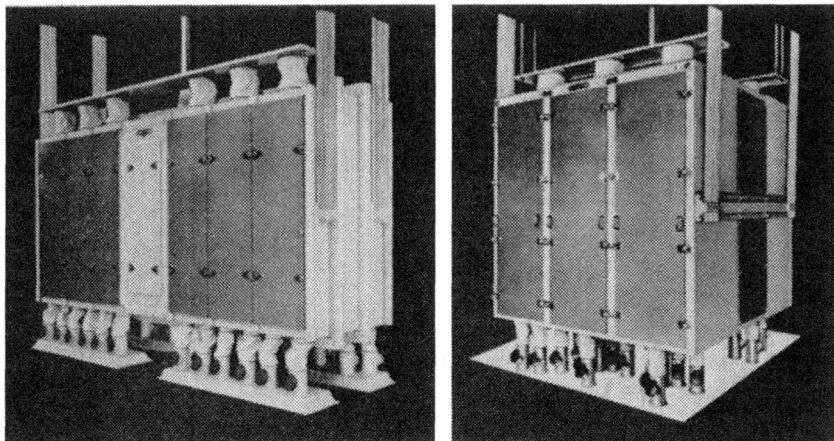

Courtesy of Simon-Carter Co.

FIG. 3. PLANSIFTERS FOR FLOUR

purpose in this case is size grading rather than purification, and purifiers are sometimes superior to plansifters for these separation requirements.

One must be careful not to overgrind the flour in the reduction system. Overgrinding damages starch granules and makes the flour unsuitable for baking (Larsen 1964). Overgrinding can be detected by a high maltose value (see wheat flour specifications).

Overgrinding should not be confused with grinding too fine. It is possible to grind flour without starch damage if the grinding is done gradually using moderate pressure increases. Heavy grinding pressures will cause starch rupture and this is to be avoided (Jones 1940).

The Scratch System

In addition to the break system and the reduction system, some mills have a standby system called the scratch system. If the mill is operating properly, that is the tempering is such that good release of endosperm is obtained on the break, the scratch system can be bypassed. If not, the scratch system is employed to maintain proper release of endosperm from bran. Therefore, the scratch system is in reality an extension of the break system.

Air Classifiers and Grinders

Recently, some flour mills have added a system to their mill stream which further processes flour through additional grinding and particle size separation. The system depends upon air classification of flour particles in the micron size range. The minimum size separation on a sieve is about 60μ. Air classifiers work in the range of 10 to 40μ (subsieve size).

The research papers leading to this development appeared in the 1950's (Hess 1953; Elias and Scott 1957; Elias 1958; Gellrich 1958). Patents on a practical milling method soon issued (Pillsbury Mills 1959). The purpose of the grinders and classifiers is to separate flour particles into their constituent fractions, protein and starch, thereby producing flour fractions rich in protein and rich in starch. Then by reblending, flours suitable, for example, for cake or pastries, and flours suitable for breads can be designed. This reduces the miller's dependence on wheat mixes for flour quality.

In actual practice, a problem exists in finding a market for the large volume of low protein or starch fractions at prices adequate to justify the installation and operation of the special equipment.

If the air classification technique can be perfected to a point where starch can be concentrated without protein (Wichser 1958), then industrial outlets for the starch might be found.

In this high velocity process, finished flour from the roller mill is further reduced in size in special grinders. Impact mills rotating at about 350–400 ft per sec are useful, but at this speed there remains considerable proteinaceous material still adhering to starch. At higher speeds, the flinging of the endosperm chunks against the rotating pins in these pin disk mills causes considerable starch fracture so millers are turning instead to attrition grinding (Graham 1965). Turbo grinders are examples (Pillsbury Mills 1959).

In these grinders, the flour particles spin in air vortices created behind rotating blades and become smaller by rubbing one on the other. This method breaks the protein matrix without undue starch damage. Following this, the product is separated in air classifiers using flow-dynamic properties (size, shape, and specific gravity) to separate protein particles, starch granules, and endosperm chunks. In such separators, centrifugal force keeps the heavy particles being conveyed in the air stream to the outside of the

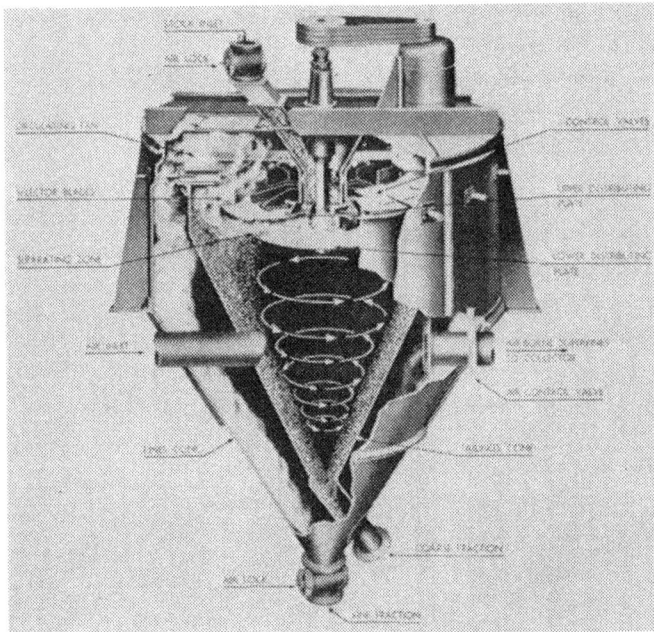

Courtesy of Sturtevant Mill Co.

FIG. 4. DIAGRAM OF AN AIR CLASSIFIER

chamber enabling a "cut-point" separation to be made by a valve located at the proper position in the air stream.

At the present state of the art, a medium protein flour (10–12%) can be fractionated commercially, into about 20–30% of a flour suitable for cakes and pastries. From 5 to 15% of the original flour becomes a fraction containing about 20% protein and is suitable for blending with bread flours while remaining 55 to 75% of the original flour is largely unchanged. Adding back the high protein fraction to the unchanged portion of the flour can, for example, raise the protein content from 10 to 12%. Thus, whereas, the original flour of 10% protein may have been suitable for biscuits, high velocity reprocessing of the flour could result in 30% of the flour suitable for cake and 70% suitable for bread baking.

The Conveying System

Some means must be provided for the transport of flour stock from machine to machine. Older mills depended upon gravity, and this is why flour mills were built so high. The wheat and flour were moved to the top by bucket elevators (that is, endless belts having attached buckets which pick up product at the bottom and dump it at the top) and then the flour flowed by spouts to the rolls and to the sifters.

Bucket elevators have two serious disadvantages: they are dusty, and they provide a place for insects to grow. Consequently, flour mills are rapidly converting to the air conveying of flour and are abandoning bucket elevators and gravity spouts.

Pneumatic conveying is the term applied to air conveying in flour mills. The higher power costs of pneumatic conveying are offset by cheaper construction, less mill clean up, less infestation, cooler flour stock, and metal rather than wood spouting. Pneumatic conveying depends upon the fluidizing effect of air flowing through a bed of particulate matter. The air force may be positive pressure but in flour mills vacuum is normally used. Powerful suction fans provide the negative pressure. The air intake is through the roller mills, and this air movement has the desirable effect of keeping the rolls and the flour cool during grinding.

Pneumatic conveying was an important advance for the milling industry (Shellenberger 1965). It made it possible to abandon the classic arrangement of a mill on 4 or 5 floors to arrangements on 1 or 2 floors. This allowed for new alternatives in the placement of machines. Thus, because of air handling systems, great variety of mill layouts became technically possible.

Courtesy of Simon-Carter Co.

FIG. 5. PNEUMATIC CONVEYING SYSTEM
FOR FLOUR

THE STORAGE OF FLOUR

The bulk handling of flour has been made possible by pneumatic conveying systems. Bulk storage has a number of advantages. Paramount is the ability to blend flour after milling to achieve optimum quality. Another advantage is that adequate storage of flours allows greater milling flexibility. Different grades can be milled and held until needed. This increases running time on a mix and avoids costly machine setting changes that are required when wheat blends are changed.

Capacity varies widely but most mills have bulk flour storage for from 2 to 4 days of production. The practice of milling directly into these tanks results from the fact that commercial users now buy more than three-quarters of the flour milled in the United States. Since this flour is of a standard grade for a single customer, long runs are possible. Flour types are blended to meet the special need of a customer on a "quality assured" basis. Special bin designs permit the discharge of product without particle segregation by density or size (Weber 1966).

The movement of flour in bulk is by special railroad cars called Air-Slide cars (Pratt 1957). Trucks of similar design are also used.

The pneumatic movement of flour through the mill, into blending and storage bins, thence to Air-Slide cars or trucks and from cars into bakeries has brought about profound changes in handling and distribution techniques. Whereas a short time ago, most flour moved in packaged form, today the reverse is true.

Packaged flour is put up in bags holding from 2 to 100 lb. The larger bags are for bakery use, the smaller for the consumer market. Since the techniques for packing and crating are not unique to the dry milling industry, they will not be discussed here.

Agglomerated Flour

In 1963 and 1964, a special grade of flour was introduced in which the particle size and bulk density had been altered through agglomeration (Black and Bushuk 1967). In the flour agglomeration process, the surface of flour particles is wet with water, the individual particles allowed to stick one to another and the resulting agglomeration of a number of individual particles dried. The resulting particle when viewed under a microscope has the appearance of a grape cluster. Similar techniques have been used for some time previously to prepare agglomerated dried milks and similar products.

These agglomerates of flour have characteristics of commercial significance. The product is dust free, flowable, requires no sifting because it does not pack, and disperses quickly in cold water. It has been suggested that the baking quality is not altered by this treatment (Claus and Brooks 1965) but in practice some problems have been found.

Agglomerated flour has been designated as a special grade of flour under government standards. If the problem of changing hydration capacity and baking variation can be overcome, the product will be undoubtedly see considerable use.

THE MILLING OF CORN

There are still mills, particularly in southern United States, where corn milling is done on millstones. These mills make whole corn meal. Such stone-ground, whole corn meal is an unbolted product. Rancidity problems that result from the presence of the corn germ in this meal limit the area of distribution. The characteristic flavor of "old-fashioned stone-ground corn meal" is due to incipient rancidity.

Before the technique of roller milling of corn could be used for corn, a successful way of removing the corn germ had to be found. In corn, the germ makes up about 12% of the seed compared to about 3% in wheat. Roller mills tend to break this rather large entity into small pieces rather than flatten it. These small pieces are hard to remove by sifting. Thus, a new system of degermination was needed. This need was answered when the Beall corn degerminator was introduced to the dry corn milling industry.

The Beall degerminator is an attrition device. It is composed of a cone shaped shell rotating around a stationary inner cone. Corn is fed into the smaller end of the cone and works its way down to the large end. During the passage, the corn is rubbed between the stationary cone and the rotating cone, both cones having special knob-like and augerlike surfaces. During the rubbing, the corn is dehulled and the germ loosened and knocked out.

Entoleters can also be used for dehulling and degerming. The Entoleter differs from the Beall in that it is an impact rather than an attrition machine. Corn is fed through a center opening in the Entoleter and falls on a rapidly rotating disk containing pins on the surface. The pins of the rotor throw the corn against a stationary wall thereby dehulling and degerming it.

As with other types of dry cereal milling, the object of dry corn milling is to make the cleanest separation possible of endosperm, bran, and germ. With the exception of degermination, dry corn milling, mechanically, is similar to wheat milling (Kaiser and Ferguson 1967).

Upon receipt, the corn first goes through a cleaning process. Both dry cleaning and wet cleaning techniques are used. The use of an electrostatic separator is of particular importance here. Because of the relationship of the size and weight of corn kernels to the size and weight of rat pellets, separation is difficult. Fortunately, electrostatic separators work very well. Usually, the electrostatic separator is added at the tail end of reels and disk separators and just before the wet stoner and washer.

After cleaning comes conditioning. Normally, the moisture content of the corn is raised not to 17% as with wheat but to about

21%. This is done because the germ of corn tends to be more friable than the germ of wheat and if it is too dry it will break into small flour sized pieces during degerming. If enough water is added, not only is the bran toughened but so is the germ. Conveyors and whizzers are both used to add moisture to the corn.

Degerming follows the conditioning according to procedures already discussed. Following the degerming (and dehulling), the corn must be dried so it can be handled on roller mills and in sieves. The moisture is brought down to about 15%. Inclined barrel shaped, rotary driers are used and the product is air cooled after drying in louvered coolers. At this point, the product is aspirated to remove the bran, and then it is ready for the main part of the milling system.

The milling system consists of grinding, sifting and classifying, purifying, aspirating, and in some cases, final drying. The normal flow is through break rolls and then to plansifters. The fines go to the next break roll and the coarse particles to purifiers and then to the germ rolls. The germ rolls flatten the germ so that it is easily removed by sifting. The break rolls are followed by reduction rolls which produce the final fine flour.

In a dry corn mill, the break system is longer than the reduction system. Since corn flour is not as important a product as corn grits or corn meal, a long reduction system is not needed to reduce the size of the endosperm particles (see corn product specifications).

Tip caps found at the end of the corn germ are a problem peculiar to corn. Since they are an intense black color, the presence in corn flour badly discolors the product. Tip caps are to be avoided in dry corn milling and are removed with the corn bran by aspiration.

Careful treatment of the germ in the dry corn milling industry is accomplished by a special system of rolls which is set aside for the cleaning of corn germ. Since corn germ is 12% of the seed and contains about 34% corn oil, it is economical to extract the oil from corn germ and sell it as corn oil. This requires clean germ and hence the emphasis on this system.

Every dry corn milling plant has an oil extraction plant. Normally, an oil press is used to extract the oil. The extraction and refining of corn oil is discussed in a subsequent chapter and will not be considered further here. Several references are available (Stimmel 1941; Striver 1955; Neenan 1951; Gehle 1937).

THE MILLING OF RYE

There are many similarities between the milling of rye and the milling of wheat (Schopmeyer 1962; Blank 1958; Zwilingenieur 1956; Nottin 1945; Prochazka 1938; Mayer 1936). In either case, the same basic type of machinery is employed to produce a powdery or granular material from a cereal grain by careful pulverizing of the seed. In both instances, the purpose is to make the flour substantially free of bran and of germ. The novice might have trouble distinguishing a rye mill from a wheat flour mill although the differences would be immediately recognizable to one skilled in the trade.

A special problem in rye milling occurs during cleaning. Rye is normally graded for size as well as dockage (impurities) and moisture. For example, US No. 1 plump rye has no more than 5% thin kernels, whereas No. 3 has over 25% thin kernels. Because of these differences, special precautions must be taken during cleaning to remove only irregular-sized rye kernels. Gravity tables are used to advantage here.

Rye is ground on a series of corrugated break and reduction rolls. Smooth reduction rolls are not used. If rye is ground on smooth rolls, the rye flour has a tendency to flake and the endosperm is not properly released. Break rolls are generally cut somewhat finer than for wheat grinding and reduction rolls are cut to approximately the corrugations of the scratch rolls used in wheat milling. The length of the break system and the reduction system is about the same in a rye mill.

Because rye endosperm is soft, it does not break into chunks (middlings) as readily on the break rolls as does wheat endosperm. The consequence is that so much of the flour is released on the break rolls that purifiers are not needed for middling classification as is the case in wheat milling. On the other hand, rye flour is more difficult to sift so more sifting surface is required in the plansifters than for a corresponding production of hard wheat flour.

Ash and protein are not important specifications in the sale of rye flours. Color is a much more critical specification. The granulation of the rye flour (this is generally called "dress") is a second important criterion. Flavor is a third.

The lack of importance of ash and protein is understandable. Rye protein in contrast to wheat protein will not readily form a gluten. Thus, rye bread is normally a blend of wheat flour and rye flour. As a consequence, the baker is more interested in appearance and taste qualities of the rye than he is in the baking quality.

THE MILLING OF DURUM

The method of milling durum is similar to the milling of wheat but the purpose is quite different. In the milling of wheat, flour is the desired end product. In the milling of durum, middlings are wanted. Consequently, in a durum milling system, the break system—where middlings are formed—is emphasized, and the part of the reduction system where flour is formed is deemphasized (Paige 1936; Ager 1912). Reduction for middling sizing rather than flour production is the desired action.

Since the natural yellow color of durum is wanted in the final middlings (middlings are called semolina in durum milling) the job of the durum buyer is important. It is he who is responsible for maintaining constant yellow color of desirable quality in the durum mix.

Durum goes through the normal cleaning house and tempering operations used for wheats. The tempering of durum while using the same equipment as wheat is different in that the time required for tempering is less than for wheat. This is because the object of durum milling is to produce middlings and not flour. Excessive tempering softens the endosperm and results in higher percentages of flour.

The break system in a flour mill normally numbers four. In a durum mill at least five breaks are used. Such a system provides very gradual reduction of the stock so necessary for good middlings production while still avoiding large amounts of break flour.

The rolls in a reduction system are used as sizing rolls only. They function the same as the sizing rolls in a flour mill, reducing the coarse middlings to a uniform particle size. In a flour mill, the sizing is done to produce a uniform product for further grinding on the reduction rolls. In a durum mill, however, sizing is done to make a uniform product size.

Normally, the sizing rolls in a durum mill are corrugated on their surface. If a mill has 5 breaks, it may have 5 sizing rolls. Two rolls handle the large chunks, the rest smaller endosperm particles.

Some durum mills have several rolls set aside for the production of durum flour. Since this is not the primary product of durum milling, these rolls are usually set apart from the main milling system. Durum flour is particularly desired for noodles.

The sifting system in a durum mill relies heavily on purifiers. Plansifters are little used. The reason for the predominance of purifiers in a durum mill has been outlined in the discussion of the function of purifiers in the section on wheat millings. Because of the layering effect of upward rising air currents, purifiers stratify middling chunks according to size better than plansifters. Since middlings are the end product of durum mill, the heart of the mill is the purifier system.

GENERAL TESTS FOR CEREAL FLOURS

Introduction

Each class of cereal flours has its own peculiar specifications, but there are other specifications common to all. These more general tests are moisture, protein, ash, color, fiber, and particle size.

Moisture

Air ovens are commonly used to determine the moisture in flour. The usual conditions are 266°F for 1 hr with constant air movement.

In recent years, a number of semiautomatic ovens have been developed using a variety of heat sources. The Carter-Simons moisture tester and the Brabender moisture tester are two of this type.

Protein

There is only one method of importance for measuring the protein content of cereal products. This is the time-tested Kjeldahl technique. In this method, the product is digested by the application of heat and strong sulfuric acid, usually in the presence of a catalyst. The nitrogen is converted to ammonia which forms a salt with the sulfuric acid. After the solution has cooled, the ammonia is released by adding a strong base, the freed ammonia distilled into a standard acid solution and the excess acid determined by back titration.

In this way, the amount of nitrogen in a cereal is determined. By using a factor, the per cent nitrogen is translated into percent protein.

Ash

The importance of ash determination will be discussed in the flour specifications part of this chapter. The higher the ash, the more bran particles there are in the flour. Therefore, ash is used as an indication of flour grade or extraction.

The principle of this test is to burn the cereal flour until all organic material is converted to gases and driven off along with water. The residual metallic oxides are weighed and the percent of ash is calculated.

Normally, this determination is made in a muffle oven at temperatures between 425° and 600°F. The holding time varies from a minimum of 2 hr to overnight.

Color

The color of cereal flour can be measured either by comparison with standards or by suitable electronic instrumentation.

One visual method is called the Pekar Color Test. Samples of the flour are placed side by side on a flat glass or metal plate. The surface of the flour is smoothed using a spatula. The slicks are either examined unwetted (durum) or immersed in water and then air- or oven-dried (wheat, rye, or corn flour). The wet Pekar process intensifies the dark color of off-grade flours and bran specks are easily seen. The flours are graded visually by comparison to a standard product.

Electronic color graders depend upon measurement of reflected light, usually in the yellow region of the spectrum. The cereal product is made into a paste with water, and light of a specific wavelength is allowed to impinge on this paste (Kent-Jones and Brown 1966). The whiter the flour, the higher the reflectance value.

Fiber

To determine fiber, it is necessary to extract the sample with ether and digest it, first with dilute sulfuric acid and then with dilute sodium hydroxide. After each digestion, the solubilized carbohydrates and proteins are removed by filtration and washing. The residue is dried and weighed directly.

Particle Size

There are four general methods of measuring particle size adaptable to cereal products. These are sieve analysis, microscopic measurements, sedimentation techniques, and air permeation techniques.

Sieve analysis depends upon the passage of cereal products through standard mesh screens of graduated size under standard conditions. The percentage of material on each sieve is determined. A machine called a Ro-tap is widely used, which, through mechanized agitation, assures standardized test conditions for the sieving. This type of analysis is used for cereal grains and for some cereal products.

For smaller particles, measurement using a microscope having a calibrated eyepiece can be used. A representative sample is brought into the microscopic field and the number of particles in each linear range is counted.

Sedimentation techniques depend upon Stokes Law which says that in a given liquid or gas, larger particles of equal density settle more rapidly. Therefore, the amount of a sample which precipitates in a known time is measured and by application of Stokes Law, the dimensions of the particle can be calculated. From these data, a size distribution curve can be drawn.

Air permeation techniques can be used to measure the particle size of cereal products. As the size of the particles becomes smaller, the average interstitial space becomes less. Thus, a given volume of small particles will give more resistance to the passage of a liquid or a gas than will the same volume of large particles. The time required for passage of a given volume of gas at constant pressure, or the pressure required to pass a given volume of gas in a given time, is measured.

Fat

The fat content of cereal products is estimated by solvent extraction procedures. A sample of well dried material is extracted with anhydrous ether for from 4 to 16 hr. The ether is driven from the fat by heat and the ether-free fat sample weighed. The percentage of fat in the original material is calculated.

SPECIAL TESTS FOR WHEAT PRODUCTS

Introduction

In addition to the general tests just discussed, there are several more specific tests that are applied primarily to the products of wheat milling. These are protein quality, starch quality, hydrogen ion concentration and baking tests.

Gluten Quality

Besides the amount of protein (Kjeldahl test), flour millers must take into account the quality of the protein. This quality is referred to as strength and is related to the fact that strong glutens in a dough (gluten refers to the complex of proteins in flour) resist extension and that the gluten dough strands break before the extension is very great. Such gluten is wanted for bread. Weak glutens are easily extended and can stretch long distances (i.e., are plasto-elastic). These glutens are better for cookies, pastries, and cakes. The gluten quality of biscuit flour is intermediate between these two extremes.

In Europe, the gluten quantity and gluten quality are often determined at the same time. This is done by washing the gluten from the flour using an automatic device (Theby Test) and weighing the resulting gluten dough ball. Since water is present, the weight of the dough ball is greater than for the calculated Kjeldahl protein. The dough ball gluten results when divided by three roughly approximate Kjeldahl protein values. The gluten is then suspended in a lactic acid solution and shaken. The weaker the gluten, the more goes into colloidal solution and the greater the cloudiness of the solution. Turbidity measurements are made on this lactic acid-gluten suspension. The values obtained are related to gluten strength (Biechy 1957).

Gluten strength is measured in other types of special machines. The resistance to mixing is one factor that can be measured. In such devices, the flour is mixed into a dough under standard conditions and the resistance to this mixing procedure over time is

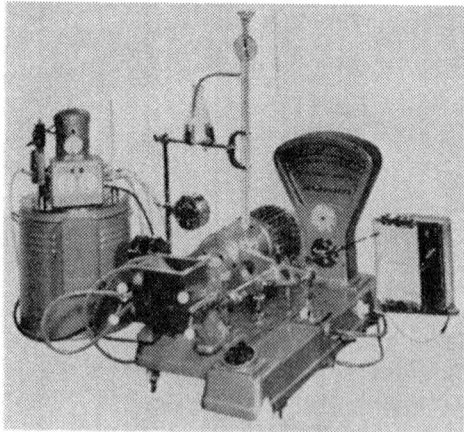

Courtesy of C. W. Brabender Instrument Co.

FIG. 6. THE FARINOGRAPH, AN INSTRUMENT USED TO MEASURE THE MIXING PROPERTIES OF A FLOUR

measured. The resulting curve is a measure of flour strength. Generally speaking, the resistance increases rather rapidly to a peak and then slowly drops as the mixing continues. The stronger the flour, the longer it takes for these changes to take place.

For the measurement of gluten strength, extensometers appear to be most sensitive. For example, the effect of dough strengthening agents (oxidizing compounds such as bromates and iodates) and dough weakening agents (chemicals such as sulfites or sulfhydryls, or proteolytic enzymes) are easily detected (Sullivan 1948).

A third method of measuring gluten strength, the MacMichael Test, is related to the Theby Method. Flour is suspended in a weak lactic acid solution and the gluten allowed to swell. Then, the resistance of the swelled gluten to the shear between a suspended bob and a spinning bowl is measured in a special viscometer. The results are reported in degrees MacMichael. Strong glutens have greater resistance to shear.

Starch Quality

Starch makes up in excess of 80% of the flour. The quality of this starch, particularly the degree of structural damage that has occurred in milling, is of particular importance. If there is not enough damage, enzymes cannot convert the starch to sugars so necessary for the metabolism of yeast during fermentation. Thus, in bread flours, a degree of starch damage is desirable. If too much starch damage occurs, the baking quality of the flour, particularly in cakes, cookies and biscuits, is harmed. Consequently, tests of starch damage are particularly important.

The amount of maltose freed from starch during the milling process is determined in one such test. The flour is buffered and held at 86°F for 1 hr. The amount of reducing sugar is determined with thiosulfate and the results expressed as milligrams of maltose in 10 gm of flour. Results may vary from 100 to beyond 500.

Another method of measuring starch damage is to allow yeast to ferment the flour under standard conditions and to measure the amount of gas produced. The amount of maltose is the limiting factor. This test is called gassing power.

An enzymatic test for starch quality devised by Sandstedt and Mattern (1960) is widely used. It depends upon the breakdown of starch by amylase enzyme under standard conditions. The degree of starch damage is measured by the amount of sugar released from the starch.

Finally, the amylograph, an automatic physical testing unit, is now in wide use as a measure of starch damage. In the device, a starch-water paste is heated under controlled conditions at a rate of about a 1°F rise in temperature each minute. During this time, the viscosity of the paste is being measured and automatically recorded. At some certain point, the crystallinity of the starch will begin to be disrupted and the granules begin to swell. This swelling increases the viscosity and is so recorded. If the starch is mechanically or enzymatically broken, gelatinization will begin at an earlier temperature and the peak viscosity will not be as great.

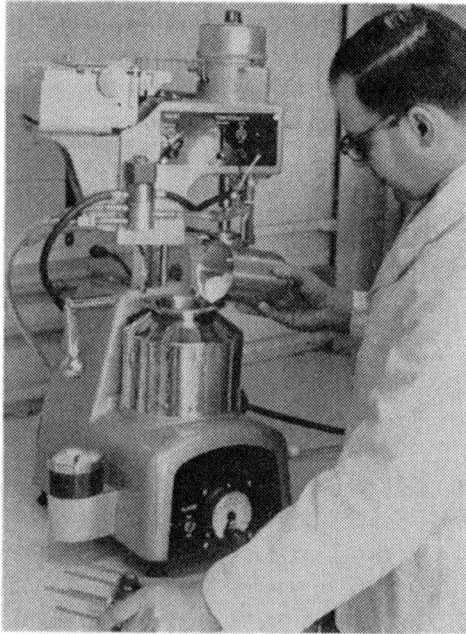

Courtesy of C. W. Brabender Instrument Co.

FIG. 7. THE AMYLOGRAPH, AN INSTRUMENT USED TO MEASURE THE QUALITY OF STARCH IN FLOUR

Hydrogen Ion Concentration

Some soft wheat flours, particularly cake flours, are treated with gaseous chlorine to weaken the gluten. In this reaction, hydrochloric acid is formed. Therefore, the pH of a flour suspension can be used as a measure of chlorine treatment.

Baking Tests

The tests just described are an attempt to predict by chemical and physical techniques the baking quality of a flour. In the last analysis, however, a baking test must be applied.

It is virtually impossible to standardize on a single baking procedure. The types of products made in different lands and in different shops within a country vary greatly. Each producer has his own peculiar set of conditions. For each baker, there is a baking test which suits his conditions and the miller serving that baker must adjust his baking test accordingly. There is no uniformity in these tests.

If the reader wishes some reference as a starting point for a baking procedure, the standard baking tests in the Cereal Laboratory Methods (Anon. 1962) are recommended.

SPECIAL TESTS FOR CORN PRODUCTS

Introduction

Moisture, particle size, fat content, and visual appearance are important corn flour specifications. In addition, acidity and malt extract are special tests applied to this type of product.

Acidity

The acidity of corn flour products is important. This test, which is not generally applied to other cereal flours, is used as a measurement of undesirable microbiological action. If the acidity is high, the microbiological spoilage is high (Zeleny 1940).

The first step in the acidity test is to leach the free fatty acids from the flour. This may be done by shaking the corn meal with benzene or sometimes water to extract the acidic materials. The liquid is decanted or filtered from the corn product and the fatty acid content of the filtrate determined by titrating with standard potassium hydroxide solution using phenolphthalein indicator. The results are calculated as percent lactic acid.

Malt Extract

Since some corn flours are used in the brewing industry, it is necessary to get a measure of the degree of conversion of starch to soluble solids as an indication of the amount of fermentable carbohydrates. The determination in question is called the malt extract test.

In this test, the corn product is added to water and gelatinized. The gelatinized slurry is heated, ground malt is added, and the mixture is held at 140°F for a period of time to allow the malt enzyme to act on the carbohydrate. The temperature is then raised to boiling to inactivate the malt enzyme. The solution is cooled, filtered, and the specific gravity of the filtrate measured. From tables on the relationship of specific gravity to sugar content, the amount of soluble solids is estimated and reported as percent conversion.

SPECIFICATIONS FOR WHEAT PRODUCTS

Introduction

There are two general classes of products coming from a wheat mill. These are the flour and the millfeed. While percentages vary somewhat, they hover around 70% flour and 30% millfeed. In turn, these two general classes are subdivided into products depending upon the degree of purity desired. A list of these subclasses follows:

70% Straight Flour
  Patent flour (less than 70% of wheat)
  Clear flour (residue left when a patent flour is removed from a straight flour)
30% Millfeed
  14% bran (seed coat material left after milling flour)
   2% germ (wheat seed embryo)
  14% shorts (everything left after the bran and germ have been removed from millfeed)

These terms apply only to milling in the United States. Other terms are found in other countries. Lockwood (1952) has an excellent comparison of terms for similar products as they exist in different lands and so does the Miag milling dictionary (Kessler 1945).

The terminology for flour is confusing to the milling novice. Names such as short patent and the first patent or first clear and second clear are commonly used. One needs only to remember that all of the flour coming from a mill is called straight flour. If it is further purified, it is separated into two fractions—a patent flour and a clear flour. The better of the two is the patent flour. The flour remaining after the removal of a patent from a straight flour is a clear flour. This terminology applies to all types of flours whether they be cake, pastry, cookie, biscuit, or bread flour.

The best measure of the degree of refinement of a flour is either ash content or color. The most refined flour comes from the center of the wheat berry and the least refined next to the bran. As one approaches the bran coat during milling, bran specks begin to appear, causing a discoloration of the white flour; hence, the validity of color measurement as an indication of quality (Kent-Jones and Brown 1966). Also as one approaches the bran coat, the innermost layer (the aleurone layer) is scraped away. Since the aleurone layer contains about 60% of the ash of wheat (Shetlar *et al.* 1947), the ash of the resulting flour rises very sharply. Thus, low grade flours coming from near the bran are high in ash and ash becomes a measure of flour quality.

Although it might be natural to assume that flours of the same degree of milling purity would have the same uses for baking, this is not necessarily true. Some wheat flours are better for bread and some are better for cakes. Intermediate stages lie between these two extremes. Most importantly, these flours differ in protein content (Atkins and Geddes 1939; Finney and Barmore 1948). In bread, one wants lots of tough elastic gluten since gluten is the structural component of the bread loaf. In cakes, this type of gluten is not wanted. Less gluten and gluten having a soft characteristic is the objective here.

Particle size is becoming of more importance as a quality measurement of wheat flours (Shellenberger *et al.* 1950). While the effect of particle size is not clearly understood, it is probably related to the starch gelatinization conditions found in the particular product in which the flour is to be used.

Bread Flour

Miller and Johnson (1954) list some 27 different tests for determining the quality of flour for bread baking. Only certain of these many tests will be described. The reader should remember that other tests are available and commonly used. Here, bread flour is specified in terms of six constants. These are moisture content, protein content, ash content, starch quality, protein quality, and particle size.

These constants are based upon conditions in the United States where highly mechanized bakeries are commonly found. Workers in other nations would take issue particularly with the minimum protein level since the wheats available to them are sometimes of lower protein and they have been able to compensate for this fact by gluten conditioning in milling or by formulating around this flour to produce an acceptable loaf of bread.

Moisture contents as high as 16% are quite common in Europe, particularly Northern Europe. This is because the combination of quick usage and cool climate minimizes the possibility of mold growth. But, where storage conditions are a problem, and this

TABLE 1

SPECIFICATIONS FOR BREAD FLOUR

Type of Measurement	Test Used	Values	Units of Measurement
Moisture	air oven	14.5 maximum	%
Protein	Kjeldahl	11.5 minimum	%
Ash	. . .	0.50 maximum	%
Starch quality	maltose	450 maximum	mg maltose/10 gm flour
Protein quality	Farinograph	7 C dimension, minimum	Brabender units
Particle size	Fisher	20 minimum	Fisher units

includes most of the United States, it is clearly established that 14.5% is the maximum moisture that can be tolerated (Anderson and Alcock 1954).

Cookie Flours

In Europe, biscuit flour means cookie flour whereas in North American, biscuit flour refers to a flour for a chemically leavened bread. This specification referring to cookies, therefore, also refers to European biscuits. A rather weak gluten is essential for good cookie production. Strong glutens prevent good cookie spread and hamper the molding of cookies to a specific shape.

TABLE 2

SPECIFICATIONS FOR COOKIE FLOURS

Type of Measurement	Test Used	Values	Units of Measurement
Moisture	air oven	14.5 maximum	%
Protein	Kjeldahl	9.5 maximum	%
Ash	. . .	0.44 maximum	%
Starch quality	maltose	200 maximum	mg maltose/10 gm flour
Protein quality	MacMichael	50 maximum	MacMichael units
Particle size	Fisher	12–18	Fisher units

Pastry Flours

Pastry flours (such as pie crust flours) are related more to cake flours than to cookie flour in the sense that the tendency is towards a lower protein. A weak gluten is wanted.

TABLE 3

SPECIFICATIONS FOR PASTRY FLOURS

Type of Measurement	Test Used	Values	Units of Measurement
Moisture	air oven	14.5 maximum	%
Protein	Kjeldahl	8.5 maximum	%
Ash	. . .	0.44 maximum	%
Starch quality	maltose	200 maximum	mg maltose/10 gm flour
Protein quality	MacMichael	40–50	MacMichael units
Particle size	Fisher	12–20	Fisher units

Cake Flours

Cake flours are characterized by fine particle size, low protein content and weak gluten. A special problem in producing cake flours is to get sufficiently fine flour without appreciable starch damage.

TABLE 4

SPECIFICATIONS FOR CAKE FLOURS

Type of Measurement	Test Used	Values	Units of Measurement
Moisture	air oven	14.5 maximum	%
Protein	Kjeldahl	8.5 maximum	%
Ash	. . .	0.36 maximum	%
Starch quality	maltose	150 maximum	mg maltose/10 gm flour
Protein quality	MacMichael	30–40	MacMichael units
Particle size	Fisher	12.5 maximum	Fisher units

Cracker Flours

Soda crackers are made from low protein flours having relatively strong gluten. Actually, somewhat different flour types are used for the cracker sponge and cracker dough. The flour specifications given here are a compromise but are suitable for either.

TABLE 5

SPECIFICATIONS FOR CRACKER FLOURS

Type of Measurement	Test Used	Values	Units of Measurement
Moisture	air oven	14.5 maximum	%
Protein	Kjeldahl	9.5 maximum	%
Ash	. . .	0.43 maximum	%
Starch quality	maltose	200 maximum	mg maltose/10 gm flour
Protein quality	MacMichael	65–85	MacMichael units
Particle size	Fisher	15–20	Fisher units

Biscuit Flours

Biscuit flours are used for making the baking powder leavened biscuit common to the United States. These flours are higher protein products having relatively strong gluten characteristics.

TABLE 6

SPECIFICATIONS FOR BISCUIT FLOURS

Type of Measurement	Test Used	Values	Units of Measurement
Moisture	air oven	14.5 maximum	%
Protein	Kjeldahl	8.0–10.5	%
Ash	. . .	0.40 maximum	%
Starch quality	maltose	250 maximum	mg maltose/10 gm flour
Protein quality	MacMichael	40–70	MacMichael units
Particle size	Fisher	12–18	Fisher units

Wheat Millfeed

The remaining products besides flour coming from a wheat flour mill are the bran, shorts, and germ. A detailed description of the specifications for millfeed is available (Anon. 1958). Table 7 gives the general specifications for these products.

TABLE 7

SPECIFICATIONS FOR WHEAT MILLFEED

Product	Moisture	Protein	Fat	Fiber
	%	%	%	%
Bran	15.0 maximum	14 minimum	4.0 minimum	12 maximum
Shorts	15.0 maximum	16 minimum	3.5 minimum	6 maximum
Germ	15.0 maximum	25 minimum	8.0 minimum	4 maximum

Bleaching and Maturing Agents

Chemical bleaching agents and chemical gluten maturing agents are widely used to improve the properties of wheat flour. While most chemicals appear to have both a bleaching and maturing action, the terms refer to quite different effects. The bleaching of flour suggests the oxidation of the yellow pigment of flour so that the finished flour is whiter than untreated flour. Maturing of flour refers to the oxidation of flour gluten so that its baking characteristics are improved in that the dough is drier, machines better, and bakes better.

Since both changes are oxidative in nature, most bleaching agents have some effect on gluten properties and maturing agents have some bleaching action. Chlorine compounds are widely used for these purposes as are peroxides. Bleaching is normally done by the flour miller while maturing agents are most often added to flour in the bake shop. Bleaching agents are usually applied in gaseous form. For an excellent discussion of bleaching and maturing agents, an article by Harrell (1952) is suggested.

SPECIFICATIONS FOR CORN PRODUCTS

Introduction

The corn kernel is different from other cereal seeds in that the endosperm opposite the tip of the kernel is quite soft while the other parts of the corn seed endosperm are hard. When ground, the soft portion easily breaks into flour (called soft flour) while the rest of the corn endosperm remains in chunks. These larger pieces of endosperm are classified according to size or ground into flour (called sharp flour).

The yield of endosperm products in corn milling varies between 65 and 70%. The residual is bran, germ, and shorts. These are combined to produce a product for animal feeding called hominy feed.

Before the germ is added to the hominy feed, however, the valuable corn oil in it is removed by expression and the use of filter presses (Striver 1955). Thus, the germ in corn millfeed is a partially defatted product.

Classified by size, the products of corn milling are these:

Grits		6	to	16	USBS Sieve
Meals					
Coarse		16	to	24	USBS Sieve
Medium		24	to	40	USBS Sieve
Cones		40	to	70	USBS Sieve
Flours					
Sharp		70	to	100	USBS Sieve
Soft		70 maximum USBS Sieve			

Soft Corn Flour

This flour comes mostly from the end opposite the tip of the kernel and lies between the flinty endosperm and the hull. It is separated from the meals and grits by sieving the early break products.

TABLE 8
SPECIFICATIONS FOR SOFT CORN FLOUR

Measurement	Test Used	Values	Units of Measurement
Moisture	air oven	14.5 maximum	%
Particle size	screens	70.0 maximum	USBS[1] sieves
Fat content	ether extract	2.0 maximum	%
Acidity	fatty acids	1.0 maximum	% as lactic acid
Appearance	visual	normal	. . .

[1] US Bur. Std.

Sharp Corn Flour

This flour is produced by grinding to the proper size the flinty endosperm particles that come from the early break rolls and bolting the product to take out the flour-like material. The specifications are similar to those of the soft corn flours, but the feel of the product is harsh rather than soft. This is caused by the sharp edges of the flour particles in contrast to the rounded edges of the soft flour particles. Hence, the name sharp flour.

TABLE 9
SPECIFICATIONS FOR SHARP CORN FLOUR

Type of Measurement	Test Used	Values	Units of Measurement
Moisture	air oven	14.5 maximum	%
Particle size	screens	70 to 100	USBS[1] sieves
Fat content	ether extract	1.25 maximum	%
Acidity	fatty acids	1.0 maximum	% as lactic acid
Appearance	visual	normal	. . .

[1] US Bur. Std.

Cones

This product is similar to sharp flour except that it is coarser. It feels much the same as sharp flour. Cones are often sold to the brewing industries and are used in other food industries where flour dust cannot be tolerated.

TABLE 10
SPECIFICATIONS FOR CONES

Type of Measurement	Test Used	Values	Units of Measurement
Moisture	air oven	14.5 maximum	%
Particle size	screens	40–70	USBS[1] sieves
Fat content	ether extract	1.25 maximum	%
Acidity	fatty acids	1.0 maximum	% as lactic acid
Appearance	visual	normal	. . .

[1] US Bur. Std.

Corn Meal

Corn meal is a product somewhat smaller than corn grits but still much coarser than a corn flour. Its chief use is for table consumption, and in this case, white corn products are most often sold. Corn meals are also used to make gelatinized corn flour. Such flours find their way into industrial as well as food uses.

TABLE 11

SPECIFICATIONS FOR CORN MEAL

Type of Measurement	Test Used	Values	Units of Measurement
Moisture	air oven	14.5 maximum	%
Particle size	screens	16 to 40	USBS[1] sieves
Fat content	ether extract	1.5 maximum	%
Appearance	visual	normal	. . .

[1] US Bur. Std.

Corn Grits

Grits of screen sizes in the area of 8 to 12 are made. These products are used by breweries as a fermentation substrate. Brewing conditions depend upon definite cooking times and temperatures and so the particle size of the grit is important and must be controlled. Thus, close attention is paid to particle size.

Visual inspection is important in these grit products since in such coarse materials it is hard to remove adventitious materials by sifting techniques. While formerly, only white corn was used in the production of corn grits, the brewers preference for this product has largely disappeared.

TABLE 12

SPECIFICATIONS FOR CORN GRITS

Type of Measurement	Test Used	Values	Units of Measurement
Moisture	air oven	14.5 maximum	%
Particle size	screens	4 to 16	USBS[1] sieves
Fat content	ether extract	1.0 maximum	%
Starch conversion	extract	8.0 minimum	. . .
Appearance	visual	less than 1% hull	

[1] US Bur. Std.

Pearl Hominy

This product is sometimes called number four grits because it will pass through a No. 4 screen but will remain on top of any finer screen. Its use is for the production of breakfast cereals. Normally, this product is made from white corn and not from yellow corn.

Visual inspection is of great importance. Freedom from foreign particles such as yellow corn, soybeans, and insect refuse is a necessity before further processing to breakfast cereals is possible. Many of these foreign materials are of the same size and concentrate with the number four grits during sizing. Thus, avoidance depends upon good milling practice and quality control by visual appearance.

TABLE 13

SPECIFICATIONS FOR PEARL HOMINY

Type of Measurement	Test Used	Values	Units of Measurement
Moisture	air oven	14.5 maximum	%
Particle size	screens	3½-4	USBS[1] sieves
Fat content	ether extract	0.75 maximum	%
Appearance	visual	less than 1% hulls	...

[1] US Bur. Std.

Hominy Feed

Corn millfeed is sold for animal feeding (hominy feed). Hominy feed is a combination of corn bran, corn shorts, and defatted corn germ.

TABLE 14

SPECIFICATIONS FOR HOMINY FEED

	Moisture	Protein	Ash	Fiber	Fat
Hominy feed	% 15.0 max.	% 10.0 min.	% 2.9 max.	% 6.0 max.	% 5.5 min.

SPECIFICATIONS FOR RYE PRODUCTS

Introduction

The protein content of rye flour is not of great importance even though rye flour is used for bread baking. Rye protein does not form a gluten in the sense that wheat protein does. For this reason, most rye breads contain a mixture of rye flour and wheat flour, the latter for structure and the former for taste and appearance. Thus, the important factors in the milling of rye are dress (granulation), color, and flavor rather than protein content.

Ergot contamination is a special problem in rye products. Ergot is quite poisonous and rye is a natural habitat for this fungus.

Therefore, ergot must be avoided and is commonly mentioned in the specification for buying rye grain.

When rye is milled, the end products are about 80% endosperm products and 20% millfeed. The endosperm product is commonly sold as a rye meal or as a rye flour.

White, Medium, and Dark Rye Flour

While it is difficult to generalize about rye flour specifications, an analogy to wheat flour milling is useful. White rye flour is a patent grade, medium rye flour is a straight grade, and dark rye flour is a clear grade. In contrast to wheat flour, the higher ash products are not necessarily less desirable. As ash rises, so does rye color and rye flavor.

The data in the following tables are typical for the United States.

TABLE 15

SPECIFICATIONS FOR WHITE RYE FLOUR

Type of Measurement	Test Used	Values	Units of Measurement
Moisture	air oven	14.5 maximum	%
Ash	muffle oven	0.58–0.78	%
Color	Pekar	white	comparison
Protein	Kjeldahl	7.0–9.1	%

TABLE 16

SPECIFICATIONS FOR MEDIUM RYE FLOUR

Type of Measurement	Test Used	Values	Units of Measurement
Moisture	air oven	14.5 maximum	%
Ash	muffle oven	1.11–1.39	%
Color	Pekar	medium white	comparison
Protein	Kjeldahl	10.1–12.8	%

TABLE 17

SPECIFICATIONS FOR DARK RYE FLOUR

Type of Measurement	Test Used	Values	Units of Measurement
Moisture	air oven	14.5 maximum	%
Ash	muffle oven	2.05–2.83	%
Color	Pekar	dark	comparison
Protein	Kjeldahl	13.7–16.2	%

Other Rye Products

Rye meal is sold for pumpernickel and other food uses. There are numerous varieties which differ from each other according to granulation. Examples of the size distribution are shown in two extremes, fine and coarse, in the next table.

TABLE 18
SIZE CLASSIFICATIONS FOR COARSE AND FINE RYE MEAL

USBS[1] Sieves	Pumpernickel Rye Meal	Fine Rye Meal
On 8	30	10
On 20	46	40
On 40	14	20
On 60	5	5
Thru 60	5	25

[1] US Bur. Std.

Rye millfeed does not contain bran. Rather, it is an impure middling product.

TABLE 19
SPECIFICATIONS FOR RYE MILLFEED

Product	Moisture	Fat	Fiber	Ash
Middlings	% 15.0 max	% 4.5 max	% 8.2 max	% 6.0 max

SPECIFICATIONS FOR DURUM PRODUCTS

Introduction

Durum, like rye and in contrast to wheat, is milled to color and size specifications and with little regard for protein considerations. The yield of endosperm in milling is 70–75%, the remainder being millfeed products.

Normally, a coarse product is desired. Hence, durum is tempered lightly to give coarse granulation during milling. The result is that the ash content of durum products is unusually high compared to wheat products of similar extraction.

The end products of durum milling are classified according to particle size. The coarsest middling products are called semolina, the next smaller material is called granular product, and the finest material is known as durum flour.

Semolina

Semolina quality is specified according to color, ash, and granulation. Flour is any product that passes through a 100-mesh screen, and flour should not exceed 3% by weight.

TABLE 20

SPECIFICATIONS FOR DURUM SEMOLINA

Type of Measurement	Test Used	Values	Unit of Measurement
Color	Pekar	yellow[1]	comparison
Ash	. . .	0.69 maximum	%
Granulation	USBS[2] sieves	on 30; 4	%
		on 50; 20 to 40	%
		on 70; 20 to 30	%
		on 100; 10 to 20	%
		through 100; 3 maximum	%

[1] Absence of bran specks is important.
[2] US Bur. Std.

Granular Product

There are many grades of granular product. The only restriction is that the product contain not more than 15% durum flour. Granulations vary. The ash generally runs higher than semolina, being about 0.72% maximum.

Durum Flour

Semolina and granular product go into macaroni and spaghetti goods. Durum flour is used for noodles. Color and granulation properties are most important. Ash specifications are secondary. Methods for measuring the color of durum flour have been described by Matz and Larsen (1954).

TABLE 21

SPECIFICATIONS FOR DURUM FLOUR

Type of Measurement	Test Used	Values	Units of Measurement
Color	Pekar	light yellow[1]	comparison
Ash	. . .	0.60 to 0.72	%
Granulation	USBS[2] sieves	on 100; 25 maximum	%
		through 100; 75 maximum	%

[1] Absence of bran specks is important.
[2] US Bur. Std.

Durum Millfeed

There is a striking similarity in the specifications for durum millfeed and wheat millfeed. This is but another indication of the close botanical relationship of wheat and durum. The recent report of a cross of durum and wheat to produce grain having good bread baking potential (Kaltsikes *et al.* 1968) is an important technological advance and heralds interesting new developments ahead.

TABLE 22

SPECIFICATIONS FOR DURUM MILLFEED

Product	Moisture	Protein	Fat	Fiber
	%	%	%	%
Bran	15.0 max	14.0 min	4.0 min	12.0 max
Shorts	15.0	16.0 min	3.5 min	8.0 max
Germ	15.0 max	25.0 min	8.0 min	4.0 max

BIBLIOGRAPHY

AGER, J. J. 1912. A modern Italian semolina mill. Am. Miller *15*, 130–134.

ANDERSON, J. A., and ALCOCK, A. W. 1954. Storage of Cereal Grains and Their Products. Am. Assoc. Cereal Chemists, St. Paul, Minnesota.

ANON. 1940. Development and use of baking powder and baking chemicals. U.S. Dept. Agr. Circ. *138*.

ANON. 1958. Flour and Feedstuffs-Laws and Regulations. Millers National Federation, Chicago, Ill.

ANON. 1962. Cereal Laboratory Methods. Am. Assoc. Cereal Chemists, St. Paul. Minnesota.

ATKINS, T. A., and GEDDES, W. F. 1939. The relationship between protein content and strength of flours in gluten enriched flours. Cereal Chem. *16*, 223–231.

BIECHY, T. 1957. Theby gluten washing machine. Mühle *23*, No. 13, 306–307.

BLACK, H. C., and BUSHUK, W. 1967. A laboratory flour agglomerator. Cereal Sci. Today *12*, 517–518.

BLANK, R. B. 1958. Basic differences between rye and wheat flour milling operations. Am. Miller *56*, No. 8, 18–23.

CLAUS, W. S., and BROOKS, E. M. 1965. Some physical, chemical and baking characteristics of instantized wheat flour. Cereal Sci. Today *10*, 41–43.

ELIAS, D. G. 1958. The protein displacement process. Am. Miller *86*, No. 8, 15–19.

ELIAS, D. G., and SCOTT, R. A. 1957. British flour milling technology. Cereal Sci. Today *7*, 180–184.

FINNEY, K. F., and BARMORE, M. A. 1948. Loaf volume and protein content of hard winter and spring wheats. Cereal Chem. *25*, 291–311.

GEHLE, H. 1937. Conditioning and grinding of corn. Mühle *74*, 361–362.

GEHLE, H. 1952. Wheat Conditioning. Miag, Braunschweig, Germany.

GELLRICH, W. H. 1958. Adaptation of air classification. Milling Prod. *4*, 12–13.

GRAHAM, J. C. 1965. The use of air classifiers in the flour milling industry. Northwestern Miller *273*, 25–32.

HARRELL, C. G. 1952. Maturing and bleaching agents used in producing flours. Ind. Eng. Chem. *44*, 75–100.

HESS, K. 1953. The protein and lipoid differentiation in flour and gluten. Getreide W. Mehl *3*, 81–85.

JONES, C. R. 1940. The production of mechanically damaged starch in milling as a governing factor in the diastatic activity of flour. Cereal Chem. *17*, 133–169.

KAISER, F. J., and FERGUSON, H. K. 1967. Mammoth corn mill comes of age. Am. Miller Proc., Aug., 20–23.

KALTSIKES, P. J., EVANS, L. E., and BUSHUK, W. 1968. Durum-type wheat with high bread making quality. Science *159*, 211–213.

KENT-JONES, D. W., and AMOS, A. J. 1957. Modern Cereal Chemistry. 5th Ed. Northern Publishing Co., Ltd., Liverpool, England.

KENT-JONES, D. W., and BROWN, T. R. 1966. Automatic and continuous monitoring of flour quality and production. Cereal Sci. Today *13*, 80–90.

KESSLER, G. T. 1954. Milling Phrases and Definitions. Miag, Braunschweig, Germany.

LARSEN, R. A. 1964. Hydration as a factor in bread flour quality. Cereal Chem. *41*, 181–188.

LOCKWOOD, J. F. 1952. Flour Milling, 3rd Edition. Northern Publishing Co., Liverpool, England.

MATZ, S. A., and LARSEN, R. A. 1954. Evaluating semolina color with photoelectric reflectometers. Cereal Chem. *31*, 73–86.

MAYER, L. 1936. Rye conditioning. Mühle *73*, 1330–1332.

MILLER, B. S., and JOHNSON, J. A. 1954. A review of methods for determining the quality of wheat and flour for bread making. Kansas State College Tech. Bull. *7*.

NATH, N., SINGH, S., and NATH, H.P. 1957. Studies on the changes of the soluble carbohydrates of chapaties during aging. Food Res. *22*, 25–31.

NEENAN, J. L. 1951. The degerminating corn mill. Am. Miller *79*, 44–45.

NOTTIN, P. 1945. Mixed milling of wheat and rye. Bull. Assoc. Miller *62*, 309–311.

OXLEY, T. A. 1948. The Scientific Principles of Grain Storage. Northern Publishing Co., Liverpool, England.

PAIGE, W. A. 1936. Durum semolina milling. Am. Miller *64*, No. 8, 48–50.

PILLSBURY MILLS. 1959. Improvements in or relating to dry process of fractionating milled cereal flour stocks. British Pat. 815,597.

PRATT, D. 1957. Chemical and baking changes which occur in bulk flour during short term storage. Cereal Sci. Today *2*, 191–195.

PROCHAZKA, F. 1938. Practical experiences in the conditioning of wheat. Mühle *75*, 599–602.

SANDSTEDT, R. M., and MATTERN, P. J. 1960. Damaged starch. Quantitative determination in flour. Cereal Chem. *37*, 379–390.

SCHOPMEYER, H. H. 1962. Rye and rye milling. Cereal Sci. Today *7*, 138–144.

SHELLENBERGER, J. A. 1965. Fifty years of milling advances. Cereal Sci. Today *10*, 260–262.

SHELLENBERGER, J. A., WICHSER, F. W., PENCE, R. O., and LAKAMP, R. C. 1950. Flour granulation studies. Kansas State College Tech. Bull. *4*.

SHETLAR, M. R., RANKIN, G. T., LYMAN, J. F., and FRANCIS, W. G. 1947. Investigation of the proximate chemical composition of the separate bran layers of wheat. Cereal Chem. *24*, 111–122.

STIMMEL, E. P. 1941. Dry corn milling. Am. Miller *69*, No. 10, 30–33.

STRIVER, T. E. 1955. American corn milling systems for de-germed products. Assoc. Operative Millers Bull.

SUGDEN, G. H. 1956. Various aspects of wheat conditioning. Cereal Sci. Today *1*, 136–142.

SULLIVAN, B. 1948. The Mechanism of the Oxidation and Reduction of Flour. Am. Assoc. Cereal Chemists. St. Paul, Minn.

THORPE, T. E. 1927. A Dictionary of Applied Chemistry, 2nd Edition. Northern Publishing Co., Liverpool, England.

WEBER, K. 1966. Trends in bulk flour storage design. Northwestern Miller *276*, 37–39.

WICHSER, F. W. 1958. Baking properties of air classified flour fractions. Cereal Sci. Today *3*, 123–126.

ZELENY, L. 1940. Fat acidity in relation to heating of corn in storage. Cereal Chem. *17*, 29–37.

ZWILINGENIEUR, G. F. 1956. Milling procedure of a combined wheat and rye mill. Muellerei *9*, 697–698.

CHAPTER 2

Samuel A. Matz | # Baking

INTRODUCTION

The baking of leavened cereal dough is a very ancient practice, its origin antedating the historical era. Refinements and improvements in the art came very slowly over the millenniums until the acceleration of scientific and technological advances which began in the nineteenth century started to affect baking as well as most other food processing techniques. Within the last 2 or 3 decades great strides have been made toward putting baking practices on a firm scientific footing. It is true, however, that some of the procedures currently used in the baking industry are based on traditional concepts and customs which are not technologically sound. Art still plays a role even in the most modern plant.

Bakery products are an important part of the diets of nearly everyone in the United States and of a majority of the world's inhabitants. In the United States, however, the amount of these products consumed per capita has been in a declining trend for many years. This slow but steady trend has sometimes been attributed to the increasing predominance of standardized bread and rolls made by highly mechanized bakeries, but a far more likely explanation is that it is a part of the overall decrease in consumption of cheap, starchy foods as greater numbers of people become financially able to obtain their calories from more expensive sources, such as the animal-derived foods. Notwithstanding the trend, the baking industry is, and will continue to be, a very important factor in our economy.

In this chapter, an attempt will be made to acquaint those readers who are not specialists in baking technology with some of the most important processing techniques, equipment, and special ingredients used in bakeries. Space limitations obviously make it impossible to cover in detail an industry and a technology which embrace thousands of different products extremely diverse in their characteristics and in the methods used to produce them. Several products representative of major classes have been selected and described in moderate detail. General procedures and equipment used for numerous commercially important products are also covered. Readers needing further information will find a more thorough survey in *Bakery Technology and Engineering* (Matz 1960) and *Baking Science and Technology* (Pyler 1952) or they can refer to the original references listed in the bibliography.

SPECIAL INGREDIENTS USED IN BAKING

Eggs, milk, salt, shortening, water and sugars are ingredients commonly used in bakery products. In some cases, there are special requirements for these ingredients when they are to be used in certain bakery products. For these specifications, the reader is advised to consult a more specialized text.

On the other hand, flour and leavening agents are ingredients which are responsible for the characteristic appearance, texture, and flavor of most bakery products and they will be discussed in some detail in the following sections. To these raw materials may be added some specially compounded mixtures, such as yeast foods and dough improvers, which do not have any applications outside the baking industry.

Flour

Wheat flour is unique among cereal flours in that, when mixed with water in the correct proportions, its protein component will form an elastic network which is capable of holding gas and which will set to a rather firm spongy structure when heated in the oven. It is this behavior which makes possible the production of bread as we know it.

In order to secure batter and doughs which will handle satisfactorily and yield finished products of good eating quality, the proper flour must be used. Suitability of a flour for a particular purpose is governed primarily by the variety of wheat from which it was milled, the protein content of the flour, and the milling conditions.

For the purposes of this discussion, wheats may be divided into soft and hard, with the latter group subdivided into winter and spring wheats. Hard wheats are indispensable for the production of flour intended for use in bakery foods of low density, such as ordinary white bread. Flours made from hard wheats can yield doughs which are elastic, have excellent gas-holding properties, and respond well to the usual bake-shop processing techniques. Spring wheats often yield flours which are somewhat "stronger" than those obtained from winter wheats.

Flours from soft wheats are used for cakes, cookies, pie crusts, and other products when a high specific volume is not essential and a tender texture is desired. They are also usually whiter in color and blander in flavor than flours from winter wheats.

Purpose of Milling.—The purpose of milling is to separate the wheat endosperm (the source of flour) from the germ and the bran layers, and to reduce the endosperm chunks to a fine powder. In conventional milling processes this is accomplished by passing the wheat kernels and their products through a series of pairs of rolls and sieves. The initial sets of rolls break open the seed coat and strip out the friable endosperm. Subsequent sets of rolls grind the products finer, or perform other essential functions such as flattening the germ (which facilitates its separation from the endosperm). At each stage, stacks of screens each having a different mesh size separate the ground material into several streams. These streams may be further processed by grinding, they may be drawn off for feed, or they may be combined to yield flours of different qualities.

By selecting the proper streams, the miller can furnish flours of widely varying quality from the same wheat blend. Of course, the inherent quality of the wheat places an upper limit on the excellence of flour which can be obtained from it. Ordinarily, about 70% of the wheat kernel emerges as flour of some sort. Usually the first endosperm fractions obtained from the wheat kernel, constituting perhaps 50 to 90% of the total flours, are combined to make a so-called "patent" flour. The remaining flour would then be called a "clear" flour, and is considered to be of inferior quality because it is of darker color and usually exhibits poorer baking performance. If all of the flour streams are combined to make only one product, the latter is called a "straight" flour. Figure 8 demonstrates some of these relationships.

The complex nature of flour and the many kinds of products in which it is used make it impossible to prescribe a simple chemical or physical test for overall quality. The most successful tests have been based upon the comparison of baked products made from the experimental flour and a control sample. In most cases, the conditions presumed to exist during commercial processing are duplicated in the test, so far as feasible. Nearly all specifications for flours intended to be used in bakery products include some test of this type. Many such tests are based on the "standard" bread baking and cake baking tests of the American Association of Cereal Chemists (Anon. 1962). Test baking procedures

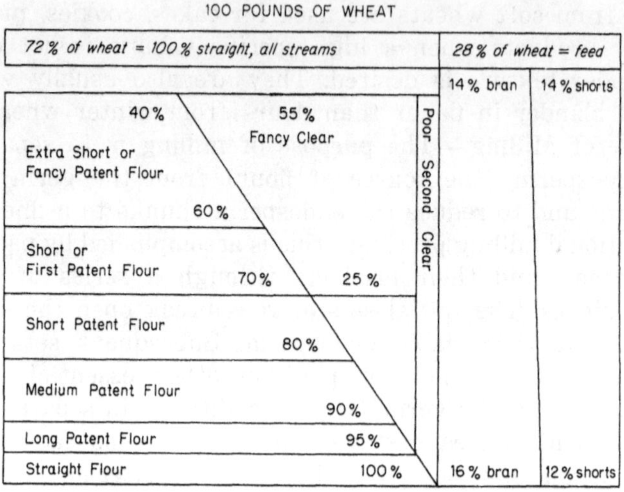

100 POUNDS OF WHEAT

72 % of wheat = 100 % straight, all streams		28 % of wheat = feed	
		14 % bran	14 % shorts

```
40 %                         55 %              Poor
                         Fancy Clear           Second
Extra Short or                                 Clear
Fancy Patent Flour
                    60 %

Short or
First Patent Flour       70 %      25 %

Short Patent Flour
                               80 %

Medium Patent Flour
                          90 %

Long Patent Flour            95 %

Straight Flour              100 %     16 % bran    12 % shorts
```

Diagram by Dr. C. O. Swanson;
Courtesy of Burgess Publishing Co.

FIG. 8. RELATIONSHIP OF FLOUR GRADES TO EXTRACTION
PERCENTAGES

for cookies, crackers, biscuits, and several other products have also been published.

Of the chemical tests for flour quality, protein, ash, and moisture determinations are in widest use. The moisture content has the obvious value of indicating the amount of an inactive substance—or diluent—which is present, and, in addition, flour containing more than about 14.5% moisture does not store well. The protein content (actually a Kjeldahl determination of total nitrogen is the most common test) is closely related to the baking performance of the flour. High protein content is desirable for bread flours while cakes, cookies, and pie crusts require flours of relatively low protein content.

The ash content is an indirect indication of the protein quality of the flour. Patent flours have lower ash contents than straight or clear flours from the same wheat. Much of the higher ash content of the latter two flours results from the inclusion of a larger amount of the bran and adjacent layers which also contribute higher percentages of nongluten protein. Higher ash content is almost always associated with darker color in the flour and this effect carries over into the baked product. Fat, fiber, starch, reducing sugars, and amylase activity are other chemical determinations sometimes applied to flours.

Many empirical physical tests have been developed for estimating flour quality. Several of these measure the resistance of doughs to some type of mixing action. The Brabender farinograph is probably the best known of this type. Other devices determine the resistance of dough pieces to stretching or tearing, etc. It is often difficult to establish the relationship of values obtained with these machines to the baking quality of a flour.

The color of flour is an important quality factor and may be measured subjectively in the Pekar test, which requires the comparison of smooth wetted surfaces of the unknown and control flours, or it may be determined electronically by a reflectance colorimeter. The importance of particle size in flour behavior has only recently been recognized. Generally, the fragments of flour are too small to be accurately classified by sieves, and more accurate methods must be used. Microscopic measurements are precise but very tedious. Techniques involving determination of sedimentation rates and resistance to permeation by streams of gases have been developed and are described in the literature.

When the rate of viscosity increase in heated aqueous suspensions of flour is followed, valuable information about the gelatinization characteristics of the starch can be derived. These data can in turn be used to predict some aspects of the performance of the flour in certain baked products. Consequently, some specifications for cake, pie and pastry flour include requirements for minimum or maximum viscosity development under standardized conditions. By including a time variable, the rate of action of the starch digesting enzymes can be estimated. This is a factor of importance in the fermentation of bread doughs.

Leavening Agents

Yeast.—Bakers' yeast is composed of the living cells of *Saccharomyces cerevisiae*. There may be a small amount of diluent included in some brands to improve their dispersibility in water, but most of the weight of a fresh block of yeast is composed of active cells.

Bakers' yeast may be purchased in the compressed form as cakes of 1 oz, 2 oz, 1 lb, and 5 lb or as active dry yeast in the form of small granules. Compressed yeast contains about 70% water and must be stored at refrigerator temperatures while active dry yeast contains approximately 8% water and can be kept at higher temperatures if necessary. Compressed yeast can be used simply by mixing it into a dough along with the other ingredients but active dry yeast must first be rehydrated. The temperature of the rehydration water is very critical and the yeast will be completely inactivated if the water is too hot or too cold. Automatic devices for rehydrating yeast under optimum conditions have been made available by some manufacturers of the product.

Yeast performs its leavening function by fermenting carbohydrates such as glucose, fructose, maltose, and sucrose. It can not use lactose, the predominant carbohydrate of milk. The principal products of fermentation are ethanol and carbon dioxide, the latter being the leavening gas. Ethanol is important in the aroma of baking bread, a fact which is not always recognized. There are various other products of the yeast's activity which flavor the baked product and change the dough's physical properties.

Quality tests for yeast include baking techniques similar in principle to those used for estimating flour quality, and gas production measurements. Some rather complex instruments have been devised for determining the amount of carbon dioxide given off by yeast which has been mixed in a dough of standard formulation. In this country, simple Blish-Standstedt pressuremeters (shown in Fig. 9) are widely used. They consist of a cylindrical brass bowl coupled to a removable head which bears a pressure-relief valve and an aneroid gage or mercury manometer type of pressure sensing device. Such pressuremeters are maintained in constant temperature baths during the measuring period.

Chemical tests are not of value in determining the fermentation capabilities of bakers' yeast.

Photo by Dr. H. J. Peppler

FIG. 9. BLISH-SANDSTEDT PRESSUREMETERS
IN WATER BATH

Chemical Leavening Agents.—Layer cakes, cookies, biscuits and many other bakery products are leavened by carbon dioxide originating from added sodium bicarbonate (baking soda). When soda is added alone it tends to make the dough alkaline, leading to flavor deterioration and discoloration, and the carbon dioxide is released very slowly. Addition of an acid along with the soda promotes a vigorous evolution of gas and keeps the dough pH near neutrality.

The rate of gas release from solution controls the size of the bubbles in the dough and consequently influences the grain, volume, and texture of the finished product. Much research has been devoted to developing leavening acids which will maintain the rate of gas release within the desired range. Acids such as acetic (as from vinegar) or lactic (as from sour milk) usually act much too fast. The most successful compounds have been cream of tartar (potassium acid tartrate), sodium aluminum sulfate (alum), sodium acid pyrophosphate, and various forms of calcium phosphate. The chemical formulas of these compounds do not adequately define their function in doughs. Small amounts of additives included during manufacture can have a profound effect on the rate of reaction of the compound. Granule size and form also have important modifying actions. For example, companies specializing in leavening acids usually offer several types of sodium acid pyrophosphate. Although the chemical formulas of the main constituent are the same and analysis will reveal only variations in the trace elements, the slowest reacting member of the sodium acid pyrophosphate series will cause an initial rate of gas evolution many times slower than the rate promoted by the fastest reacting member.

Most bakers use baking powder instead of measuring soda and leavening acids into the dough in separate additions. This substance is a mixture of soda and acid in correct proportions plus diluents which simplify measuring and improve stability. The end products of baking powder reaction are carbon dioxide and harmless salts having more or less bland flavors. Baking powders from all reliable firms have about the same amount of "available carbon dioxide" and differ, if at all, in the speed of reaction. Most commercial baking powders are of the double-action type, i.e., they give off a small amount of the available carbon dioxide during the mixing and make-up stages and then remain relatively quiescent until the temperature begins to rise after the batter is placed in the oven. This type of action insures that excessive loss of leavening gas will not occur if the baker finds it necessary to leave the batter in an unbaked condition for long periods.

Quality testing of baking powders involves a bake test made under actual conditions of use. It is difficult to predict commercial performance from the results of standardized laboratory tests because one of the attributes of a good baking powder is its lack of sensitivity to changes in conditions. There is a common laboratory technique for determining available carbon dioxide which involves measurement of the volume of gas released from baking powder which has been treated with a strong acid. This test gives no indication of the speed of reaction of the baking powder when it is mixed into a dough or batter.

Special Minor Ingredients

There are a few kinds of specialized compositions which can be added to a dough in small amounts in order to modify its properties. These include enzymes, yeast foods, and dough improvers. Enzymes which have been recommended for addition to doughs are proteases, which act on the gluten to alter dough extensibility, amylases, which digest some of the starch so as to supply fermentable carbohydrates to the yeast, and lipoxidases, which whiten the dough by destroying flour pigments and perhaps affect the flavor.

Yeast foods are usually composed of ammonium salts, phosphates, and sulfates. These yeast nutrients presumably encourage yeast growth and may indirectly accelerate gas production. If the water supply is unusually soft, the salts in yeast foods may have a direct beneficial effect on the physical properties of the dough.

Oxidizing agents can be very effective in improving the handling characteristics of dough and the specific volume and texture (indirectly) of the finished products. The most common oxidizer is potassium bromate, but potassium iodate and calcium peroxide are also used to some extent. These substances exert their effect on the mechanical properties of dough by causing the formation of additional cross-bonds between gluten molecules.

Not all flours benefit from the additional of oxidizers. Well-aged flour, or flour which has been treated with optimum quantities of oxidizers at the mill, may not show any beneficial results from the addition of dough improvers and may, in fact, give doughs and bread of inferior quality when supplemented in this manner. Flours which react best to supplementation are likely to be long extraction, freshly milled flours from wheats of generally good baking quality. Short extraction flours from poor varieties of wheat may not be improved at all by the addition of oxidizers.

THE SPECIAL EQUIPMENT OF BAKING

Much of the equipment used in preparing bakery foods is highly specialized, often having been developed especially for some particular task which was formerly performed by hand. This is true partly because some of the processing steps in bakery work are not closely analogous to any other operation in the food field, but in other cases (e.g., mixers), special equipment is necessary because the materials being handled possess unusual characteristics and respond poorly to the machines used for less sensitive foods.

Materials Handling Equipment

Many bakeries, including those of moderate size, have found that bulk handling of flour is more economical, convenient, and sanitary than the century-old steps of bag transfer and dumping. Most large flour mills are now equipped to ship flour in bulk trailers or freight cars. The bakery must install a number of receiving bins and a pneumatic conveying system. The conveying system may be of the positive pressure or negative pressure type. The former is easier to engineer and is probably less expensive but it may be more hazardous since a break in the conveyor will result in ejection of large amounts of explosive flour dust into the atmosphere.

For batch mixing, the flour charge is usually weighed in a scaling mechanism located immediately above the mixer bowl. A "loss-in-weight" procedure based on measurements at the bin is also possible. It is advisable, but not essential, to have some sort of sifting device in the delivery end of the system.

Bulk shortening installations are fairly common. These are based, of course, on storing and conveying fat in the melted state. For some bakery products, a plastic fat must be added to the mixer in order to obtain the desired texture and appearance in the finished product. This can be accomplished by running the liquid fat through a chilling unit, such as a Votator with a refrigerated jacket, and extruding the solidified fat directly into the mixer bowl. Many excellent liquid metering devices are available for measuring liquid shortening.

Bulk sugar is used by some bakeries. Although granular sugar can be handled by pneumatic conveying, it is more common to use and store sugar syrups. Cane sugar syrups, some special syrups and many different kinds of corn syrups can be delivered in bulk. Generally, the economics of these systems are favorable only for the larger plants. Many preparation techniques can be adapted to allow the use of liquid sugars although problems may be encountered if there is a critical "creaming" step in the process.

Mixers

In many food manufacturing processes, mixing is a step conducted merely to assure a uniform distribution of the separate ingredients throughout the mass of the finished batch. This function is also important in the mixing of doughs and batters, but here two other purposes are evident. These are the incorporation of minute bubbles of air to serve as foci for the evolution of carbon dioxide, and the "development" of the gluten. The latter function is critical in the preparation of bread doughs and the like but is of no significance in mixing layer cake batters, pie doughs, etc.

Nearly all bread doughs (batch process) made in this country are mixed in equipment similar to that shown in Fig. 10. These horizontal dough mixers are very effective in developing the gluten because the essentially unidirectional action of the agitator bars rapidly orients and aligns the protein fibrils. The agitator is composed of 3 cylindrical members, 2, 3, or 4 in. in diameter, disposed parallel to the long axis of the bowl. The mixer arms are mounted on two spiders connected to a powered shaft which passes axially through the mixer. In some cases, a swing bar replaces one or more of the fixed arms. Other less common designs are also available for the agitators.

The mixer bowl may have a smooth interior or it can be provided with baffles or stationary bars mounted parallel to the agitator shaft. Since a great deal of heat is generated during high speed mixing, the bowls of horizontal mixers are jacketed so that cooling can be accomplished by circulation of ethylene glycol or by direct expansion of a refrigerant gas.

Two mixing speeds are usually available. High speed (about 70 rpm) is used for dough development while low speed (about 35 rpm) is used primarily for the initial incorporation of water. To open the mixing chamber, the bowl is rotated through about 90° by the power unit. Horizontal dough mixers are available in capacities of about 1,000 lb to about 2,000 lb from 7 different manu-

Courtesy of Baker Perkins, Inc.

FIG. 10. HORIZONTAL DOUGH MIXERS SHOWING BULK FLOUR WEIGH-
ING SYSTEM AND A DOUGH TROUGH BEING ELEVATED

facturers in the United States. Laboratory size units or larger mixers can be obtained on special order.

Batch mixing of batters is accomplished with equipment such as that shown in Fig. 11. These vertical planetary mixers have bowl capacities of up to 340 qt. Various agitator designs are available. For example, wire whips are used for egg white foams, batter beaters for the usual layer cake mixes, and pastry cutters for pie doughs. Provision is usually made for three different beater speeds.

There are specialized mixers which can be mentioned only briefly here. The Artofex mixer, of European design, combines two agitators traveling through intersecting elliptical paths and a constant revolving bowl to give an action resembling that of hand kneading. These devices have attained some popularity with pie manufacturers and are also used for some specialty breads. The Oakes continuous mixer can give high aeration of certain types of cake batters. The Morton "Whisk" permits batters to be mixed

Courtesy of American Machine and Foundry Co.

FIG. 11. BATCH MIXER FOR CAKE BATTERS SHOWN HERE HAS
A BOWL CAPACITY OF 340 QT OF BATTER

in a pressurized chamber. This is very beneficial to the volume
of air leavened products such as angel food cakes.

Dough Forming Equipment

Dough Dividers.—At some stage in the processing, and frequently immediately after mixing, it is necessary to divide a large
mass of dough into pieces corresponding to single units of the
finished product. The machines which perform this function
operate on a volume displacement system rather than on a weight
basis. The dough is rammed into a chamber having adjustable
volume. and the measured piece is then severed from the main
body.

Figure 12 illustrates the principal operating elements of a

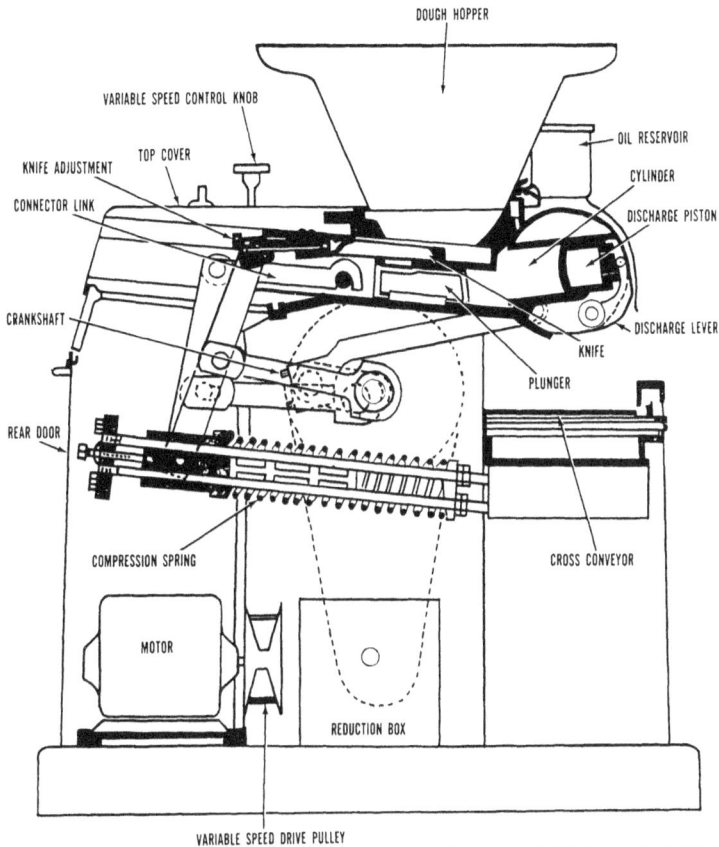

DOUGH HOPPER

VARIABLE SPEED CONTROL KNOB

OIL RESERVOIR

KNIFE ADJUSTMENT   TOP COVER

CYLINDER

CONNECTOR LINK

DISCHARGE PISTON

CRANKSHAFT

DISCHARGE LEVER

KNIFE

PLUNGER

REAR DOOR

COMPRESSION SPRING

CROSS CONVEYOR

MOTOR

REDUCTION BOX

VARIABLE SPEED DRIVE PULLEY

Courtesy of US Dept. of Defense

FIG. 12. DIAGRAM OF A DOUGH DIVIDER

typical dough divider. The dough flows from the hopper into the underlying compression chamber. At the start of the cycle, a knife moves horizontally to cut off a piece of dough near the hopper bottom. Next, the ram or piston moves forward pressing the severed dough piece into a chamber contained in a rotatable cylinder. At the end of the ram stroke, the cylinder turns, cutting off the excess dough, and, finally, the discharge lever ejects the measured dough piece. In the return cycle, the emptied cylinder is turned back so that the cavities face the compression chamber, and the knife and compression piston withdraw, allowing more dough to be drawn into the chamber by gravity and suction.

Commercial models (see Fig. 13) have from 2 to 8 pockets in the cylinder and operate at speeds up to 25 strokes per minute.

Scaling range is from 6 to 36 oz. Motors with ratings of up to 7.5 hp are used. The volume of the pockets which scale the dough pieces can be adjusted by changes of the piston depth.

Anything which affects the dough density will change the weight of the scaled pieces. Since the dough continues to ferment in the hopper, with the gas production contributing to a lower density, a slight decrease in scale weight can be expected as each

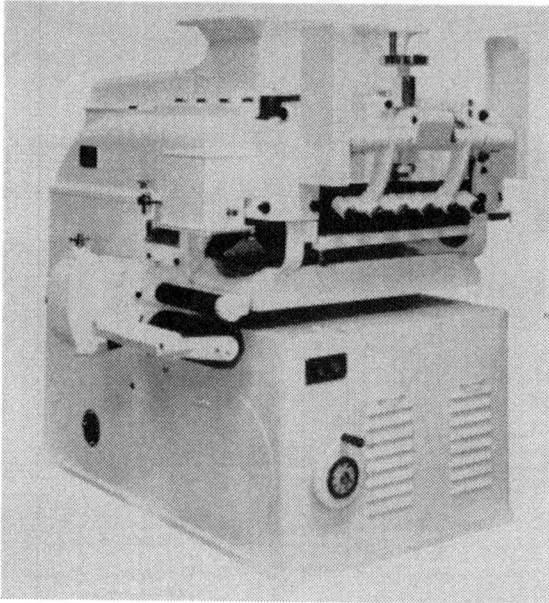

FIG. 13. SIX-POCKET DOUGH DIVIDER WITH VARI-
ABLE SPEED DRIVE

batch of dough is processed, with a rise in weight as a new batch starts to flow into the compression chamber. If the batches are small enough and divider operation is rapid enough, the changes in piece weight will probably be within limits that can be tolerated. If the divider has to be shut down for an appreciable length of time while there is dough in the hopper, a considerable error in scaling weight must be expected.

Rounders.—When the dough piece leaves the divider, it is irregular in shape with sticky cut surfaces from which the gas can readily diffuse. The gluten structure is disoriented and so not in a condition suitable for molding. It is the function of the rounder to close these cut surfaces, giving the dough piece a smooth and dry exterior, to make a relatively thick and continuous skin around the dough piece, to re-orient the gluten fibrils, and to form the dough into a ball for easier handling in the subsequent processing steps. It performs these functions by rolling the well-floured dough piece around the surface of a drum or cone while moving it upward or downward along this surface by means of a spiral track. As a result of this action, the surface is dried by an even distribution of dusting flour as well as by the dehydration occurring because of its exposure to the air, the gas cells near the surface of the ball are collapsed forming a thick layer which inhibits the diffusion of gases from the dough, and the dough piece assumes an approximately spherical shape.

Rounders may be conveniently classified as bowl-, drum-, or umbrella-type. The conical or bowl variety consists of a rotatable cone-shaped bowl around the interior of which is placed a stationary spiral track or "race." Figure 14 schematically illustrates such

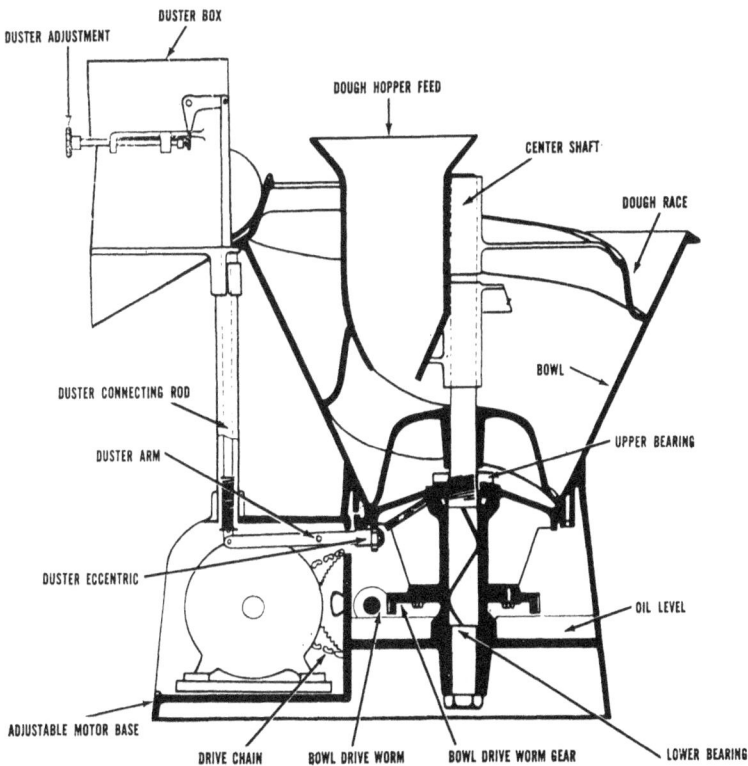

Courtesy of US Dept. of Defense

FIG. 14. SCHEMATIC DIAGRAM OF A BOWL-TYPE ROUNDER

a device. From the conveyor leading from the divider the dough pieces fall into the feed hopper of the rounder and then drop to the bottom of the rotating bowl. The pieces are tumbled and rolled along the dough race until they emerge from the top of the bowl and fall onto the belt leading to the intermediate proofer.

A second popular type of rounder is the so-called umbrella or inverted cone variety. These machines differ from the preceding type in that the dough piece is carried along the outside surface of a cone which has its apex facing upward. Figure 15 shows such a rounder.

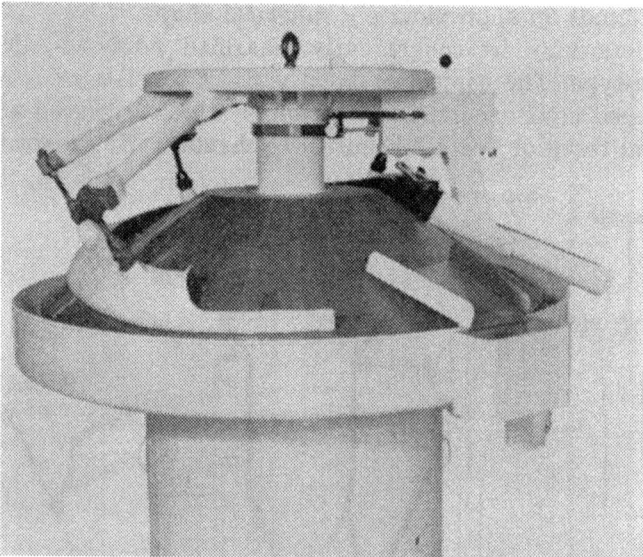

Courtesy of Baker Perkins, Inc.

FIG. 15. UMBRELLA-TYPE LOAF ROUNDER

The third type of rounder is the drum rounder. This machine differs from the bowl and umbrella styles in that the cone segment has very little slope to its sides, i.e., the sides are almost vertical. The dough piece enters near the bottom of the drum and rolls upward.

In addition to their form, rounding machines may vary in the texture or composition of the rotating surface, in the means provided for adjusting the relationship of the dough race to the drum or cone, in the method of applying dusting flour, etc. The rotating surface is usually corrugated vertically or horizontally, but the design and the size of the ribs vary considerably from one manufacturer to another. The surface may be waxed or it may be coated with a plastic such as Teflon to reduce sticking. Frequently, a device to shunt aside oversize dough pieces (doubles) is fixed to the exit chute.

Fermentation Rooms and Proofing Equipment

In all conventional processing methods for yeast-leavened products, there are one or more rest stages during which the dough is allowed to ferment so as to accumulate carbon dioxide and flavoring constituents. In the usual bread-making procedure there are 3 or 4 of these steps: the fermentation, which occurs shortly after the dough leaves the mixer; the intermediate proof which takes place after the dough has been divided and rounded; and the pan proof during which the shaped and panned dough pieces are allowed to accumulate gas just prior to being placed in the oven. In most plants, the fermentation stage is divided into two sections separated by a remix step.

It is highly desirable that the doughs be held under conditions of controlled temperature and humidity during these steps, since the rate (and, to some extent, the quality) of the fermentation is strongly dependent upon the temperature, and the handling properties of the dough surface depend upon the extent to which drying occurs. Therefore, all modern bakeries have fermentation rooms and intermediate proofing cabinets with controlled atmospheres for storing the dough during the rest periods.

A fermentation room is essentially a well-insulated box provided with means for humidifying and warming the inclosed air. Figure 16 indicates construction principles of a fermentation room.

When fermenting dough is transferred in bulk, dough troughs on casters are used as the containers. When shaped pieces are proofed, they are placed on trays, the trays are placed on racks, and the racks are wheeled into the proof room. In some plants, racks or troughs may be moved in and through the holding room by a powered arrangement of one sort or another. Air conditioning systems for fermentation rooms should be capable of maintaining a dry bulk temperature of 80°F $\pm$ 1°, and 76% RH $\pm$ 1%, regardless of the temperature and relative humidity of the surrounding atmosphere.

In the manufacture of bread, there is a specialized piece of machinery called the intermediate proofer which receives the dough pieces as they come from the rounder. Gas is generated as

allow for standard
trough — 30" width

allow for jumbo
trough — 36" width

Overall
trough length

Overall trough
length plus 12"

Overall
trough length

Floor or
wall bumpers

Sound, heavy, floor base
to anchor door frame

6"

6"

Door frame, two 2"x 6"

6'

5'-10"

6'

Floor or roof above

8'
min.

Wall bumper, or
2"x 2"x ¼" angle
floor bumper

7'-3"

Flooring
Insulation
Ceiling

SECOND FLOOR

GROUND FLOOR

Earth

Concrete
2" Insulation
Concrete

Vapor seal inside of wall
with asphalt paint

Fill with rock wool

rock
wool

Hardboard,
plywood, or
tile

2"x 6"

INSIDE WALL DETAIL

OUTSIDE WALL DETAIL

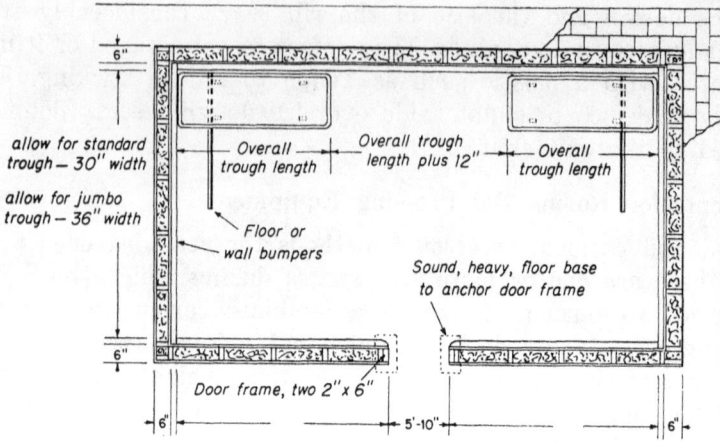

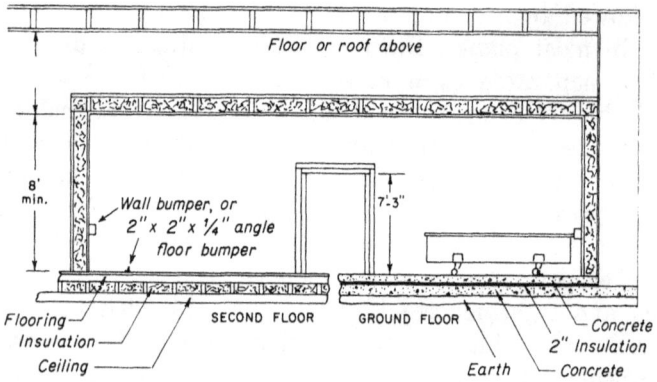

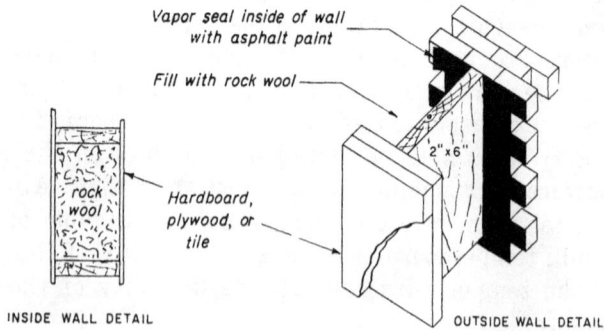

Courtesy of Fred D. Pfening Co.

FIG. 16. CONSTRUCTION DETAILS OF A FERMENTATION ROOM

the dough pieces are carried through the intermediate proofer and the dough "relaxes" so that it is in better condition for molding. No large capacity (over 3,000 lb/hr) intermediate proofers are available with air conditioning equipment built-in. The humidity and temperature of the inclosed space are controlled to some extent by the dough pieces themselves, and the dough is not very sensitive to changes in these factors during this stage in its processing.

Most intermediate proofers used at the present time are the overhead type in which the principal part of the cabinet is raised several feet above the floor. When overhead space is not adequate, one of the floor-level models can be installed. Intermediate proofers may be conveniently classified as belt-types and tray-types, the latter variety having many subtypes. Belt proofers consist essentially of endless belts running in a closed cabinet. The dough pieces are carried forward to the end of the cabinet, then dropped down on the next lower belt traveling in the opposite direction, and so on until they reach the exit conveyor.

Tray-type conveyors include those proofers which have segmented conveyors of any design. The dough pieces may be moved in metal pans, troughs or buckets, wooden trays or canvas loops. Figure 17 illustrates schematically an overhead tray-type proofer which has perforated metal troughs or tubes in which the dough pieces are carried. This diagram also shows details of the loading mechanism, a vital part of the intermediate proofer. The loading mechanism receives the dough pieces which come along the conveyor belt in single file and arranges them in rows in the receptacle of the intermediate proofer. The proofer trays may contain spaces for 2 to 8 dough pieces placed across the tray. Figure 18 shows an intermediate proofer with related equipment.

Dough Molding and Shaping Equipment.—In the bread-making plant, the molder receives pieces of dough from the intermediate proofer and shapes them into cylinders (loaves) ready to be placed in the pans. There are several types of molders, but all have four functions in common: sheeting, curling, rolling, and sealing. Some writers consider the last two as one function, since they are performed simultaneously. Figure 19 is a schematic diagram of a simple type of molder in which these operations are illustrated.

The dough as it comes from the intermediate proofer is a flattened spheroid. The first operation of the molder is to flatten this spheroid still more to form a thick sheet which can be

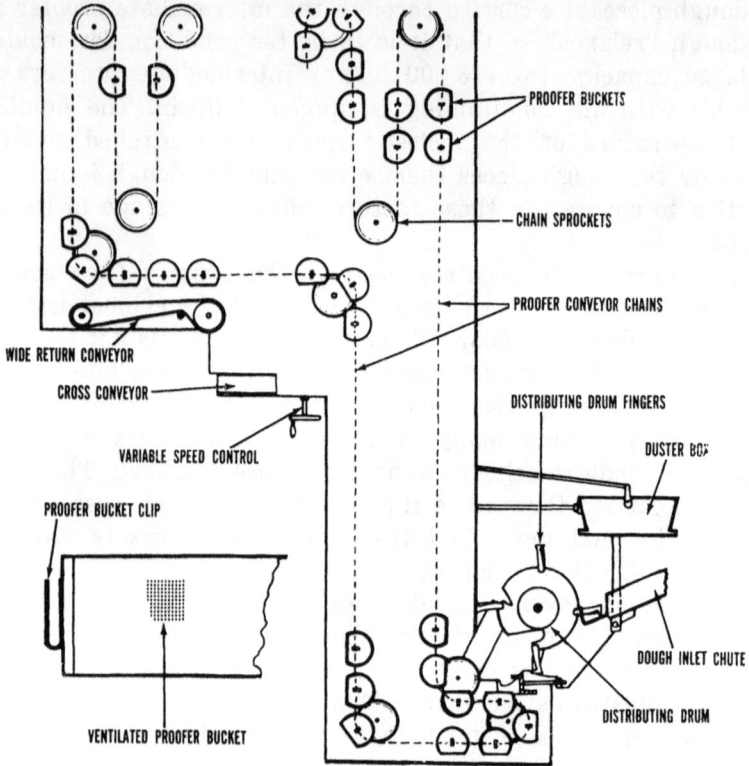

PROOFER BUCKETS

CHAIN SPROCKETS

PROOFER CONVEYOR CHAINS

WIDE RETURN CONVEYOR

CROSS CONVEYOR

DISTRIBUTING DRUM FINGERS

DUSTER BOX

VARIABLE SPEED CONTROL

PROOFER BUCKET CLIP

DOUGH INLET CHUTE

DISTRIBUTING DRUM

VENTILATED PROOFER BUCKET

Courtesy of US Dept. of Defense

FIG. 17. DIAGRAM OF A TRAY-TYPE INTERMEDIATE PROOFER SHOWING LOADING MECHANISM

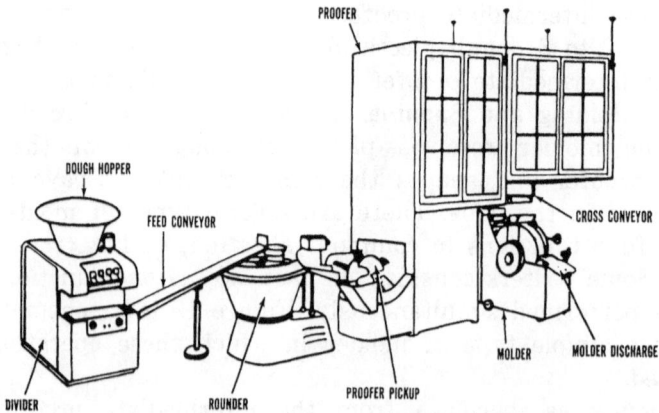

PROOFER

DOUGH HOPPER

FEED CONVEYOR

CROSS CONVEYOR

MOLDER

MOLDER DISCHARGE

DIVIDER

ROUNDER

PROOFER PICKUP

Courtesy of US Dept. of Defense

FIG. 18. MAKE-UP EQUIPMENT, SHOWING POSITION OF INTERMEDIATE PROOFER

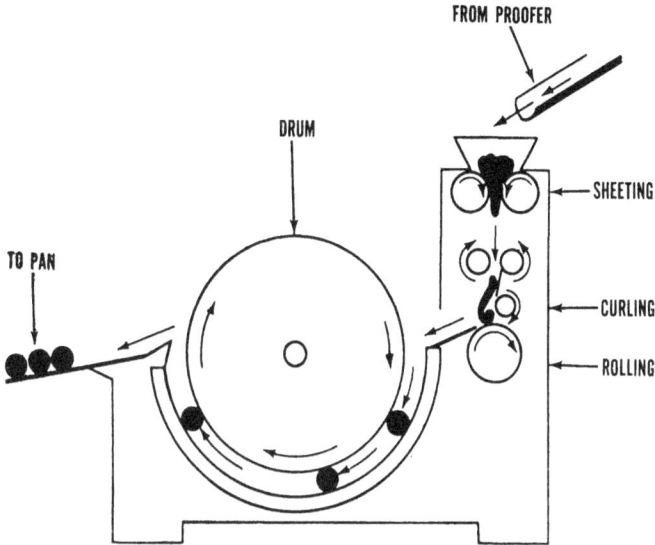

Courtesy of US Dept. of Defense

FIG. 19. DIAGRAM OF A SIMPLE DRUM MOLDER

properly manipulated in the later stages of molder operation. This effect is usually achieved by 2 or more (usually 3) consecutive pairs of rollers, each succeeding pair being set closer together than the ones which preceded it. The first pair of rolls, called the head rolls, exerts only a relatively slight pressure on the dough piece. The second set of rolls, the center rolls, operates at an intermediate pressure. The last set, which may be called either the sheeting rolls or the lower rolls exerts the maximum pressure on the dough sheet. The gradual reduction in thickness caused by this multiple roller system minimizes the punishment received by the dough, so that tearing and similar problems are reduced.

After the dough has been sheeted out, it is curled up into a loose cylinder. This operation is conventionally performed by a special set of rolls, but it may also be done by a pair of canvas belts. The lower conveyor moves the dough piece forward until the upper curling belt or mat engages the front end of the piece, brings it back, and curls it up into a loose cylinder. A more advanced development substitutes a short length of woven metal mat or linked metal bars for the upper curling belt.

The layers in the cylinder of dough are not tightly adherent when it leaves the curling section. The next function of the molder

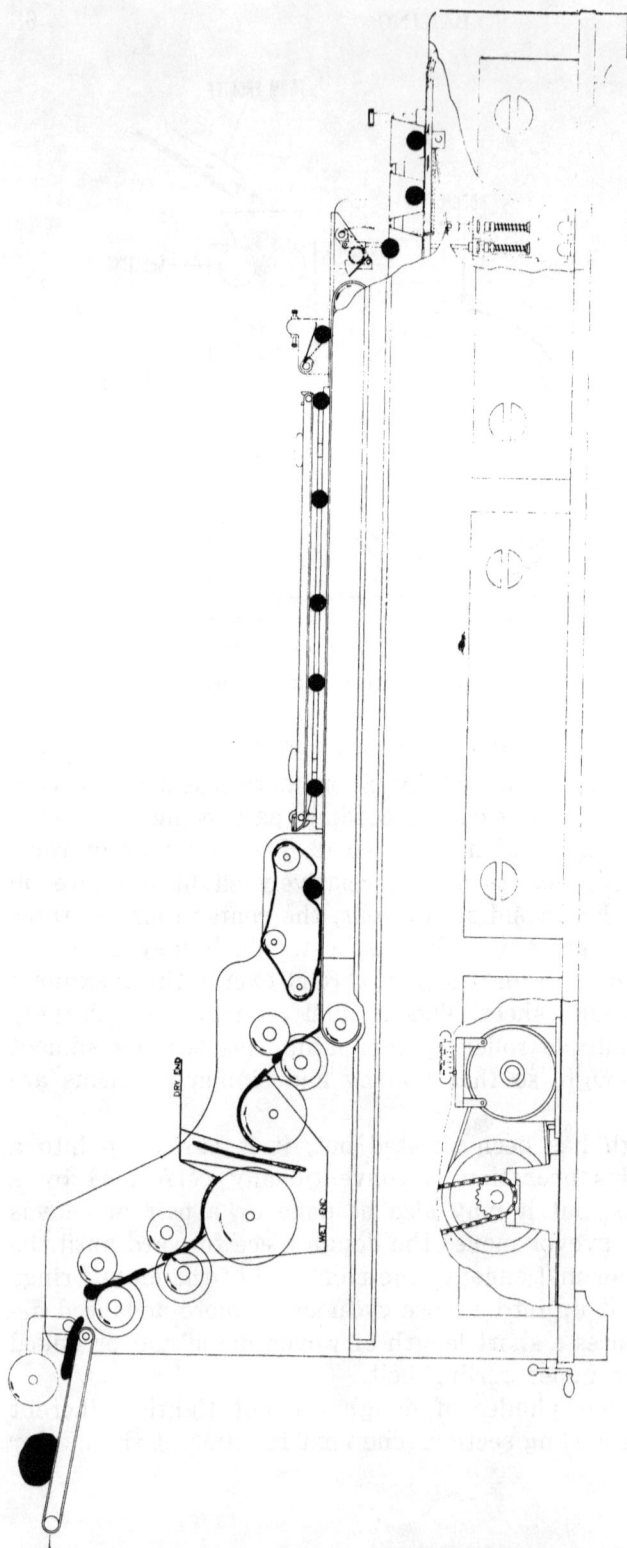

Fig. 20. A Reverse-Sheeting Molder and Automatic Panner

is to seal the dough piece so that it will not unroll when it expands in the oven. In this step, the cylinder of dough is also lengthened so that its axial dimension is somewhat greater than the length of the pan. Entrapped air is expelled from between the spirals in this operation. The conventional molder achieves the desired results by rolling the dough cylinder between a large drum surfaced with canvas and a semicircular compression "board" having a smooth surface. Clearance between the drum and the board is gradually reduced along the route of dough travel so that the piece is continuously compressed.

Many modifications in the basic steps have been made in an attempt to improve the uniformity and texture of the finished product. The cross-grain molder curls the dough sheet at right angles to its direction of travel through the sheeter rolls. The reverse-sheeting molder was devised to curl the sheet of dough so that the wet end of the piece would be folded into the center of the loaf. The dough piece is turned over or reversed between the second or third set of rolls, thus placing the original trailing end in the leading position. Figure 20 is a diagram of a molder-panner which includes these operations. Another type of molder, developed primarily to give loaves with more uniform cell structure, twists the dough pieces after they have been rolled into cylinders.

In general, the operations involved in shaping and forming sweet goods have not been mechanized as completely as those for making bread loaves. However, mechanical benches such as the one illustrated in Fig. 21 can automatically sheet sweet doughs, apply liquid and solid flavoring agents, roll the sheet into an endless

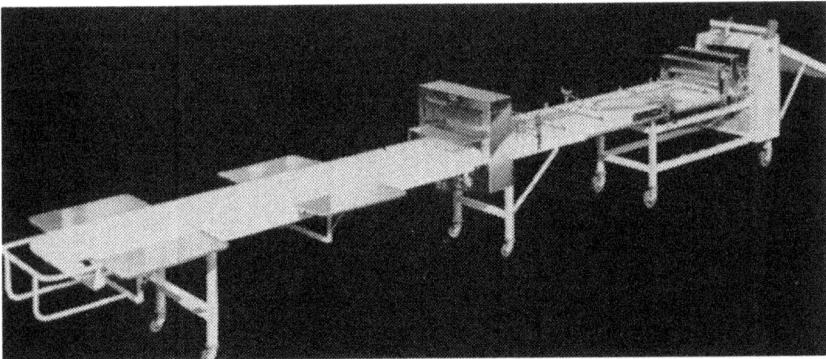

Courtesy of Moline, Inc.

FIG. 21. A MECHANICAL BENCH FOR SWEET GOODS PRODUCTION

helical tube, seal the edge, and cut off segments of the desired size.

Continuous Bread Making Processes

The conventional bread-making process was regarded as difficult to mechanize because of the problems involved in handling and holding large quantities of dough which could not be pumped or otherwise transferred in a continuous manner. In recent years, continuous processing has been achieved by mixing the dough just prior to depositing it in the pans. The fermentation flavor is obtained, not by holding large masses of dough for long times at controlled temperatures and humidities, but by fermenting liquid concentrates which can be pumped directly into the mixer. There is some fairly recent evidence which indicates that the pan fermentation alone is sufficient to give the finished bread a typical aroma and taste, without the use of the so-called liquid sponges.

Continuous bread making plants have been sold by Baker Perkins, the Baker Process Division of Wallace and Tiernan, and American Machine and Foundry Co. The latter's Amflow system requires a "broth" which is a blend of yeast, sugar, milk, salt, yeast food, vitamins, and mold inhibitor in water. This mixture is fermented in large tanks for 2 to 2.5 hr before being transferred through a transfer tank, constant level tank, and heat exchanger (the latter reduces the temperature which has risen due to fermentation) to the incorporator. In the incorporator, oxidant, shortening, and flour are mixed with the sponge by a high speed agitator. The very fluid dough which results is extruded in loaf size pieces which drop into pans. The pans pass through a proofing area where the dough increases in size by a predetermined amount before it is sent to the ovens and baked in a conventional manner. The flowsheet in Fig. 22 summarizes these steps. The other two continuous processes differ in significant details from the operation described, and each of the three has certain advantages and disadvantages which cannot be described here because of space limitations.

Ovens

Early types of bakery ovens were essentially chambers of brick or stone which were heated by building fires in them. Technical refinements over the years have been directed mostly toward improving the temperature control and minimizing the labor required to get the products into and out of the oven. Modern ovens suitable

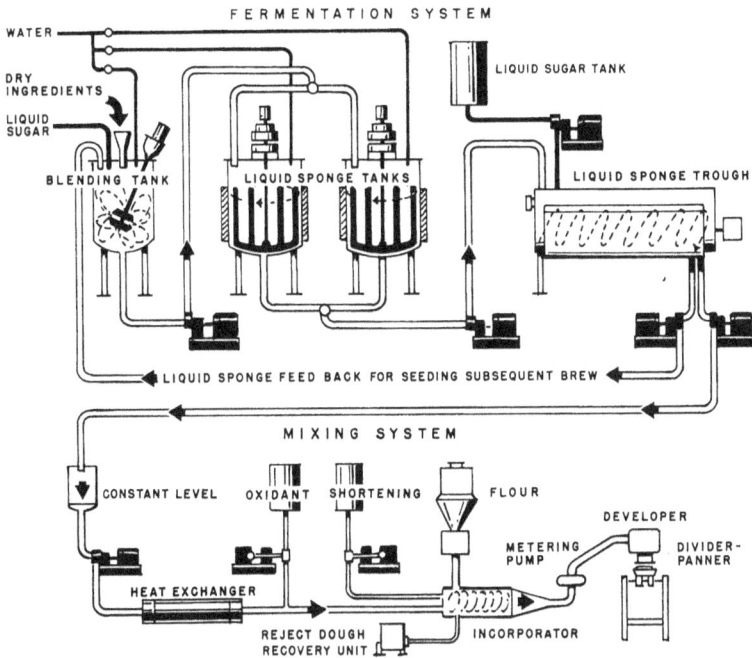

Courtesy of American Machine and Foundry Co.

Fig. 22. Flow Diagram of the Amflow Fermentation and Mixing System

for quantity production can be classified as reel ovens, single-lap ovens, double-lap ovens, and traveling-hearth ovens. In small capacity plants or for special purposes, peel ovens or rotary-hearth ovens are still used.

The reel oven consists essentially of a series of shelves each suspended between arms radially disposed from powered axles located in the approximate center of the sides of the oven. The arms are slowly rotated during the baking cycle so that the shelves describe a cylindrical path. Rotation of the product means that top-to-bottom differences in temperature will not cause product irregularities. Pans can be placed on the shelves as the latter pass a narrow opening in the front of the oven. Baked loaves are removed through the same opening. The reel oven is not suited for production of more than about 1,000 to 2,000 lb of bread per hr.

Single-lap and double-lap ovens also use traveling trays, but, unlike the reel oven, they use a system of chains, sprockets, and curved tracks to move the trays back and forth within the chamber. As a result of these design features, they have more capacity per unit area of floor space than the reel ovens have. The single lap oven has two horizontal arrays of pans while the double-lap oven has four horizontal runs. While the single lap oven has 2 pairs of sprockets like any ordinary conveyor, the double-lap has 5 pairs of sprockets, 2 of which are used to carry a vertical or inclined run of trays. Double-lap ovens have been designed to bake as much as 10,000 lb of bread per hr.

Tunnel ovens (traveling hearth ovens) are the most satisfactory type for high capacity plants and are the only kind suitable for continuous processing systems. The hearth is a band of solid steel or steel mesh which carries the pans through a long heating chamber open at both ends (see Fig. 23). The product is loaded on the

Courtesy of Baker Perkins, Inc.

FIG. 23. A VIEW OF AN AUTOMATIC LOADING AND UNLOADING 120-FT TRAVELING HEARTH BAKERY OVEN AND AUTOMATIC BREAD DEPANNING MACHINE

band at one end and taken off at the same level from the other end. Temperature can be varied in different sections of the tunnel to give a flexibility not attainable in any other type of oven.

The integration of equipment into a plant for manufacturing many kinds of bakery foods is shown in Fig. 24.

THE REACTIONS OF BAKING

Many of the techniques applied in the bakery are unlike those of any other technology. For this reason it is important to consider

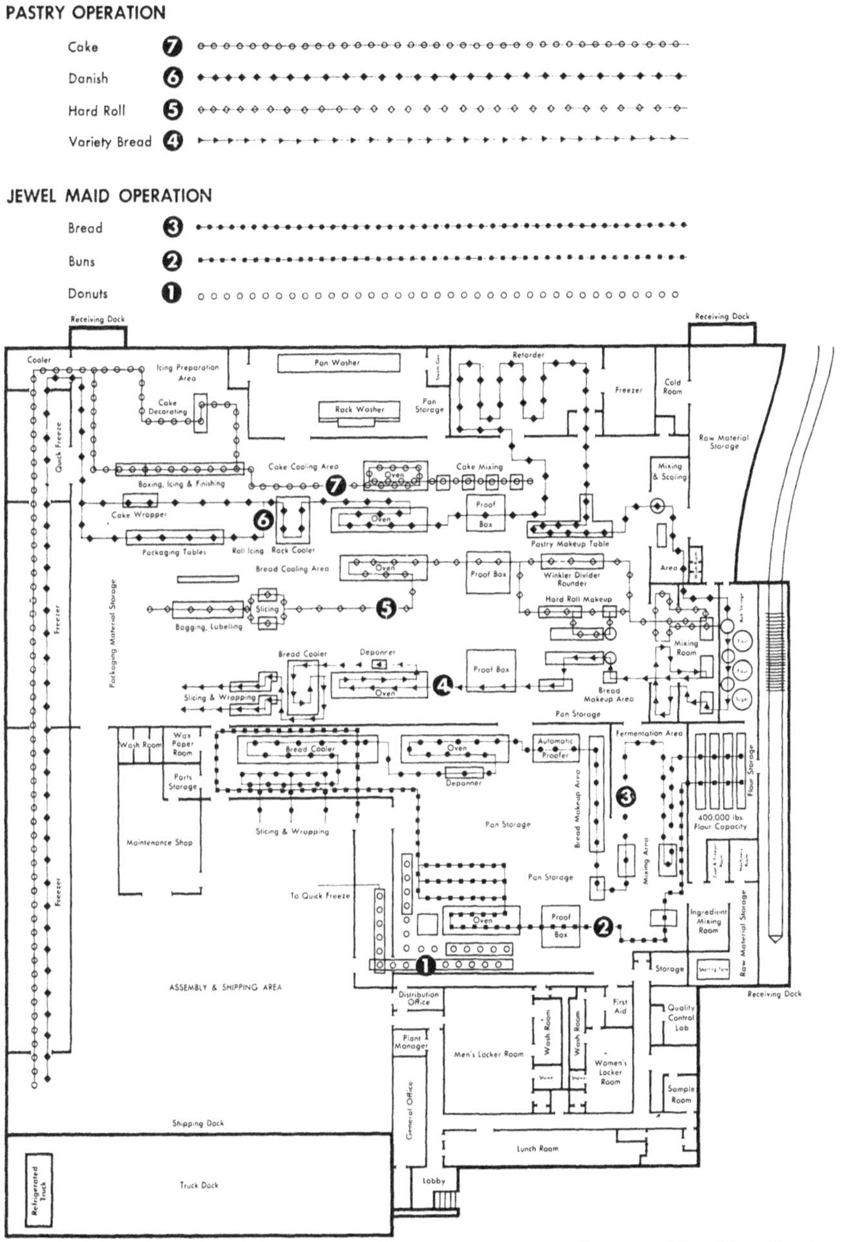

PASTRY OPERATION

Cake **7**

Danish **6**

Hard Roll **5**

Variety Bread **4**

JEWEL MAID OPERATION

Bread **3**

Buns **2**

Donuts **1**

Courtesy of Jewel Tea Co., Inc.

FIG. 24. A COMPLETE PLANT FOR MANUFACTURING SEVERAL KINDS OF BAKERY FOODS

them here even though limitations of space preclude a detailed discussion of all of the specialized processes. There will be no attempt to include sample formulations or "how-to" instructions for performing specific operations.

Mixing

The major objective of all mixing processes is, of course, the effecting of a uniform distribution of the separate ingredients. Bakery mixing also has the special purpose of introducing into the dough or batter bubbles of air which not only leaven the product to some extent but serve as foci for the evolution of leavening gas from dissolved carbon dioxide. In addition, dough mixing has the very important function of developing the gluten so that the vesicle walls will possess maximum elasticity.

The order in which ingredients are added to the mixer may be critical, especially in batter processing. Generally, however, sugar, salt, vitamins, yeast foods, mineral supplements, malt, and milk can be added to doughs or batters in any sequence. They are sometimes dissolved in the water, but the water used for dispersing the yeast should not contain any of these ingredients. It is considered desirable to add shortening last when mixing doughs since the lubricating effect of the fat may interfere with the early stages of gluten development. The initial step in some cake preparation methods is a "creaming" process in which a fluffy mass is formed by beating the shortening and sugar. The air bubbles which are incorporated during creaming have a beneficial effect on the grain of the cake.

Bread doughs are mixed either by the sponge dough method or by the straight dough method. The latter is the simpler process, all of the ingredients being added at the start. Sponge dough mixing requires two separate additions of ingredients interrupted by a lengthy fermentation period. All of the yeast, the yeast nutrients, the malt, 50 to 75% of the flour, and enough water to yield a moderately stiff dough are mixed together in the first stage. This "sponge", as it is called, is fermented for a few hours (e.g., 3 to 4) and then is mixed with milk, salt, sugar, shortening, and the remainder of the flour and water.

A very important function of the mixer in bread processing is the development of the dough. Development, in this special sense, can be defined as the use of the mixer to yield a dough having satisfactory handling properties and capable of being processed into a finished product of good quality. If a dough has not been developed properly, it will be difficult or impossible to handle by ordinary means during subsequent processing steps and will yield a product of inferior quality regardless of the corrective measures which may be applied.

Macroscopically, development is accompanied by an increase in extensibility of the dough, a decrease in the apparent wetness and stickiness, and other changes. When well-developed, the dough can be stretched manually into a thin film. On the molecular basis, development appears to be due to increases in the orderly orientation of the gluten molecules, in the number of intermolecular bonds, and in the degree of hydration of the proteins. Oxidation may also play a part. If mixing is carried beyond the optimum stage, the dough starts to break down, eventually becoming sticky and "short."

Development must be carefully avoided in the mixing of batters for cakes and the like. Increasing the elasticity of the gluten network in such mixtures leads to defects in the texture of the finished product, especially to toughness. Pie doughs are also harmed by development; the ideal mixing treatment for these products is the minimum action necessary to give a uniform mass.

Leavening

The gas primarily responsible for the volume increase in chemically- or yeast-leavened baked goods is carbon dioxide, although steam, air and ethanol may also contribute to the expansion. Carbon dioxide may originate from added sodium bicarbonate or from the action of yeast on fermentable sugars. In either case, it is originally present in the aqueous phase of the dough as an equilibrium mixture of carbonate ion, bicarbonate ion and dissolved carbon dioxide. If the source of the gas is the added bicarbonate, the amount which goes over into the gaseous phase is determined by the pH, the temperature, the concentration of the bicarbonate, and some other poorly understood factors.

When yeast is present, the situation is considerably more complex. The potential supply of gas is then limited by the amount of fermentable sugars, but these compounds may be, and usually are, replenished continually by amylolytic enzymes acting on the starch of the dough. The rate at which the sugars are supplied may not be fast enough to permit optimum yeast activity. The rate at which the yeast uses available carbohydrates depends upon the temperature, the pH, the concentration of certain co-factors, the osmolality of the aqueous phase, the concentration of specific inhibitors, the condition of the yeast cells, and the ratio of yeast to sugar, as well as other conditions. Few of these factors are subject to change at will, and practical control of yeast action is obtained by varying the temperature and the concentration of sugars.

In chemically-leavened doughs and batters, all of the soda is dissolved soon after it is mixed with the other components. Control of the rate of gas evolution by varying the solubility of the soda has been found to be impractical, and leavener manufacturers have resorted to manipulating the speed with which acid is released into the aqueous phase. In some cases, the solubility of the leavening acid determines the rate at which it reacts with bicarbonate. Sodium acid pyrophosphate, a common leavening acid, is hydrolyzed to give the more effective orthophosphate. The rate of this hydrolysis is determined by the activity of enzymes in the dough, the pH, etc.

Not all of the gas given off by the yeast or chemical leavening systems is retained within the dough. Carbon dioxide is constantly diffusing through the vesicle walls and much of it is ultimately lost to the atmosphere. The stretching, cutting, compression, and twisting undergone during processing drives out considerable quantities of the gas. The amount of carbon dioxide retained at the time the dough goes into the oven is the useful amount and is affected by the composition of the dough (including the quality of the flour proteins) and the severity of the treatment accorded the dough pieces.

Baking

Some of the important overall phenomena occurring in the oven are expansion, coagulation of proteins, gelatinization of starches, and evaporation of water. Expansion occurs not only as the result of increases in the volume of gases already present in the vesicles, but as the consequence of further evolution of carbon dioxide, the increased vapor pressure of water, and the volatilization of ethanol-water azeotrope.

Coagulation of protein and gelatinization of starch alter the characteristics of the vesicle walls causing them to become more permeable to carbon dioxide. The protein becomes less elastic, retarding expansion of the loaf. There is not enough water in the dough to permit complete gelatinization of the starch, and, as a result, the carbohydrate remains in a state of incomplete hydration. The crumb of the baked bread is thus dry and elastic (as compared to the dough) rather than sticky and gummy.

Loss of water is continuous during the baking period but is much more rapid during the last few minutes. The dehydration, in combination with protein denaturation and starch gelatinization, fixes or sets the crumb so that it does not collapse when cooled. However, some shrinkage always occurs in the later stages of baking and after removal from the oven because the structure is not strong enough to withstand completely the forces exerted on it.

A layer of collapsed and thickened cells, i.e., a crust, surrounds the dough piece even before it is placed in the oven. This film becomes more elastic as it is heated and exposed to the atmosphere of steam in the oven, permitting the desired expansion to take place. Ultimately, however, denaturation and dehydration take place and limit further expansion. Since the crust is heated much more efficiently than the interior of the loaf, it sets before expansion of the interior ceases. This results in formation of the "break and shred," areas along the sides of the loaf where the first crust has broken and separated, allowing the exposure of under layers which then form a secondary crust of somewhat different appearance.

Nonenzymatic browning and perhaps caramelization of the outer layers of dough are responsible for the color of the crust and for some of the odor and taste of bread. It is important that radiant energy be properly balanced with the thermal energy transferred by convection and conduction to prevent burning of the crust before the interior has completely baked.

PRETZELS

The feature which distinguishes pretzels from all other baked goods is the glossy, dark brown, hard, characteristically flavored crust which results from a dip into hot caustic solution before the pieces are baked. Pretzels are generally formed in a traditional shape, and have a small cross section, a low moisture content, and a coating of salt crystals. Some exceptions to these generalities are soft pretzels, which are relatively high in moisture content and have a cross-sectional area not much different from certain pastries, and pretzels in shapes such as circles, nuggets, and sticks. The following discussion is restricted to the hard or crisp type of pretzel.

The efficient production of pretzels requires highly specialized equipment, at least for the cooking and forming stages. The two main manufacturers of pretzel equipment are the Reading Pretzel

Machinery Co., and the American Machine and Foundry Co. Extrusion equipment is also made by the Baking Specialty Machine Co., of Pennsauken, N. J.

In the Reading Pretzel Machinery equipment, a helix forces dough from a hopper through a slot in the face plate of the extruder. The dough is cut into small strips as it is extruded. The dough strip is rolled to the predetermined thickness by passing it between two canvas belts separated by a gradually decreasing height. At the end of this rolling process, the length is adjusted by clipping the ends from the strip of dough. The dough strip then enters the twister. As the shaped pretzel dough leaves the twister, it passes under a roller which exerts a slight pressure to set the knots.

A reciprocating conveyor places the raw pretzels across a proofing belt approximately 40 ft long. After about 2 to 10 min proof, the pieces drop on to a wire mesh belt which carries them through the cooking solution.

The caustic section consists of a relatively small tank in which the raw pretzels are cooked, and a larger make-up tank. The caustic solution is pumped (or flows by gravity, if the make-up tank is at the higher level) to the immersion or cooking tank. The liquid level in the cooker is maintained by adjusting an overflow pipe.

The caustic solution contains from 0.5 to 1.5% sodium hydroxide or sodium carbonate. A mixture of these two chemicals can also be used. Temperature is maintained at 180° to 200°F by any convenient heating method. Pieces remain in the solution for 10 to 25 sec, the time depending upon the temperature and the concentration of the alkali.

If the lye concentration becomes excessive for any reason, there may not be complete conversion of the chemical to sodium bicarbonate in the baking and drying cycles, and the pretzels will have a hot taste due to the residual sodium hydroxide. There appears to be no FDA regulation governing the concentration of sodium hydroxide in the cooking solution, or the amount remaining on the baked pretzels.

The cooked pretzels are transferred to the oven belt for salting and baking. The salter, located about 15 sec from the cooker, consists of a supply hopper from which the salt is dispensed by a grooved roller. The salt slides down a chute until it is approximately 2 in. above the belt and then drops the rest of the way. The usual aim is to have 2% salt on the finished product, but it is necessary to apply it at the rate of 8 to 10% due to its failure to adhere to the pieces. Excess salt falls through the oven belt into a tray and is recovered.

The types of salt most commonly used are evaporated salt and rock salt. Crystal size is dependent on the size and shape of the product as well as market preference, but in all cases is relatively coarse. Evaporated salt has a whiter color and cleaner taste, but is more apt to dissolve during the baking process. Rock salt is opaque and has a somewhat bitter taste. Uniformity of crystal size is very important.

After salting, the pretzels are ready for baking. The oven in the Reading Pretzel Machinery line is usually 50 ft long. The bake section is the top part of the oven, and it is heated by burners above and below the band. Depending upon the dimensions of the pieces, and other factors, baking times will vary from 4 to 10 min. Temperatures are usually high during the first half of baking, say in the range 475° to 600°F, and lower during the last half, perhaps 375° to 475°F. The final moisture content of the piece as it leaves the oven section will be between 12 and 18%, and usually near 15%.

The baked pretzels fall from the oven belt on to a slide which directs them to the drying belt. The drying section is underneath the baking compartment and separated from it by heavy insulation. Pretzels which cling to the baking belt are removed by a doctor blade. The belt in the drying section travels in a direction opposite to the movement of the belt in the bake section. The speed of the drying belt is variable, but it runs much slower than the belt in the baking section so that the pretzels form a bed 2 to 4 in. deep. Here they are dried for 25 to 90 min at temperatures from 225° to 250°F, and reach a moisture content of 2 to 4%.

In addition to drying the pretzel, the final heating step tends to allow the moisture to equilibrate within the pieces so that checking will be minimized. Slow cooling in the absence of drafts is also desirable for the same reason. Most pretzel companies try to keep breakage below 15% in the packaging and preceding steps. The relatively large amount of checking in twisted pretzels is due to the moisture gradients between the thicker knotted parts and the rest of the piece.

Although the usual pretzel oven does not exceed 90 ft in length, some technologists have suggested that longer ovens would have a beneficial effect on checking by reducing the rate of baking and drying, thereby allowing the moisture content to equilibrate more completely.

Stick pretzels are extruded in a manner somewhat similar to that used for macaroni products. A common type of extruder contains a group of 5 heads containing 10 to 12 holes per head. The dough is forced through the extruding head by a helix and flows on to the proofing belt. As the dough nears the end of the proofing belt, it is cut into the desired lengths by disk-shaped blades which travel across the belt. The stick pretzel pieces pass through a caustic and salting operation similar to that used for twist pretzels. The oven temperature is usually kept at a constant 420°F, and the time of baking is 4 to 5 min. The drying section is held at 225° to 250°F and the sticks are exposed for approximately 55 min. The oven will be about 25 ft long.

Logs and nugget pretzels are extruded like the sticks, but are cut off into lengths at the extruder head.

On the AMF equipment, the twisters work differently (see Fig. 25) than the Reading twisters. AMF has ovens similar to the

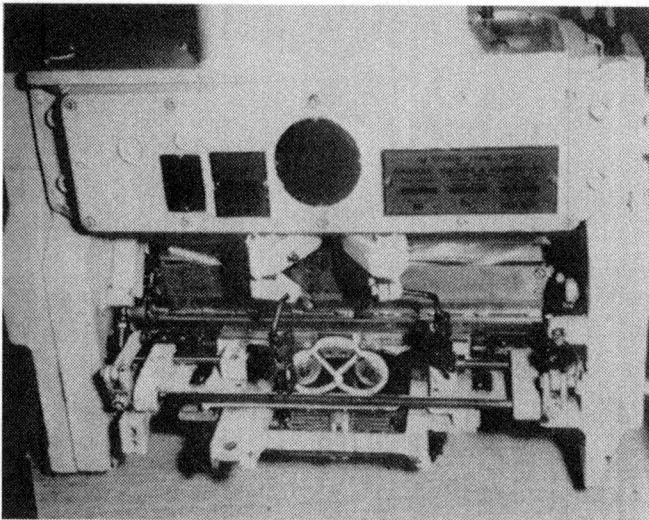

Courtesy of American Machine and Foundry Co.

FIG. 25. DOUGH TYING HEAD OF PRETZEL FORMING MACHINE

Reading type, but they also make or distribute a single pass oven that is approximately 90 ft long. In this oven, the cooked pretzel pieces enter a baking section about 30 ft long held at 450°F and then pass through a drying section held at 225° to 250°F.

Pretzel dough is a stiff, lean, straight dough. The following formulas are representative:

	Twist Lb	Stick Lb
Flour	160	160
Shortening	2	4
Malt syrup (nondiastatic)	2	4
Yeast	0.4	0.4
Ammonium bicarbonate	0.06	0.25
Water	67	62

Yeast food is added as required, and absorption is adjusted to meet flour requirements and mixing conditions.

The kind of flour used in twisted pretzels is very critical. A soft winter wheat flour with a protein content of 8.5 to 9.5%, an ash content of 0.40 to 0.50%, and a viscosity of 65 to 80 might be suitable for many operations (Reisman 1969). Flour is less critical for sticks, logs, and nuggets. Corn syrup is sometimes used in the place of malt syrup, primarily for the purpose of accelerating fermentation. In some cases, the doughs are entirely chemically leavened.

Mixing is done in horizontal single- or double-armed mixers. The dough is brought out at 80° to 90°F. It is necessary that the dough for twist pretzels be dry and tough in order to handle properly in the tying machines. Stickiness causes many problems in machining.

Scrap dough must be handled judiciously. If used in fresh dough in too great a proportion, fermentation and flavor may be adversely affected by the high salt content.

The pH of the entire pretzel, determined on a slurry in the usual manner, will be in the range of 7.5 to 9.0, but the interior "crumb" will ordinarily be in the acid range (below 7.0).

COOKIES AND CRACKERS

The most critical factor in the manufacture of crackers and cookies is the interaction of the doughs and the forming equipment. To a very considerable degree, the appearance and texture of the finished products are limited by the kind of equipment used in processing. For example, a deposit cookie will almost always be more open in texture and have a more irregular surface than rotary molded pieces. A logical basis is thus provided for categorizing cookie varieties and the formulas and procedures used in making them according to the type of equipment used to form the pieces. A useful classification on this basis might have four categories:

(1) **Rotary Molders.**—Dough is forced into shaped cavities in a rotating cylinder. Complex shapes can be formed, and the cookies are usually fairly dense and very uniform.

(2) **Cutting Machines.**—Pieces are cut from a sheet of dough. Rotary cutters or reciprocating cutters (stamping machines) are both common. The shaping is usually restricted to outline cuts and docking pin holes leaving the surface flat and featureless although simple designs such as names or initials can be impressed on the cookie either before, or at the same time as, the outline cut. These products are fairly uniform and often tend to have a more or less stratified internal structure. They may be quite dense or very open.

(3) **Extruders.**—Dough is extruded through an orifice, usually in a continuous stream. Pieces may be cut off with a wire, a guillotine cutter, or disk-shaped knives, as in the cases of wire-cut cookies, some bar cookies, and fig bars, respectively, or the separation may occur as the result of adhesion to the retractible oven band, as in the case of deposit cookies. Extruded cookies are less uniform, and cover a wider range of textural qualities than those made by other methods. They are generally more open in structure than molded cookies.

(4) **Miscellaneous.**—A catchall category set up to contain those preparation methods which do not fit into one of the preceding classes. Sugar wafers and puffed products can be included in this category.

The principal features of these processing methods and of the products made by them will be discussed briefly in the following paragraphs.

Rotary Molded Cookies

In formulating a dough for rotary molded cookies, the consistency must be such that it will feed uniformly and fill all of the crevices of the die cavity under the pressure existing in the feeding hopper. The dough blank must be capable of being extracted from the cavity without undergoing distortion or forming tails of considerable size, but it must adhere to the die roll enough to prevent the dough from falling out before it reaches the extraction roller. The blank must have sufficient cohesion to hold together and not break up at any of the transfer points before or after baking. The dough must flow very slightly or smooth out during forming and baking so that woodiness or undesirable irregularities and cracks in the surface are not apparent in the finished cookie.

Usually, the spread and rise should be minimized so as not to blur or distort the design.

Doughs formulated to these requirements are usually fairly high in sugar and shortening, and low in moisture. The typical dough is crumbly, lumpy, and stiff, with virtually no elasticity. What cohesiveness exists is due primarily to the fat content. The development of gluten is definitely to be avoided. Table 23 gives some comparative formulas for plain rotary cookies. To these may be added flavoring ingredients such as butter, spices (e.g., mace or

TABLE 23
COMPARATIVE FORMULAS FOR PLAIN ROTARY MOLDED COOKIES

	Sandwich Base Cookies			Butter Cookies Lb
	Butterscotch Lb	Vanilla Lb	Chocolate Lb	
Flour	100	100	100	100
Powdered sugar	20	32	33	23
Shortening	25	27	24	22
Acid cream	0.5	—	—	0.3
Invert syrup	—	—	5	—
Sodium bicarbonate	0.5	0.5	0.8	0.3
Salt	1.5	1.5	1.5	1.5
Ammonium bicarbonate	—	—	0.25	0.35
Frozen whole eggs	3.5	3	1	3
Sweetened condensed skim milk	6	6	3	6
Butter	1.2	—	—	4
Cultured butter flavor	0.15 to 0.30	—	—	0.5 to 1.0
Vanillin	—	0.05	0.05	0.05
Lecithin	0.3	0.25	0.35	0.4
Malt	1.4	—	0.4	1.2
Water, variable	10	8	13	8
Dutched cocoa	—	—	5	—
Chocolate liquor, natural	—	—	8	—

nutmeg), pure vanilla, or traces of orange or lemon oil. The desirable volatiles of pure vanilla are largely lost in a typical rotary cookie bake. Yellow color will be required in butter and vanilla varieties, and carbon black in the chocolate. About 2 oz of dissolved gelatin added per 100 lb of flour improves the gloss of chocolate rotary doughs. Table 24 summarizes the formulas for a few specialty items.

Most manufacturers use flour of about 8.1 to 8.2% protein for rotary base cakes, although a range of 7.1 to 9.2% has been reported. Ash should be about 0.415, with a range of 0.33 to 0.47% being satisfactory. Oleo shortening added in the liquid condition is suitable for most of these doughs, but vegetable shortening can also be used. Powdered sugar and sugar syrups are the preferred

TABLE 24

FORMULAS FOR SPECIALTY ROTARY COOKIES

	Molasses Lb	Variety Almond Short Lb	Coconut Lb	Sugar Lb
Flour	100	100	100	100
Shortening	35	20	15	35
Sugar	30	28	13	48
Invert syrup	4	2	3	7.5
Frozen whole eggs	—	4	—	2
Sweetened condensed skim milk	10	3	6	6
Brown sugar	—	—	20	—
Molasses	10	—	—	—
Acid cream	—	0.2	—	—
Sodium bicarbonate	1	0.5	0.62	0.35
Ammonium bicarbonate	—	0.25	—	0.5
Salt	1.5	1	0.75	1.5
Butter	—	15	20	—
Water	5	5	6	7
Almond slices	—	6	—	—
Extra fine coconut	—	—	5	—

sweetening ingredients. Nonfat milk solids are often added, but it is thought that condensed milk is preferable since the liquid ingredient removes any possibility of lumps appearing in the finished cookie. Lecithin at about the 0.4% level will improve machineability.

One stage mixing is often perfectly satisfactory, but a creaming operation with most of the minor ingredients added before the flour goes in gives added assurance that lumps of undistributed ingredients will not appear in the cookie. Dough temperatures from 72° to 90°F are being used for rotary sandwich bases.

Cutting Machine Biscuits

Cutting machine goods vary in type from the soda cracker to the hard sweet cookie. The essential characteristic of the dough is that it must be sufficiently cohesive to form the continuous sheet from which the pieces will be cut and to hold the scrap web, if any, together, so that it can be lifted from between the pieces. Properties of the finished products can be made to vary over a wide range, but they are generally thin and crisp or hard.

Soda Crackers.—Soda crackers or saltines, and the many variants of this product, such as oyster (soup) crackers, club crackers, cheese crackers, etc., are made from a sponge dough that has been fermented for many hours. An 18 hr fermentation period can be regarded as typical. A considerable amount of acid, as well as the expected alcohol and carbon dioxide, is developed by microorganisms (yeast and bacteria) during the sponge fermentation. The acid is neutralized by adding sodium bicarbonate when the dough stage is mixed, but lactic acid and some of the other fermentation by-products are carried through to the finished cracker and contribute to the characteristic flavor.

Soda crackers are formed from a continuous sheet of dough which is laminated (lapped) before being cut. The dough pieces are formed by a stamping device (reciprocating cutter) which does not entirely sever the individual crackers from the sheet. As a result, the crackers remain in a substantially continuous sheet as they travel through the band oven and are broken apart in strips after baking. The cutter also punctures the dough pieces in numerous places with "docking pins" (see Fig. 26). These holes prevent uneven or excess expansion in the oven.

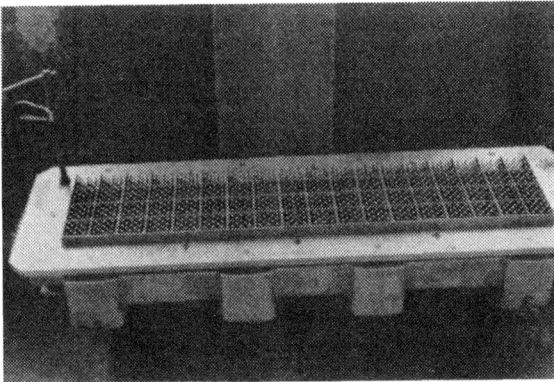

FIG. 26. A CUTTING DIE FOR SALTINES

Most of the soda cracker-type products are made from rather lean doughs. Shortening addition may be as much as 10% (FWB), but is typically only a few percent. Diastatic malt syrup is used at rates of up to 1.5%. Sugar and milk solids are not commonly added to saltine doughs, but may be used with rich specialty crackers.

The sponge flour must be moderately strong, while the dough flour can be a weaker, soft wheat type, especially if thin crackers are wanted. Salt is applied after stamping and before baking. Application rates may vary considerably, but about 2.5%, on the dough weight, is an average figure. Salt suppliers sell a crystal size and weight specifically intended for topping crackers.

Baking is on a perforated or mesh band. Temperatures are

generally high throughout the baking period, but relatively lower in the later stages.

Rich crackers, usually in a round shape, are often sprayed with a fat such as coconut oil after baking. These doughs do not go through a fermentation step. Cheese crackers contain a substantial amount of strong-flavored cheddar cheese as well as artificial flavor and color, or natural colors such as paprika. Milk crackers contain a few percent high heat nonfat dry milk. An almost infinite variety of snack crackers can be made by varying the flours (whole wheat, rye, corn, etc.), spices, colors, and flavors. Commercially available snack crackers are of both the fermented and unfermented types.

An example of a butter thin formula is:

	Lb	Oz
Soft flour	100	
Granulated sugar	4	
Butter	8	
Shortening, hydrogenated	7	
Nondiastatic malt syrup	3	
Sodium bicarbonate	2	
Acid cream		10
Water, about	26	

The sugar, salt, and shortening are mixed thoroughly, then hot water (about 200°F) is added. After all the water is in, the malt syrup is added followed by the flour and leavening ingredients, and the dough is mixed until it cleans the sides. The dough must be laminated after the initial sheeting to give the desired structure.

A milk cracker is compounded along very much the same lines:

	Lb
Soft flour	100
Shortening	11
Granulated sugar	5
Salt	0.5
Nondiastatic malt syrup	1
Dried whole milk	1
Sodium bicarbonate	0.75
Acid cream	0.75
Ammonium bicarbonate	0.15
Water, about	30

Essentially the same process is used as for the preceding variety, except that a small amount of the water (cold) is reserved for dissolving the ammonium bicarbonate and the solution is sprinkled onto the dough just as the flour becomes wetted.

Graham Crackers.—The original graham cracker, which was developed as a health food, was made by a fermentation process. Present-day varieties are strictly chemically leavened, and are popular more for their unique flavor and texture than for their dietetic properties. A well-formulated and prepared graham cracker has a slightly sweet, but well-rounded flavor with branny and caramel notes, almost nutlike. The texture should be crisp but by no means hard. The top should have a very slight gloss. A dull, grayish crust is a defect. Flavor and texture are contributed by the bran components of the graham flour, while the brown sugar, molasses, and malt also add to the flavor.

Most graham crackers sold today are called sugar-honey grahams. It is doubtful that the small amount of honey contained in these products adds much to the flavor, but the name seems to have a strong appeal to consumers. A formula which has been very successful is:

	Lb	Oz
Soft flour	50	—
Strong flour	50	—
Graham flour	25	—
Granulated sugar	25	—
Lard	18	—
Honey	11	—
Brown sugar	10	—
Dried sweet whey	4	—
Defatted cooked cottonseed meal	2	—
Sodium bicarbonate	1	4
Salt	1	4
Nondiastatic malt syrup	1	—
Acid cream	—	15
Ammonium bicarbonate	—	12
Lecithin	—	8
Vanillin	—	1
Water, variable	34	—

The sweetening ingredients and all of the water except 2 lb are heated to 165° to 170°F before addition to the flour, shortening, whey, and lecithin. After these ingredients have been mixed 5 min in the 3 spindle mixer, all the other ingredients except the ammonia and the remaining 2 lb of water are sprinkled into the dough. Immediately thereafter, the rest of the water containing the ammonium bicarbonate is added. The total mixing time must be adjusted to compensate for flour quality and mixer characteristics, but 25 min to 1 hr may be required. The dough should be about 118°F at the time it is sheeted.

The preceding formula was for a high-quality, rather expensive graham cracker. There are many cheaper versions on the market. The following formula is probably typical of low-cost examples of the product:

	Lb	Oz
Soft flour	50	—
Strong flour	50	—
Graham flour	25	—
Lard or oleo oil	10	—
Granulated sugar	15	—
Invert syrup	5	—
Molasses	5	—
Sodium bicarbonate	1	4
Acid cream	—	8
Ammonium bicarbonate	—	12
Lecithin	—	8
Salt	1	8
Vanillin	—	0.5
Water, variable	32	—

Mixing is conducted as described for the preceding formula.

Graham cracker doughs are folded and reduced one or more times after the preliminary sheeting. Formerly, this processing was performed on dough brakes, but continuous equipment is now used throughout the industry.

Baking is done on a perforated or mesh oven band to avoid cupping and other problems.

"Hard sweets" are more popular in Canada and Europe than they are in the United States. The dough for hard sweets is very stiff and it is machined so as to have little leavening action. The resulting cookie is dense and hard, but this texture is acceptable because of the thinness of the piece. The mixing time is up to 2 or 2.5 hr in a horizontal mixer at very low speeds, say 15 rpm.

Wire-cut Cookies.—Extruding equipment can be adapted to handle doughs having a wide range of physical properties. Wire-cut cookies are typical of the goods made on this type of machine. In these doughs, it is necessary to have the material sufficiently cohesive to hold together as it is extruded through an orifice and yet it must be nonsticky and short enough so that it separates cleanly as it is cut by the wire which slides back and forth over the orifices. Formulas may contain up to several hundred percent sugar based on the flour, and 100% or more of shortening, based on the flour content. Doughs may be almost as soft as cake batters or too stiff to be easily molded by hand. The very soft doughs overlap deposit doughs in consistency, while the other extreme is close to rotary molded type doughs. Advantages over rotary molded cakes are a more open texture, and over deposit goods, a more uniformly shaped cookie. Disadvantages over the rotary molded piece are the lack of design, and somewhat less uniformity.

Wittenberg (1965) classified wire-cut cookies as shown in Table 25. McGee (1955) further subdivides soft cookies into (1) drop

TABLE 25

FORMULATION OF WIRE-CUT COOKIES

	Sugar %	Flour Type[1]	Shortening %	Liquid Whole Eggs %	Final Moisture %
Low-cost promotional cookies	40 to 50	A	20 to 25	Little if any	4 to 5
Standard market shelf cookies	50	A	50	10	4 to 5
Soft cookies	up to 75	A and B	60	up to 20	12 to 15
Specialty high quality cookies	35 to 40	A and C	65 to 75	15 to 25	—

[1] Flour Types: A = soft wheat unbleached with 8 to 8.5% protein, 0.35 to 0.40% ash, 40+ viscosity, spread factor 79 to 80; B = cake flour; C = bread flour.

type, as those to be used in sandwich cookies filled with a marshmallow or imitation cream and usually having an amount of sugar equal to the flour, (2) sugar cookies, molasses cookies, coconut, raisin, date, and honey varieties, (3) shortbreads, in which the shortening is usually $\frac{1}{2}$ to $\frac{3}{4}$ as much as the flour, and (4) macaroons, with little or no flour and large proportions of sugar. Such varieties as lady fingers, based on pound cake recipes, would not be included in the above classifications.

Flick (1964) prefers to categorize soft cookies as (1) filled, (2) old fashioned sugar cookies, (3) drops, and (4) bars. For soft cookies he recommends a medium strong soft red winter wheat, lightly bleached with chlorine, with an ash content of about 0.36% and a protein content of about 9.5%, or a strong soft red winter wheat, unbleached, having an ash of about 0.41%, and a protein of about 9.5%. The shortening should be a good quality creamable fat. The sugar should be predominantly granulated with perhaps 18 to 24% invert (FWB). Eggs should be present at the 14% level, or more (FWB). Emulsifiers may be necessary.

Deposit-cookies are also made on extrusion equipment. This variety is the machine-made counterpart of the hand-bagged cookie and many of the latter formulas can be successfully adapted to automatic production. Deposit-cookies will have about 35 to 40% sugar, 65 to 75% shortening, and 15 to 25% liquid whole eggs. The flour should be from soft wheat, unbleached, with 8 to 8.5% protein and 0.35 to 0.40% ash. It should have a viscosity of 40°M or more and a spread factor of 79 to 80.

The flour must be able to carry the sugar and shortening without too much spread so that the surface configuration is preserved through baking. At the same time, the flour or other ingredients must contribute enough adhesive properties to the dough so that it will adhere to the band and pull away from the main tube of the dough in the deposit stage.

Table 26 gives typical basic formulas for three types of deposit-cookies. Modifications by adding small amounts of flavors and colors may generally be made without other adjustments. For example, orange oil, lemon oil, or butter flavors can be used to advantage in spritz and star cookies. The amount of sugar given in the formula should be divided between powdered and granulated with the proportions being chosen to give the proper dough con-

TABLE 26
FORMULAS FOR DEPOSIT COOKIES

	Peanut Butter Lb	Spritz Lb	Star Lb
Moderately strong flour	50	50	100
Bleached cake flour	50	50	—
Sugar	80	50	45
Shortening	32	40	46
Whole eggs, frozen	12	8	3
Salt	1.5	1.25	1.5
Sodium bicarbonate	0.75	0.4	0.15
Ammonia	0.25	0.125	—
Invert syrup	5	1.25	—
Butter or margarine	—	5	—
Dried sweet whey	—	1	4
Baking powder	—	—	0.5
Vanilla extract	—	0.5	0.25
Water or ice	35	6	16
Peanut butter	85	—	—

sistency and spread. A recipe for basic almond macaroon is 5 lb almond paste, 5 lb powdered sugar, 1.75 lb egg whites, and 0.75 lb white corn meal. Macaroons can be made by the cold process, hot syrup process, or cooked process (McGee 1955). They should not be made by the cold process, however, unless they are to be consumed within three days due to the possibility of soapiness developing from lipase activity.

Other types of cookies made on extrusion equipment are plain bar cookies (e.g., coconut bars) and fruit-jam-filled cookies (e.g., fig bars). Filled cookies such as fig bars are extruded through a pair of concentric orifices, the jacket dough passing through the outer ring while the filling is extruded through the inner hole.

Miscellaneous Biscuit Products and Adjuncts

Sugar wafers.—The base cake for sugar wafers is entirely different from any other cookie. It has many similarities in texture and structure to puffed snack foods. Structurally, it is a foamed dehydrated starch gel with a supporting gluten network. Most of the leavening action results from the evolution of steam during the baking phase. The batter is closely confined between heated plates so that steam cannot readily escape, and its leavening effect under these conditions is much greater than it would be if the thin sheet were exposed to the atmosphere.

The basic principles used in making the flat base cake for sugar wafers are also used in manufacturing ice cream cones (waffle cones) and novelty cream-filled shapes such as circles, hemispheres, peanut shells, etc. Some of these novelty shapes are filled with cheese flavored, or other savory fillings, instead of sugar-shortening mixtures. All of the fillings are fat-based, however, since any appreciable amount of free moisture completely ruins the texture of the base cake, causing it to collapse.

The batter is a very fluid mixture of flour and water to which small amounts of other ingredients are added. Generally, no sugar or other sweeteners are present. See the formula below.

	Lb	Oz
Flour	100	—
Water, variable	135	—
Sodium bicarbonate	—	6
Salt	—	8
Lecithin	—	1.5
Coconut oil	1	—

To the base formula may be added colors and flavors selected to be compatible with the process. A yellow color, egg shade, is commonly added to the vanilla or plain base cake. Flavors should be added sparingly, since the predominant flavor is expected to come from the filling and any volatile compounds are lost by steam distillation during baking. Vanillin at the 0.025 to 0.05% level is a suitable flavoring. Chocolate or dark colored wafers often do not contain any cacao products and are colored brown with caramel or food dyes. For a chocolate flavor, about 10% of cocoa (preferably the high fat or breakfast type) may be added to the formula. This addition will necessitate an increase in the water.

The flour should be a short extraction soft wheat flour. Flour from white wheat is sometimes used. Both bleached and unbleached flours have been used successfully. Any speckiness shows up clearly in white flour, so a low extraction, low ash flour is recommended. The amount and quality of gluten seem to have some effect on expansion, so that a very weak flour may give a piece that is too dense or too close textured. On the other hand, gluten also affects the texture, a strong flour causing the wafer to be hard and flinty. The protein content or strength of the flour should be chosen to achieve a compromise between excessive hardness and excessive fragility or denseness. The precise specification will depend upon the kinds and amounts of other ingredients, the type of end product desired, and the conditions encountered during processing. As usual, water is varied, the amount being chosen to give a batter viscosity which will enable the batter to spread rapidly over the plate and between the plates while still retaining a substantially uniform internal structure without large voids or other defects.

Mixing is a simple procedure. Any equipment which can provide a smooth, lump-free batter is acceptable. The dry ingredients are first blended, then water is added gradually while mixing. If shortening is used, it is blended in while the mixture is still of a doughy consistency. Some experts recommend using melted shortening. As a final step, the remainder of the water is gradually mixed in.

Batters are preferably mixed at high speed although there are numerous cases where low speed horizontal mixers have been used successfully. Under certain conditions, as when overmixed, the gluten may agglomerate into strands and separate from the liquid. Settling-out is also a problem and some agitation is necessary while the batter is being held for depositing. Mixing batches of the smallest possible size and prompt usage are recommended.

Sugar-wafer-filling formulas are sugar-shortening-flavor compositions similar to those used for sandwich cookie fillings. Finely ground scrap in varying amounts is added to sugar wafer cream. The scrap consists of broken wafers and the trimmings removed at the saws. Fillings will contain from about 30 to perhaps 50% shortening, with quality types being in the upper part of this range and the economy fillings being brought as close to 30% as possible consistent with satisfactory handling properties and consumer acceptance. The 50 to 70% nonfat solids consist principally of powdered sugar and scrap. There is no doubt that scrap has a deleterious effect on the appearance and flavor of the cream, but the relatively large amounts of such material generated in the trimming operation make it essential to utilize it in some economically advantageous form.

Icings, fillings, coatings, and toppings are a necessary part of many cookie varieties but it is beyond the scope of this book to give details of their preparation. The reader interested in this information or in other details on cookie and cracker preparation should consult *Cookie and Cracker Technology* (Matz 1968).

Uniformity is a highly desirable characteristic in cookies. Although according to the usual commercial practice the container may initially be designed to fit the cookie (and even this is not always the case), ever afterwards the cookie must be made to fit the box. If the pieces coming out of the oven are too small, the container will exhibit slack fill, with consequent consumer dissatisfaction, or excess count, often leading to high weights. On the other hand, if the cookies are too large, slow packing, breakage, or illegal low weights (due to low counts) may result. The extreme importance of uniformity to processing efficiency may lead the production supervisor to make formula or processing changes which cause the desired dimensions and unit weight to be obtained at the expense of the organoleptic characteristics of the cookie.

BIBLIOGRAPHY

ANON. 1962. Cereal Laboratory Methods, 7th Edition. Am. Assoc. Cereal Chemists, St. Paul, Minn.

FLICK, H. 1964. Fundamentals of cookie production, including soft type cookies. Proc. Am. Soc. Bakery Engrs. *1964*, 286–293.

FORTMANN, K. L. 1969. Innovations in continuous processing methods. Bakers Dig. *43*, No. 6, 64–66.

KAMMAN, P. W. 1969. Quality control in continuous mix. Bakers Weekly *216*, No. 22, 28–35.

MATZ, S. A. 1960. Bakery Technology and Engineering. Avi Publishing Co., Westport, Conn.

MATZ, S. A. 1968. Cookie and Cracker Technology. Avi Publishing Co., Westport, Conn.

McGEE, O. L. 1955. Soft cookies. Proc. Am. Soc. Bakery Engrs. *1955*, 251–260.

PYLER, E. J. 1952. Baking Science and Technology. Siebel Publishing Co., Chicago.

REISMAN, H. 1969. Modern methods of pretzel production. Snack Food *57*, No. 6, 33–36, 64.

TRUM, G. W. 1969. Trouble shooting on continuous mix. Bakers Weekly *216*, No. 19, 24–27.

W_SEBLATT, L. 1961. Flavor research in the bread-baking field. Cereal Sci. Today *6*, 298–300.

WITTENBERG, H. L. 1965. Wire-cut cookies. Biscuit Bakers' Inst. Training Conference *1965*.

CHAPTER 3

Bruce W. Smith[1] | # Feed Manufacture

INTRODUCTION

An increasingly affluent American society is consuming less products of cereal grains directly and more in the form of such protein foods as meat, milk, and eggs. Of the cereal grains raised in the United States today and consumed domestically, slightly more than 86% goes for the feeding of livestock and poultry.

Feed manufacturing today is an exacting science. The feed industry combines cereal grains and mixtures of grain with other natural products and synthetics to produce what are called "formula" or "mixed" feeds. Although millions of farmers still feed some straight grain, sometimes fortified with an oil meal, to their livestock, the trend is clearly towards wider use of commercially-formulated manufactured rations.

Some large livestock production units, most notably cattle fattening operations (feedlots), have installed machinery for the processing of their feed requirements. Such equipment is not inexpensive, for it is scaled-down from similar machinery in use in commercial feed mills (see Fig. 27). Grinder-mixer combinations designed for use on small farms generally do not prove a sound investment economically.

Additions of minute quantities of expensive ingredients cannot be made precisely with the grinder-mixers designed for use on small farms.

EARLY HISTORY

The American feed manufacturing industry as a recognizable entity dates back to the start of the twentieth century. Wherry (1947) pointed out that despite resistance from some segments of the economy, typified by the large milk companies which prohibited the feeding of gluten feed to dairy cows, "Men of vision realized that a new era in poultry and livestock production could be built on the efficient utilization of these precious proteins that were wasted in an earlier day. And so it was that the golden age of better feeding began."

[1] Editorial Director, Communications Marketing, Inc., Edina, Minn.

Courtesy of Prater Pulverizer Co.

FIG. 27. EQUIPMENT FOR MILLING ON THE FARM

Larger livestock feeding units, particularly cattle
feedlots, sometimes process their own requirements
with grinder-mixer equipment like this.

It is likely that a natural catastrophe played a role in the early
development of scientific feeding. In 1888, Buffalo, New York, was
swept by floods which followed a major March snowstorm that
year. Long a major flour milling center, the Empire State city had
a number of corn processing plants. Apparently, flood waters which
invaded the mills carried out with them traces of gluten by-products when they receded.

There were many animals kept in barns and stables in Buffalo
in those days and they drank freely of the gluten-impregnated
flood waters. The result was obvious to their owners: The animals,
depending on their species, gained weight or produced milk at
record rates. By the end of 1888, Buffalo corn mills had a new
by-product business: supplying boxcar loads of wet gluten feed
throughout the dairy country of the northeastern states.

ADVENT OF DRY FEEDS

The sun and air may have been responsible for the first dry
feeds, the result of wet corn gluten feed drying naturally before

feeding. A foresighted Buckeye cereal miller, however, is believed to have been the first to process dry animal and poultry feeds deliberately.

Ferdinand Schumacher of Akron, Ohio, sold dry feed even before the flood water plagued Buffalo. In 1885, he brought out what he called "COB Feed." Contrary to the implication in its name, it was not a corn cob product. Instead, it included the milling by-products of corn, barley, and oats.

Schumacher obviously had a basic knowledge of nutrition. He reasoned correctly that if the primary cereal grains had nutritive values for humans, that the resultant by-products well could have some value remaining in them.

The Ohio miller was ahead of his time, however, and it was not until the twentieth century that a discernible feed industry began to evolve. That infant business had little resemblance to the scientific feed manufacturing industry of today.

In 1908, for example, a feed for baby chicks was advertised as being composed of wheat, corn, oats, flax, millet, kafir, "etc." Although that magic, all-encompassing abbreviation "etc." may have been all right in those days, strict state and federal government supervision today would not permit it.

Corn gluten feed is recognized as the first dry protein cereal by-product to have been fed widely to livestock. Now, as then, it consists of shelled corn from which most of the starch, gluten, and germ have been removed by wet-process milling. The major corn processors have been selling gluten feed for six decades.

First Feed Companies

The oldest feed manufacturing company in continuous operation in the United States is National Food Company, which started operations in 1885. Today a subsidiary of the Cudahy Company, it has its offices at Dundee, Illionois, and its mill at New Holstein, Wisconsin.

The firm which today is by far the largest manufacturer of animal and poultry feeds is Ralston Purina Company of St. Louis (Schaible 1970). It was founded in 1894 by the late William H. Danforth as Robinson-Danforth Commission Company. Although Ralston Purina today manufactures a wide range of human food products, its feeds (which it brand-names as "Chows") figure importantly in total company sales and profits. Ralston Purina has plants throughout the United States and Canada and in several foreign countries.

Among the larger flour millers in the feed manufacturing business are International Milling Company, Nebraska Consolidated Mills Company, Hubbard Milling Company, and Peavey Company. These firms market a substantial part of their flour milling by-products through the feeds they make. The remainder finds a ready market with other feed manufacturers which do not mill flour.

Publicly Held Companies

The following companies which manufacture formula feed are publicly held and their stock traded on the New York Stock Exchange: Allied Mills, Central Soya Company, Cudahy Company, W. R. Grace & Company, Ralston Purina, and Swift & Company.

The Carnation Company, which manufactures feeds in its Albers Milling Company division, is listed on the American Stock Exchange.

Nebraska Consolidated, International Milling, and Doughboy Industries, Inc., are publicly owned and their stock traded nationally "over the counter."

Among the diversified companies which previously manufactured formula feed but which no longer do so are Textron, Inc.; Thomas J. Lipton, Inc.; General Mills, Inc.; Corn Products Company; and Quaker Oats Company. In 1969, the latter sold its Ful-O-Pep feed division to Allied Mills, Inc.

It is the local mill that actually sells most of the feed to the livestock producer. Usually these retail establishments obtain a part or all of their feed from larger manufacturers, often combining these "factory built" rations with locally grown grains.

What they consider high freight rates has spurred the larger companies to decentralize their feed manufacturing operations. The "terminal" type plant which distributes its output over hundreds of miles has almost disappeared from the American scene. Instead, the larger companies have built smaller mills closer to the livestock production of the nation, with resultant important savings in transportation costs (see Fig. 28 and 29).

FEED PRODUCTION AND CONSUMPTION STATISTICS

Production of Feed

W. H. Strowd (1925) forecast correctly when he wrote in 1925: "We may look for a bigger, better business in commercial feeds." He commented further: "More by-products feed will be produced

Courtesy of Feed Bag Magazine

FIG. 28. MODERN US MILL

This plant typifies the feed mill being built today
in the United States; it can produce 200 tons in 8
hr and distributes within a 250 mile radius.

and less feed will be exported to meet the increased domestic demand."

Neither Strowd nor other observers of the national scene at that time dreamed of the manifold yield increases that American grain farmers would achieve in the years which followed. But Strowd was correct when he judged that increased consumption of roughages would not affect adversely the amount of grain which animals would eat. He said: "We are so far behind in applying the known principles of animal nutrition, that this offset (use of more legume hays) will hardly be felt even as a check on the increasing demand for commercial feeds."

How accurate the author of the preceding statement has proven to be is borne out by these statistics from the American Feed Manufacturers Association, the national organization of feed companies, on the consumption of formula feeds over a 20-yr span:

1948	25,500,000 tons
1958	40,000,000 tons
1968	53,600,000 tons

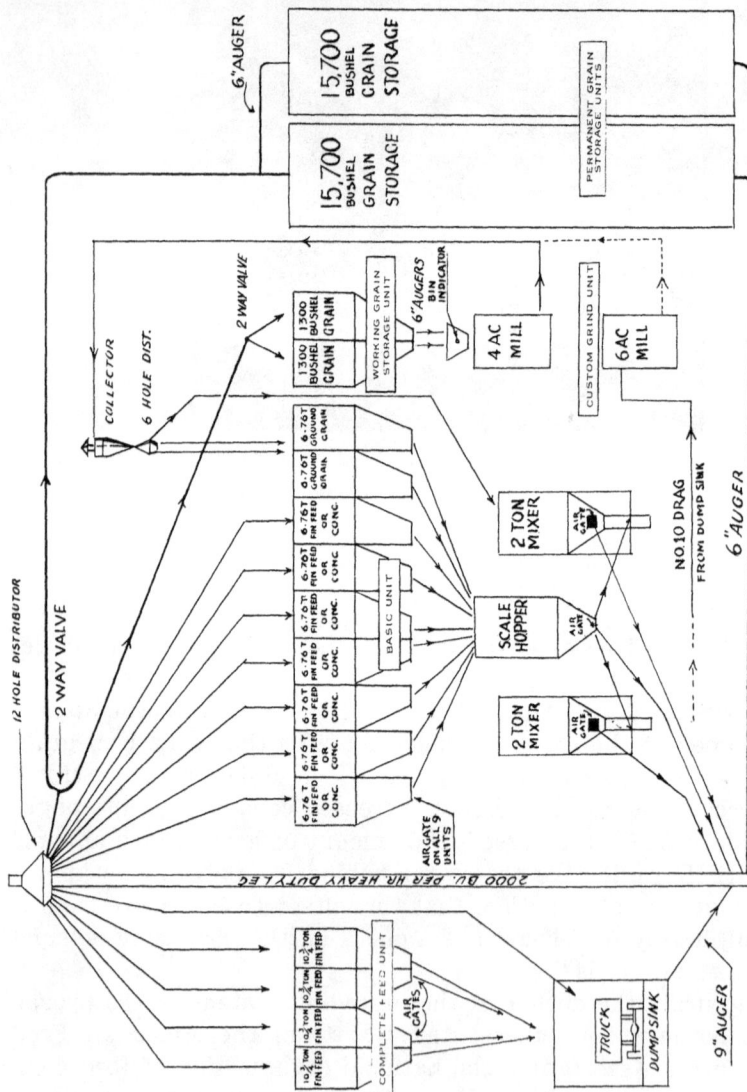

Courtesy of Prater Pulverizer Co.

Fig. 29. Flow Diagram of a Modern Local Plant

Feed companies serving areas with a radius of 15 to 50 miles operate plants typified by this diagram.

The 1959 Edition of this book carried this statement on page 407: "As higher-energy rations are developed and new additives devised to make feeds more productive, the rate of increase of total production probably will be slowed below the growth rate indicated for the years 1948 to 1958."

This proves not to have been the case. As consumer preference has sought more meat, milk, and eggs, the requirements for formula feed have continued to grow rapidly. Indeed, the higher-energy rations have been developed, tested, and accepted. But demand for the end-products of livestock production has accelerated far more rapidly than had been predicted.

The American Feed Manufacturers Association has reported (1969): "Tonnage equalled or exceeded year-earlier levels during each of the last 8 yr."

This growth pattern has been as follows with 1958 being the base period:

	%
1959	Same
1960	−1
1961	+6
1962	+10
1963	+10
1964	+11
1965	+12
1966	+25
1967	+31
1968	+34

Production by Types

In the decade from 1958 to 1968, marked changes occurred in the types of rations manufactured by the US feed industry. Here are some examples, the percentages applying to total manufactured feed production in the years cited:

For chickens, starter-grower and layer-breeder feeds, from 32% in 1958 to 23% in 1968; for boiler chickens, from 20% in 1958 to 16% in 1968; for turkeys, from 5% in 1958 to 4% in 1968; for dairy cows, up from 17% in 1958 to 19% in 1968; for swine, up from 14% in 1958 to 20% in 1968; for beef cattle and sheep, up from 6% in 1958 to 12% in 1968; and unclassified types, steady in 1958 and 1968 at 6%.

The total number of each species of livestock or poultry no longer is significant in relation to the amount of feed consumed by class. This is because the feed conversion ability of different species varies greatly and because a fair proportion of each species is not consuming any manufactured feed.

Available Feed Products

When one seeks to label a product consumed by livestock as a "feed," he becomes involved deeply in semantics. Crampton and Harris (1969) point out:

"The dictionary says, in effect, 'To name is to identify.' There are on record names of about 6,500 feedstuffs. Their names were usually given to them by the persons using them, and identified them more or less locally. These 'common names' were given with no thought of any purpose other than physical identification. It was not surprising, then, to discover that when 6,500 names for feeds were examined, about 20% of them were other names for the same products called by one name in one area but by another in a different part of the country. For example, rolled oat groats is often known as rolled oats or as table oatmeal. The milling by-product of rolled oat groats is named oat middlings by the Canada Feeds Act, but feeding oatmeal by the Association of American Feed Control Officials in the United States."

With the foregoing in mind, one can look at the feed supply picture as estimated by the US Dept. of Agr. (Clough 1969). The feed grain supply for the 1968-69 feeding year is estimated at 216 million tons. The corn supply for the same period is forecast to be slightly more than 5.5 million bushels, the grain sorghum supply at 1.03 billion bushels. About 11 million tons of soybean meal, the by-product of soybean oil refining, is expected to be fed in 1968-69.

Although other crops figure in the feeding picture, the above provides an insight into what the Department of Agriculture considers when it refers to "feed."

Allen and Hodges (1968) note that farmers are feeding more grain to their livestock today than in years past. Using Department of Agriculture terminology, they point out: "Of the total quantity fed in 1966, concentrates grain and combinations of grain with other products made up about 47%, harvested roughages 19%, and pasture 34%."

Production Abroad

The feed manufacturing industry is well-developed in a number of nations besides the United States. Western Europe has a growing scientific feed industry with US feed companies participating in more than a dozen joint ventures on the continent.

Smith (1968) notes: "The West German feed industry is following the pattern of the US firms which are its counterparts—its production capacity is growing but it is decentralizing its mills." The same is true in Italy, Spain, and France, all of which have substantial feed manufacturing industries. In other countries, the terminal mills (Fig. 30) are still predominant.

Courtesy of Feed Bag Magazine

FIG. 30. TERMINAL MILL

Economical transport, particularly by water, makes large plants practical in Europe. This mill at Malmoe, Sweden, is operated by Carl Engstrom AB.

A brake on the growth of the West German industry, however, is what is known as an "open formula" law. This requires that the exact percentages of each ingredient contained in a feed mixture be printed on the tag accompanying the sack or bulk delivery. Although the Fachverband der Futtermittelindustrie (the West German feed manufacturers' association) has been pursuing reform of the open formula law for more than a decade, it has not yet been successful.

FIELD PROCESSING METHODS

In essence, feed manufacturing is a comparatively simple operation. Cereal grains, oilseed meals, packinghouse and other by-products, sometimes molasses, and vitamin and other ingredients used in minute quantities are combined to produce rations which will permit the feeder to achieve maximum results in meat, milk, egg, or fur production.

First the cereal grains and oilseed meals must be reduced to the desired size and consistency of particle. Then the other components are added and the mixing is done.

The basic forms of dry finished feeds are mash, pellets, crumbles, and cubes. The crumbles actually are pellets which have been crushed or broken deliberately. The cubes are oversized pellets used mainly in feeding cattle and sheep. A specialty form of feed is the block, which usually measures about 10 in. in all dimensions and weighs $33\frac{1}{3}$ to 40 lb. Blocks are used mainly to provide supplemental protein and/or minerals to ruminants.

Incoming grain usually reaches the feed mill by truck or rail. In the case of the smaller local-scale plant, much of the grain requirement is filled by the farmers who purchase finished feed from the mill. Or it may come from grain farmers nearby. In areas with religious beliefs which bar the use of automobiles, grain still is hauled in by horse and wagon.

The medium-size and large mills receive ingredients by truck, rail, and sometimes by barge. Water transport is particularly popular along the Mississippi, Missouri, and Ohio rivers. Barge tows haul millions of bushels of grain annually to Gulf of Mexico ports for export shipment and to feed mills in the South.

Rail shipment of grain is still extremely important. The recent development of jumbo covered hopper cars has spurred renewed interest in rail movement of grain. Unfortunately, these modern cars are few in number compared to the demand. Boxcars still haul most of the grain moved by rail and a shortage of such cars is experienced during harvest time every year. Boxcars have a serious drawback for grain shipping because they are difficult to clean out thoroughly and frequently leak grain.

Unloading and Conveying

When a shipment of grain arrives at the feed mill, it is unloaded by gravity, by air, or manually. Frequently, power shovels are combined with hand labor to unload railroad boxcars. Hopper cars are easy to unload; their bottom gates are opened and the grain flows directly by gravity to underground conveyors.

Barges generally are unloaded by positive or negative air systems. Truck shipments usually are unloaded by opening the rear tail gate after the truck body has been elevated to 30° or 45° and dumping the cargo through a gate into an underground pit or onto an underground conveyor.

Large feed mills sometimes employ giant boxcar unloaders. These expensive machines, some of which cost $250,000, lift the entire car, tilt it to one side, and then discharge the contents when the side doors have been slid open.

The actual movement of ingredients within the mill usually is done by gravity. First, however, the grain must be elevated above the highest processing machine before the gravity process can begin. This is accomplished through the use of what commonly is called the "bucket elevator." It consists of a durable continuous belt on which are mounted cups made of metal or plastic. These elevators can move grain upwards hundreds of feet.

For horizontal movement or slight elevation, a feed mill may use a screw-type conveyor, made of mild or stainless steel; a drag conveyor, in which single or double chains haul grain along a stainless steel chute; a continuous belt, with a V-trough in its center; or an air system, in which grain is carried along in a jet-like stream of compressed air.

Storage and Conditioning

Grain is stored in concrete silos or steel tanks. An occasional wooden bin used for grain storage is still seen. Grain of comparatively high moisture content is avoided by the feed manufacturer. Generally, the moisture content of grain for feed manufacturing does not exceed 15% (12% is preferred).

Should the feed mill be unable to purchase the drier grain it requires, it may install a grain drier in a separate structure adjacent to the mill. Such a machine removes moisture from grain by treating it with hot air. Gas or oil is the fuel source for such driers.

Some feed mills, though not many, may dehydrate alfalfa or a grass such as Coastal Bermuda for use as a feed ingredient. These commodities are dried in a rotating cylinder which is heated by gas or oil flames to several hundred degrees.

Particle Reduction

The hammer mill is the basic grinding device (see Fig. 31). It is situated in the basement except in the case of small single-floor mills. The hammer mill's grinding chamber houses rows of loosely-mounted swinging hammers or plates of hardened steel. They pulverize the material by striking it sharply as they swing. The

Courtesy of Prater Pulverizer Co.

FIG. 31. HAMMER MILL

Basic grain processing machine is the hammer mill, which drops free-mounted hard-metal bars against the grain. The fineness of the product which emerges is determined by the coarseness of the screen (right center). Poultry feeds are ground finest; ruminant feeds coarsest. The separate fan (left) carries away the ground grain to holding bins or directly to the feed mixer.

processed material is ejected from the chamber when it is ground finely enough to pass through the perforations in the screen which is a part of the mill.

The screens are manufactured with perforations ranging from ⅛ in. upwards to ½ in. A screen can be removed and another installed in a matter of minutes. Some hammer mills have semiautomatic screen changers which reduce the time to 1 or 2 min. The fineness of grind is determined by the type of feed in which the ground product is to be used. Poultry rations, for example, are often finely ground; feed for cattle and sheep always is coarser.

The attrition mill, which has been in use for more than 200 yr, finds its principal use in the grinding of grain for dairy and swine feeds. Here the aim is to produce a finely textured soft product. Revolving grinding plates pulverize the grain in the attrition mill. Roller mills include pairs of corrugated metal rolls, one of which revolves at a speed 2 to 4 times that of the others. The grain is forced between the rolls and sheared by the corrugation. Stone or buhr mills are used only where a small feed business has succeeded a defunct flour milling operation.

When feed is made on a continuous basis, automatic feeders govern the flow of grain into the grinder so that it does not become overloaded. Most of these devices operate volumetrically, although some weigh the grain.

The grinder costs more to operate than any other machine in the mill. To start and stop a hammer mill repeatedly often proves extremely expensive from the power consumption standpoint. Thus the grinding cost per ton is lower in the larger continuous operations.

A hammer mill or other grinder requires a powered fan to convey away the ground material from the processing unit. Otherwise the feed mill will be extremely dusty, which usually is an indication of an inefficient operation. Furthermore, government health authorities have become increasingly concerned about the hazards resulting to workers employed in a dusty environment.

Scalpers and Separators

The cleaning process for incoming feed ingredients employs scalpers to remove coarse material from the product flow before it reaches the mixer. Separators perform a similar function and are widely used; they consist of reciprocating sieves which classify grains of different sizes and textures. Occasionally, these units are used to rough-grade grain for weight and quality.

The dairy farmer whose cow injures her intestines by ingesting a stray piece of wire or other metal terms the ailment "hardware disease." This same affliction can hurt processing machinery in the feed mill. Thus permanent or electric magnets are essential in the feed plant. They usually are installed ahead of the grinder and at other later sites in the feed processing line.

Mixing Procedure

Feed mixers consist of three basic types: (1) vertical, (2) horizontal, and (3) drum. The first is the most widely used, mainly in the smaller plants (see Fig. 32). Horizontal mixers are used in large mills and drums by some in the intermediate size range.

To make mash-type feeds, the ground grain and other ingredients, which may include a protein source, vitamins, minerals, and drugs are measured into the top or end of the mixing cylinder. The

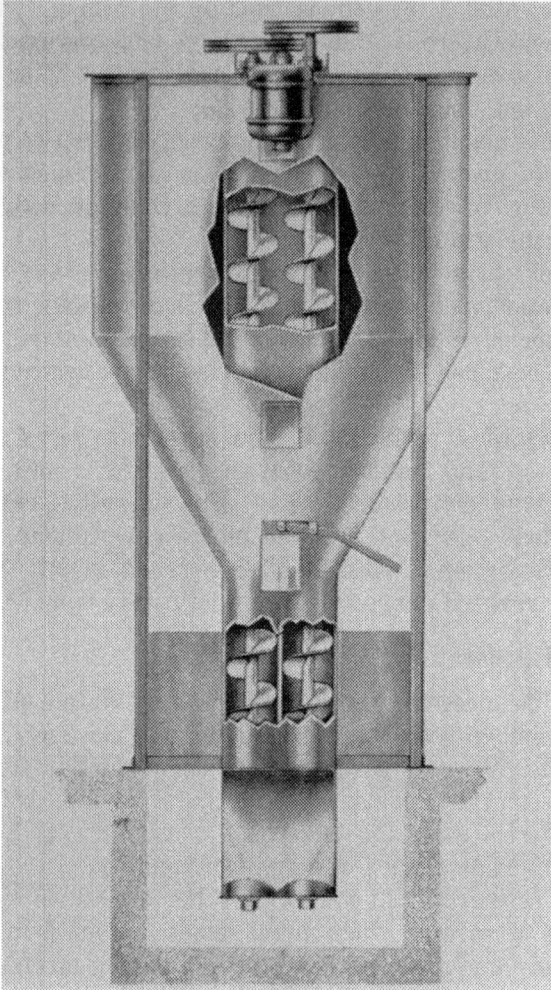

Courtesy of Prater Pulverizer Co.

FIG. 32. FEED MIXER

Second most important feed processing machine is
the mixer, which combines the ground grains with
vitamins, antibiotics, drugs, and any other additives which go to make up the complete ration.
Ingredients enter from the top, are mixed by
means of a single or double screw and then flow
downward. When thoroughly mixed, the feed can
be sacked at the bottom of the mixer or carried
away through pipes to bulk storage.

screws, paddles, or ribbons inside the cylinder then are operated for the established cycle determined to be necessary for the proper blending of the mixture. The finished feed is drawn off at the end opposite the loading entry.

When mixing a batch at a time, which usually is the case with a vertical mixer, the machine can be turned off once the cycle has been completed. The vertical mixer is most popular for smaller plants and can range in capacity from 500 lb to 7 tons. These capacity figures are misleading because a ton of fine scratch feed for chickens can be mixed properly in a unit with only 50 cu ft of capacity, while a ton of bulkier dairy cattle feed would require at least 100 cu ft.

The horizontal mixer is ideal for the larger feed mill in continuous production. Needed in such a system is a percentage feeder, to govern closely the proportions of various ingredients being introduced into the mixer. Some types of feeders measure by weight, others by volume. The former is preferred by most feed manufacturers.

A dry mash feed is ready for packaging, bulk delivery, or storage after it has been mixed. Other forms and types of feeds require additional processing. Molasses and animal and vegetable fats sometimes are added to rations. The sugar by-product must be heated to a thin consistency before it can be added. Even then, it still is sticky and usually is blended with the dry feed in a separate mixer (see Fig. 33).

The addition of fat to feeds, particularly for poultry, is a comparatively recent scientific advance. Fat usually is heated (in the case of animal fat, frequently to 170°F), before it is blended with the dry feed in a separate mixing chamber.

Pelleting

Pellet-type feed is considered essential for proper feeding of poultry, calves, rabbits, dogs, and mink. A major basic advantage of pellets is that all their components are bonded together tightly so that the bird or animal cannot separate them because of their differing degrees of flavor appeal. This means that the bird or animal is consuming the proper balance of nutrients intended for it.

The manufacture of pellets begins with a previously processed mash feed. If the pellet is to contain molasses or fat, the latter is blended with the dry mash immediately prior to pelleting. A binding agent usually is incorporated into the mixture, either sodium

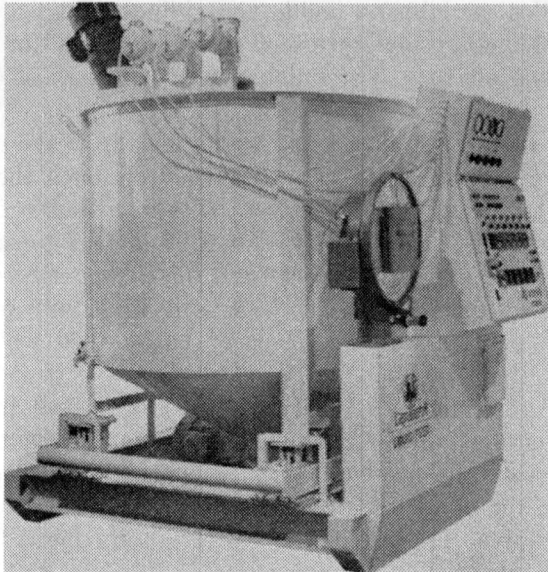

Courtesy of Barnard and Leas Mfg. Co.

FIG. 33. LIQUID FEED BLENDER

Ruminant rations in liquid form usually contain alcohol, molasses, urea, and vitamin and mineral additives.

bentonite or a lignin by-product of paper manufacturing. The latter also is claimed to be helpful in lubricating the pellet mill die for greater operating efficiency. Live steam is added to the dry mixed feed as it enters openings in the circular dies which are the heart of the pellet mill (see Fig. 34). In a few cases, a combination of water and heat is used instead of live steam.

After the pellets are formed and compressed, they are released from the dies into a conditioning unit. This is a cooling system which hardens the pellets. Broken pellets and milling fines are separated from the intact pellets and returned to the pellet mill intake for reprocessing.

In addition to the conventional pellet mill, pressure extruders which also cook the grain are in common use for specialty rations (see Fig. 35).

Making Crumbles.—Many pelleted feeds are used in the form in which they come from the pellet mill. Others have to be further reduced in size for maximum utilization by poultry or livestock. This reduction in size is called "crumblizing" and the resulting smaller product is called "crumbles." Actually, all the process does is to break or cut pellets into a smaller, irregular size.

Young chicks do particularly well on the broken pellets. It is not possible economically to manufacture intact pellets small enough for these young birds, because of the additional wear their manufacture would place on the pellet mill dies and the difficulty in handling them without breakage.

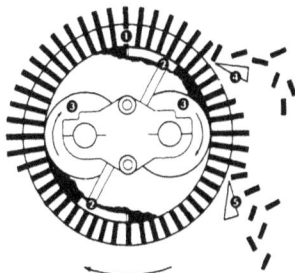

FIG. 34. HOW A PELLET MILL WORKS

Powdered or granular material is fed into rotating cylindrical die (1); then distributed evenly across the face of the die by spreader flight (2). Rolls (3) force material through die holes, forming pellets. Knife (4) cuts off pellets for further processing or packaging. Second knife (5) cuts off pellets made from material deflected to the right roll by spreader flight.

Flaking Grain

Considerable interest has been shown, particularly in the southwestern United States, in feeding flaked grain which has been conditioned by steam. Rations incorporating this flaked grain reportedly promote faster gains in beef cattle at lower feed cost. To date, only a few companies have produced units for flaking grain with steam and it is not possible at this time to describe a representative steam flaking machine.

Packaged or Not

Whether feed is packaged or sold in bulk depends to a great extent on the size of the feeding operation in terms of numbers

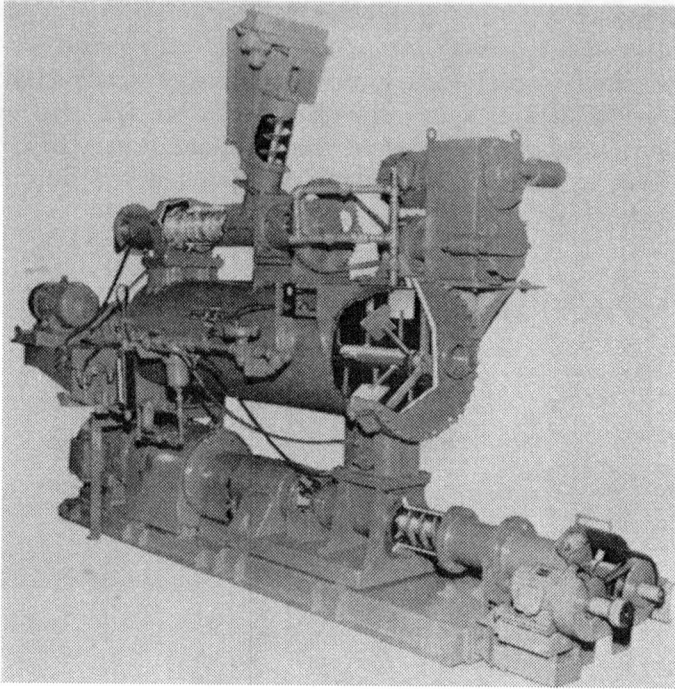

FIG. 35. COOKER-EXTRUDER

Steam pressure is utilized to cook or gelatinize cereal grains
and starches; an extruded end product is achieved by forcing
the cooked material through a comparatively thin die.

of livestock or poultry. In the early years of the US feed industry,
all feed was sacked. Currently, according to the American Feed
Manufacturers Association, 60% of the feed produced by its members is sold in bulk.

The preference for bulk versus sacked feed varies considerably
by regions. In 1968, the Pacific coast states bought 85% in bulk,
the Rocky Mountain states 83%, and the South Atlantic region
80%. In the Corn Belt, where livestock production is highly
concentrated, only 40% of the feed consumed moved in bulk.
From 1960 to 1967, bulk shipments of feed gained every year (see
Fig. 36). In 1968, the preference for bulk held steady but did not
increase.

The most popular package for sacked feeds is the 50-lb multi-
wall paper bag. Smaller sacks sometimes are used for specialty

FIG. 36. TRUCK FOR BULK DELIVERY OF FEED

Large volume users of mixed feeds usually buy their requirements in bulk. This is a typical bulk delivery truck. The vehicles often haul grains from farms to the feed mill on their return trips from feed deliveries.

items such as dog and calf feeds. Reused, recleaned burlap sacks of 100-lb capacity sometimes are used for bulky dairy feeds. The use of cotton bags, once very popular when housewives made clothes for the family from them, has virtually disappeared. Milk replacers and some other specialty rations sometimes are packed in 25-lb capacity plastic or metal pails.

Feeding Research

Every land-grant university and college in the 50 States is doing some feeds or feeding research (Schaible 1970). Nutrition specialists at public institutions appear frequently on the programs of feed manufacturers' meetings and some, where permitted, serve as consultants to the feed industry.

A degree in feed milling technology is offered by Kansas State University at Manhattan in its Grain Science & Industry Department. The university operates its own pilot feed mill (see flow diagram in Fig. 37), both as a training facility and for the production of experimental diets. In the West, Oregon State University operates a feed mill on campus for the production of both experimental rations and feeds for the university livestock and poultry populations.

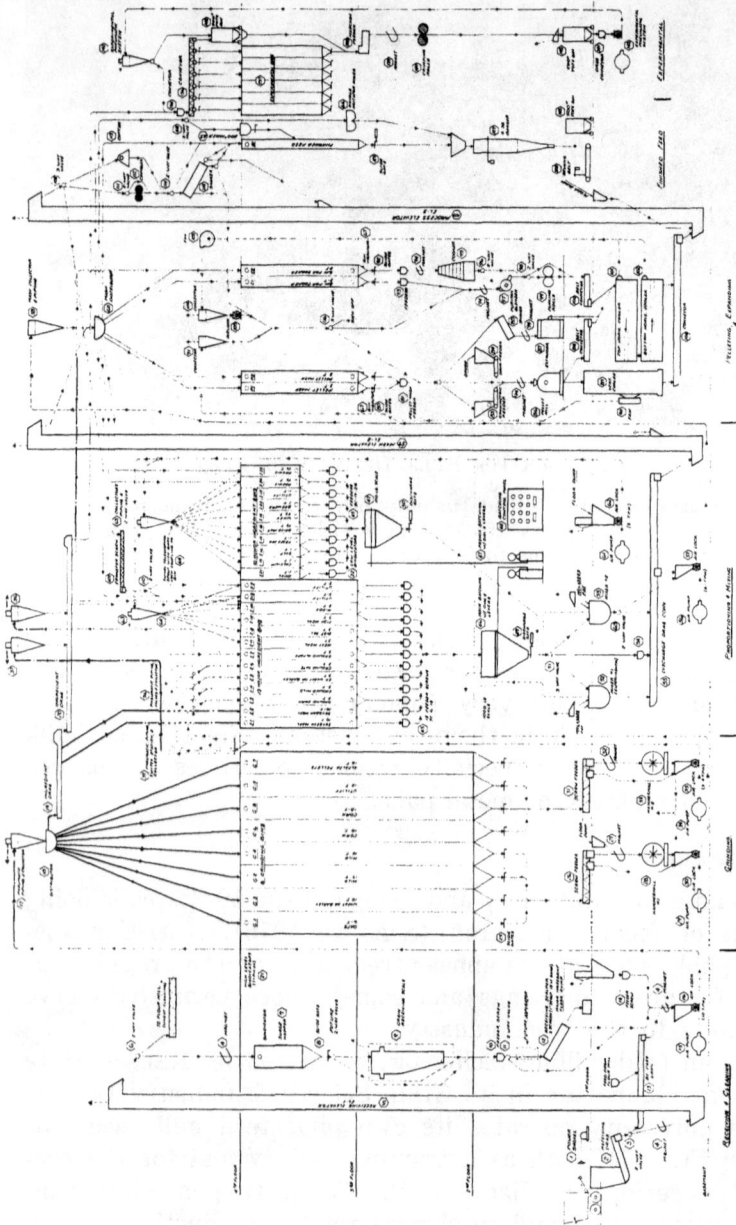

Courtesy of T. E. Stivers Co.

FIG. 37. PROCESS FLOW PLAN

Ingredient storage, material flow, and location of processing and conveying machinery in new feed mill at Kansas State University.

FEED CONSTITUENTS

Grains and cereal by-products are the bases of virtually all complete feeds, with the exception of some liquid diets for beef cattle, dairy cattle, and sheep. Still somewhat in the experimental stage, these liquid feeds usually are made up of alcohol, molasses, and urea and are fed along with ample roughages. Following is a commodity by commodity description of the basic products and by-products in broad use in modern feed formulation.

Linear programming is making inroads in feed manufacturing. This computerized procedure for least-cost formula development "is a natural for feed formulation problems. It cannot replace the nutritionist or management but it can free a nutritionist to do nutritional work and various levels of management to manage rather than punch a calculator and wonder if the job is really done." (Snyder *et al.* 1969).

In essence, linear programming selects the best protein source, for example, depending on the current market picture. It could select soybean meal in the proper percentage to replace cottonseed meal, meat scraps to replace fish meal, whatever would be the least expensive while still doing the job nutritionally.

Corn

Call it "maize" as the Europeans do, but it still is the basic grain in feed manufacturing. By itself, it is known in the feed industry and to poultry raisers as the key ingredient in "scratch" feed, along with some cracked wheat, cracked milo, steel-cut oat groats, and perhaps some other components if they are low in price and locally available.

Here are some of the most popular forms of corn and corn by-products used in the manufacture of formula feeds: (1) cob fractions—mechanically-separated portions of corn cobs; (2) bran—outer coating of the kernel; (3) feed meal—fine siftings from screened cracked corn; (4) ground corn—entire kernel ground or chopped; (5) cracked corn—same as ground corn; (6) grits—medium-sized hard flinty ground corn with little or no bran or germ; (7) flour—fine-sized hard flinty ground corn with little or no bran or germ; (8) ground ear—entire ear ground except for husks; (9) flaked corn—cracked corn run over flaking rolls; (10) hominy feed—mixture of bran, germ, and part of the kernels; (11) ground cob—the result of grinding the intact cob; and (12) gluten feed—what remains after most of the starch, gluten, and germ have been removed by the wet milling process in the manufacture of syrup or starch.

Before proceeding to descriptions of other grains and their by-products used in feeds, some definitions should be reviewed. Crampton and Harris (1969) point out:

"The term 'feed' is another whose common meaning varies. We find that 'feed barley' and 'barley feed' are not synonymous. 'Feed barley' names a grade or quality; it refers to whole barley that is rejected for use as seed or for malting. 'Barley feed' refers to a combination of barley hulls and middlings, which are the by-product obtained when the barley grain is dehulled and the groat scoured down to a rounded (pearl-shaped) particle called pearl barley. Or again, we find corn feed meal to be the siftings from making cracked corn, but corn gluten feed to be the product obtained when the bran from the wet milling of corn has been returned to the gluten part of the original grain."

Oats

Oats long has been a popular feed component. Some of the oat products used in feed formulation are: (1) groats—cleaned grain without hulls; (2) hulls—primarily the outer covering; (3) oat meal—broken rolled groats, groat chips, and floury groat fines; (4) clipped by-product—chaffy material from ends of hulls, empty hulls, light immature oats, oat dust; and (5) mill by-product—hulls and particles of groats.

Rye

Rye is not a popular feed ingredient today. If used at all, it usually is in one of these forms: (1) malt sprouts—sprouts from malted rye plus some hulls and other fractions of the malt; (2) mill run—outer covering of kernel and germ with some flour; and (3) middlings—rye feed and some other components of rye flour milling.

Wheat

This major food grain long has been used in the manufacture of feeds. Its use is greatest at those times when it is less expensive pound for pound than corn. Here are five wheat products or by-products used in feeds: (1) bran—coarse outer covering of the kernel; (2) feed flour—wheat flour with some bran and germ; (3) germ meal—mainly the germ, with some bran and middlings; (4) middlings—fine particles of bran, germ, and flour; and (5) mill run—coarse and fine bran, germ, flour, and some other minute milling products.

Rice

Rice and its by-products find their greatest use in feeds in areas where the grain is produced. Some of the more popular rice products used in feed formulas are: (1) ground rough—the whole grain and the hulls; (2) polishings—the by-product of brushing the grain to polish the kernel; (3) hulls—primarily the outer covering; (4) mill by-product—total offal including hulls, bran, polish, and broken grains; and (5) ground brown—entire product except for hulls.

Barley

Barley is the basic feed grain in the Pacific Northwest, where many rations are formulated with it as the primary ingredient. Some of the barley products are: (1) hulls—the outer covering of the grain; (2) pearl by-product—the entire end-product of pearling the grain; and (3) mill by-product—the entire residue of milling flour from clean barley.

Grain Sorghums

Milo in particular, kafir, and feterita are growing in popularity year by year in the making of formula feeds. In times past, grain sorghums were used primarily in simple scratch feeds. Now starch is being produced from grain sorghums and the resulting by-products are being fed. These are representative grain sorghum feed ingredients: (1) gluten feed—what remains after most starch and germ have been removed in wet milling; (2) rolled—whole grain run over smooth flaking rolls; (3) grits—hard flinty portions with little or no bran or germ; and (4) mill feed—mixture of bran, germ, part of the starchy portion of kernels.

NEW KNOWLEDGE

Crop scientists continually work to develop new varieties of grains that will be better than their predecessors. Currently believed to be of major importance is a corn type known as Opaque-2, which is unusually high in the essential amino acid lysine.

Opaque-2 was developed by Purdue University scientists and basically is a "modified protein" variety of corn. Scott (1968) points out: "In the United States, the greatest value of high-lysine corn will be as a replacement for both regular yellow corn and protein supplement in animal rations. High-lysine corn is expected to find its greatest domestic use in swine rations, since hogs require a high level of lysine."

Scott believes that if Opaque-2 corn is grown widely and is fed to their own swine by corn-and-pig farmers, "The most serious adjustment will be required by feed companies and soybean processors who have relied on swine supplements for a substantial volume of their business."

Similar studies are being pursued on other cereal grains used in feeds, but none has received the wide attention accorded high-lysine corn.

SELECTION OF INGREDIENTS

The market price and the delivery situation have profound effects on the decisions a nutritionist must make in the formulation of feeds. Fortunately, most formulators consider not only the economic advantages to their companies of using certain ingredients, but primarily analyze what the ingredient will contribute to the finished product in terms of economical production of meat, milk, eggs, or fur.

We now know that specific feeds must be manufactured for the several stages of an animal or bird's life, including early growth, maintenance, fattening or furring-out, and lactation. It also appears likely that the requirements of livestock differ by sex, although further study must be conducted to establish this conclusion with certainty.

Crampton and Harris (1969) acknowledge some of the complexities of feed formulation when they write: "In practice, we find it impossible to compound rations meeting exactly the quantities or proportions of nutrients set out in feeding standards, because we must deal with feeds which themselves are complex mixtures of nutrients. But we can prepare formulas for mixtures of feedstuffs whose nutrient combinations approximate the needs of specified animals closely enough so that the animal, by means of its own metabolic machinery, can make the final fitting, discarding the surpluses, and temporarily even making good minor shortages in the day's intake."

Whatever the ingredient to be selected, price cannot be the sole consideration. There formerly were companies in business which specialized in "cheap" feeds. There also were companies which made both quality and cheap rations. The cheap-only feed makers long have gone out of business and the reputable companies no longer offer cut-rate "second" lines.

Composition Analysis

The table which follows provides data used by the nutritionist in calculating the nutrient content of manufactured feeds to assure that they will do the job for livestock and poultry that they are intended to do. This information was compiled by C. W. Sievert, prominent feed industry consultant at Blue Island, Illinois, and is used with his permission.

In Table 27, average values obtained from recent research are tabulated, using round numbers as much as possible. Recommendations of the National Academy of Science—National Research Council have been considered, along with data from other sources.

Vitamin and metabolizable energy values are reported in both milligrams and kilocalories per kilogram, as well as per pound. An asterisk (*) indicates that reliable data are not available; a blank (—) indicates no value.

Feed Formulas

Now that we have data which show the composition of many feed ingredients, let us take a look at some representative feed formulas. The examples which are presented here are among the more simple. If one counted every trace element and vitamin additive individually and added them to the total number of basic ingredients such as the grains, a formula easily could contain 50 or more components.

Sievert (1969) here provides illustrations of rations and supplements for one class of swine, the brood sow. The formula under "ration" is for a complete diet, ready for the sow. If the "Supplement" formula is used, a ton of complete feed would be prepared by adding 1,100 lb of ground corn and 400 lb of ground oats to 500 lb of the supplement.

In the feeding of beef cattle, the supplement type of ration is used the most. It should provide sufficient protein to balance that which is in the grain to be fed and in the roughage which the cattle will eat. It probably also will have to provide some supplemental minerals and vitamins.

Sievert gives us 4 formulas for beef cattle supplements, 3 providing 32% protein and one with 64% protein. The major differ-

TABLE 27
NUTRIENT COMPOSITION OF FEEDSTUFFS

The vitamin columns (Thiamine, Riboflavin, Niacin, Pantothenic Acid, Choline) are expressed as **Milligrams per** Lb / Kg.

Ingredient	Protein %	Fat %	Fiber %	NFE %	Ash %	Calcium %	Phosphorus %	Thiamine Lb	Thiamine Kg	Riboflavin Lb	Riboflavin Kg	Niacin Lb	Niacin Kg	Pantothenic Acid Lb	Pantothenic Acid Kg	Choline Lb	Choline Kg	TDN %	Met. Energy Kcal Lb	Met. Energy Kcal Kg
Alfalfa (leaf) meal (20% dehydrated)	20.0	3.5	21.0	38.0	10.0	1.5	0.25	3.0	6.6	7.0	15.5	24.0	53.0	14.5	32.0	730	1600	*	*	*
Alfalfa (leaf) meal (20% sun-cured)	20.0	2.5	23.0	35.0	8.5	1.5	0.25	*	*	*	*	*	*	*	*	*	*	52	*	*
Alfalfa meal (17% dehydrated)	17.0	2.5	25.0	39.0	9.0	1.30	0.23	1.5	3.3	5.5	12.3	20.0	45.0	13.5	30.0	680	1500	47	620	1360
Alfalfa meal (17% sun-cured)	17.0	2.3	26.0	35.0	9.5	1.20	0.25	1.1	2.3	5.0	11.0	18.0	40.0	12.0	26.5	*	*	46	*	*
Alfalfa meal (15% dehydrated)	15.0	2.3	27.0	40.0	8.5	1.20	0.22	1.3	3.0	4.8	10.6	19.0	42.0	9.5	21.0	700	1540	48	*	*
Alfalfa meal (13% sun-cured)	14.0	1.8	32.0	34.0	7.2	1.10	0.20	1.5	3.3	5.0	11.0	18.0	40.0	9.0	19.9	675	1490	44	*	*
Alfalfa stem meal	10.0	1.2	36.0	34.0	7.0	0.8	0.20	*	*	*	4.5	*	*	*	*	*	*	42	*	*
Animal fat (grease)	*	96.0	—	—	—	*	—	—	—	—	—	—	—	—	—	—	—	*	3500	7700
Animal fat (tallow)	*	96.0	—	—	—	*	—	—	—	—	—	—	—	—	—	—	—		3200	7050
Apple pomace (dried)	4.0	4.5	17.0	62.0	2.0	0.10	0.10	*	*	*	*	*	*	*	*	—	—	64	*	*
Babassu meal	23.0	6.0	13.0	45.0	5.5	0.10	0.75	2.5	5.5	0.6	1.3	6.0	13.2	2.5	5.5	325	715	81	*	*
Barley	11.0	2.0	6.5	68.0	3.0	0.08	0.40	2.5	5.5	0.8	1.7	29.0	64.0	3.3	7.3	525	1160	78	1290	2840
Barley (Pacific coast)	9.0	2.0	6.5	68.0	2.3	0.05	0.40	1.9	4.2	0.6	1.3	22.0	48.5	3.6	7.9	525	1160	79	1280	2820
Barley feed	12.0	3.0	8.5	62.0	3.5	0.05	0.40	2.5	5.5	1.0	2.2	28.0	28.5	3.0	6.6	525	1160	*	*	*
Barley malt cleanings	18.0	1.5	16.0	51.0	5.5	0.10	0.45	*	*	*	*	*	*	0.6	1.3	*	*	59	*	*
Beans (cull navy)	21.0	1.0	4.5	57.0	4.5	0.10	0.10	2.1	4.6	1.4	3.1	12.5	27.5	0.6	1.3	*	*	*	610	1350
Beet pulp (dried)	9.0	0.6	20.0	58.0	3.5	0.65	0.35	0.2	0.4	0.3	0.65	7.0	15.5	0.7	1.5	375	830	78	280	620
Blood meal	80.0	1.5	1.0	2.5	5.5	0.30	0.35	0.1	0.2	1.0	2.2	13.0	28.5	2.0	4.4	300	660	62	1400	3080
Bone meal (raw)	25.0	4.0	1.5	3.0	58.0	22.0	10.0	0.1	0.2	0.6	1.3	1.9	4.2	1.1	2.4	*	*	60	*	*
Bone meal (steamed)	12.0	3.0	2.0	6.5	71.6	28.0	13.5	0.2	0.4	0.4	0.9	2.0	4.4	1.1	2.4	*	*	18	*	*
Bone meal (special steamed)	6.0	1.0	1.0	2.5	82.0	35.0	14.5	0.1	0.2	0.5	1.1	13.0	28.5	0.8	1.8	*	*	*	*	*
Bread (dried)	10.0	1.5	0.5	72.0	1.5	0.03	0.12	1.3	2.8	0.4	0.9	13.0	28.5	1.1	2.4	*	*	45	*	*
Brewers dried grains	25.0	6.0	16.0	42.0	3.8	0.25	0.50	0.3	0.6	0.5	1.1	19.0	42.0	3.9	8.6	700	1540	83	1640	3600
Buckwheat	11.0	2.0	10.0	62.0	2.0	0.10	0.25	1.8	4.0	0.9	2.0	9.0	19.8	5.0	11.0	*	*	66	765	1680
Buckwheat middlings	28.0	6.5	6.0	42.0	4.8	0.15	0.35	*	*	*	*	*	*	*	*	*	*	75	1225	2700
Buckwheat feed	16.0	3.5	15.0	54.0	4.5	*	0.50	*	*	*	*	*	*	*	*	*	*	53	*	*
Buckwheat hulls	5.0	1.0	44.0	31.0	2.5	*	*	*	*	*	*	*	*	*	*	*	*	14	*	*
Buttermilk (condensed)	10.0	2.0	0.1	12.0	3.5	0.45	0.25	6.5	14.4	*	*	*	*	*	*	*	*	26.5	*	*

Feed	1	2	3	4	5	6	7	8	9	10	11	12	13	14	15	16	17	18	19	20	21	22
Buttermilk (dried)	32.0	5.5	0.4	44.0	9.5	1.3	1.0	1.5	3.5	12.0	26.4	4.0	8.8	13.0	28.6	800	1760	85	1245	2750	*	*
Buttermilk (dried sweet cream)	32.0	4.5	0.2	45.0	8.5	1.3	1.0	1.6	3.5	13.0	28.7	4.0	8.8	14.0	30.1	800	1760	84	*	*	*	*
Charcoal	2.5	0.2	50.0	35.0	6.5	4.50	0.03	0.6	—	—	—	—	—	—	—	*	*	*	*	*	*	*
Citrus pulp (dried)	6.5	4.5	14.0	60.0	6.0	2.0	0.10	0.3	6.6	1.4	2.4	9.6	21.2	6.0	13.2	380	840	75	750	1650	*	*
Coconut meal	20.0	6.0	14.0	45.0	8.0	0.20	0.60	0.3	6.6	1.4	3.1	11.0	24.2	3.0	6.6	400	880	77	*	*	*	*
Corn (white)	8.6	3.7	2.3	68.0	1.3	0.04	0.30	2.2	4.8	1.4	1.3	8.0	17.6	2.0	4.4	*	*	80	1560	3440	*	*
Corn (yellow)	8.8	3.9	2.5	69.0	1.3	0.03	0.27	2.0	4.4	0.6	1.3	9.7	21.4	2.0	5.3	280	615	80	*	*	*	*
Corn (popcorn)	10.0	4.5	2.2	66.0	1.5	0.01	0.30	2.0	4.4	0.6	1.3	8.5	18.7	1.7	3.8	*	*	85	1320	2910	*	*
Corn and cob meal (yellow)	7.2	3.2	8.5	65.0	1.5	0.04	0.25	1.4	3.1	0.5	1.1	7.5	16.5	2.0	4.4	200	440	73	740	—	*	*
Corn bran	7.5	—	10.0	64.0	2.0	0.02	0.20	2.7	6.0	0.7	1.5	20.0	44.0	2.2	4.9	*	*	66	*	*	*	*
Corn cobs	2.5	0.5	33.0	50.0	1.9	0.10	0.04	0.8	1.7	0.2	1.1	—	—	—	—	225	495	45	*	*	*	*
Corn feed meal	9.0	4.0	3.0	69.0	2.0	0.05	0.40	1.5	5.5	0.9	1.1	10.0	22.0	2.5	5.0	600	1320	80	*	*	*	*
Corn distillers' grain (light)	24.0	9.0	13.0	44.0	2.0	0.05	0.30	0.7	1.5	1.3	2.8	12.0	26.4	2.5	5.5	1200	2640	82	1110	2140	*	*
Corn distillers' grain (dark)	27.0	9.0	11.0	41.0	2.6	0.05	0.60	0.9	5.5	6.6	6.6	20.0	44.0	2.5	11.0	2000	4400	89	1320	2910	*	*
Corn distillers' solubles (dried)	26.0	9.0	4.0	45.0	8.0	0.30	1.30	2.5	5.5	7.0	15.4	50.0	110.0	9.0	19.8	320	700	78	*	*	*	*
Cond. ferm corn extractives	24.0	0.0	0.0	19.0	10.0	0.05	1.50	1.2	2.6	—	5.7	37.0	81.0	6.7	14.8	800	1760	40	700	1540	*	*
Corn germ meal (wet milled)	21.0	1.0	12.0	52.0	2.0	0.03	0.50	2.5	5.5	1.6	3.5	18.0	39.6	1.6	3.5	650	1430	70	770	1700	*	*
Corn gluten feed	22.0	2.0	6.0	49.0	6.0	0.45	0.70	0.8	1.7	1.0	2.2	30.0	66.0	7.5	16.5	350	770	75	760	1670	*	*
Corn gluten meal (41%)	42.0	2.0	4.5	37.0	4.0	0.15	0.40	0.1	0.2	0.7	1.5	22.0	48.0	4.5	10.0	150	330	81	1170	2580	*	*
Corn gluten meal (60%)	61.0	2.5	2.0	22.0	1.5	0.01	0.40	0.1	0.2	0.9	2.0	24.0	53.0	1.3	2.9	*	*	83	1750	2860	*	*
Corn oil	—	98.0	—	—	—	—	—	—	—	—	—	—	—	—	—	—	—	—	—	4000	*	*
Cottonseed meal 43%	43.0	4.5	12.0	25.0	6.0	0.15	1.20	3.0	6.6	2.1	4.6	20.0	44.0	8.0	17.6	1300	2860	75	1000	2200	*	*
Cottonseed meal 41%	41.0	5.0	12.5	26.0	6.0	0.15	1.10	2.4	5.3	2.0	5.3	17.0	37.5	6.0	13.2	1200	2640	73	*	*	*	*
Cottonseed meal 41% solvent	41.0	1.5	13.0	28.0	6.5	0.15	1.20	3.0	6.6	2.0	4.4	20.0	41.0	6.5	14.3	1200	2340	66	830	1825	*	*
Cottonseed meal 36%	36.0	4.5	14.5	30.0	6.0	0.15	1.10	*	*	1.5	3.3	15.0	33.0	5.5	12.1	1200	3340	64	*	*	*	*
Cottonseed hulls	4.0	1.0	44.0	38.0	2.5	0.10	0.10	*	*	1.7	3.7	—	—	—	—	*	*	44	*	*	*	*
Crab meal	31.0	1.5	11.0	8.0	41.0	14.0	1.50	*	*	2.3	5.1	20.0	44.0	3.0	6.6	900	1980	27	850	1870	*	*
Egg—dried whole	46.0	42.0	—	—	3.5	0.19	0.76	1.5	3.3	4.7	10.4	1.1	2.4	—	—	*	*	*	*	*	*	*
Egg—dried yolk	30.0	58.0	—	—	3.0	0.25	1.10	2.2	4.9	3.0	6.6	0.3	0.7	—	—	*	*	*	*	*	*	*
Egg—dried white	86.0	—	—	*	—	0.13	—	*	—	—	20.4	3.0	6.6	—	—	*	*	*	*	*	*	*
Emmer (spelt)	13.0	2.0	10.0	62.0	3.5	0.04	0.35	—	—	—	—	—	—	—	—	*	*	72	*	*	*	*
Feterita	11.0	3.0	2.5	70.0	2.0	0.02	0.30	*	*	0.9	2.0	7.5	16.5	3.5	7.7	600	1320	79	1050	2300	*	*
Feathers, hydrolyzed poultry	85.0	2.5	2.0	2.0	4.0	0.20	0.20	*	*	0.9	2.0	16.5	55.0	4.0	8.8	*	*	63	*	*	*	*
Fish meal (menhaden)	60.0	7.0	1.0	2.0	20.0	5.4	2.8	0.3	0.7	2.2	4.9	25.0	55.0	4.0	8.8	1360	3000	58	1300	2870	*	*

TABLE 27 (*Continued*)
NUTRIENT COMPOSITION OF FEEDSTUFFS

Ingredient	Protein %	Fat %	Fiber %	NFE %	Ash %	Calcium %	Phosphorus %	Thiamine Lb	Thiamine Kg	Riboflavin Lb	Riboflavin Kg	Niacin Lb	Niacin Kg	Pantothenic Acid Lb	Pantothenic Acid Kg	Choline Lb	Choline Kg	TDN %	Met. Energy Kcal per Lb	Met. Energy Kcal per Kg
Fish meal (sardine)	65.0	5.5	1.0	4.0	14.0	4.0	2.8	0.2	0.5	2.5	5.5	26.0	57.0	4.0	8.8	1300	2860	61	995	2190
Fish meal (Peruvian)	66.0	4.5	1.0	1.0	15.0	4.2	2.8	0.2	0.4	4.0	8.8	43.0	95.0	4.2	9.3	1700	3800	73	1020	2250
Fish meal (herring)	70.0	7.0	1.0	2.0	11.0	3.0	2.2	*	*	4.0	8.8	40.0	88.0	5.0	11.0	1800	3960	74	1360	3000
Fish meal (white)	63.0	3.0	1.0	2.0	21.0	7.5	3.5	0.8	1.7	4.0	8.8	30.0	66.0	4.0	8.8	4000	8800	62	*	*
Fish meal (red)	57.0	8.0	1.0	1.0	26.0	7.5	3.8	*	*	3.0	6.6	27.0	60.0	3.8	8.4	1550	3400	74	1385	3060
Fish solubles condensed (50%)	31.0	5.0	0.5	2.0	10.0	0.6	0.7	2.5	5.5	6.5	14.2	75.0	165.0	16.0	35.0	1800	4000	42	*	*
Fish homogenized (condensed)	29.0	8.0	0.1	2.0	15.0	1.6	1.0	*	*	1.3	2.8	17.0	37.5	4.3	9.5	800	1760	*	650	1430
Fish oil	—	98.0	—	—	—	—	—	—	—	—	—	—	—	—	—	—	—	*	3600	8000
Flaxseed	24.0	35.0	6.5	24.0	4.0	0.25	0.55	*	*	*	*	—	—	*	*	*	*	108	*	*
Flaxseed screenings	15.0	9.0	12.0	46.0	7.0	0.35	0.45	*	*	*	*	*	*	*	*	*	*	56	*	*
Grain screenings	15.0	4.5	10.0	—	6.5	0.15	0.35	4.0	8.8	1.0	2.2	20.0	44.0	3.4	7.5	*	*	61	*	*
Hominy feed, white	10.5	5.5	5.0	—	3.0	0.02	0.50	*	8.8	1.0	2.2	20.0	44.0	3.4	7.5	*	*	83	1300	2860
Hominy feed, yellow	10.5	5.5	5.0	—	2.8	0.05	0.50	3.5	7.7	0.9	2.0	22.0	49.0	3.4	7.5	430	950	84	1365	3000
Kafir corn	10.5	3.0	2.5	70.0	2.0	0.04	0.30	1.8	4.0	0.7	1.5	18.0	40.0	6.0	13.2	300	660	81	1500	3300
Kafir head chops	9.5	2.5	8.0	64.0	4.0	0.08	0.25	*	*	*	*	25.0	55.0	1.0	2.2	*	*	68	*	*
Kelp (dried)	6.0	0.5	8.0	44.0	35.0	1.0	0.25	*	*	2.0	4.4	*	*	1.0	2.2	150	330	29	*	*
Kudzu meal	13.0	2.0	30.0	41.0	7.0	2.75	0.20	*	*	3.4	7.5	*	*	*	*	*	*	49	*	*
Lespedeza meal	13.0	2.5	30.0	46.0	5.5	0.90	0.20	*	*	3.9	8.6	*	*	*	*	*	*	46	*	*
Linseed meal (old process)	33	5.0	9.0	37.0	5.6	0.40	0.80	2.2	5.0	1.6	3.5	15.0	33.0	7.2	16.0	810	1800	76	*	*
Linseed meal (solvent)	35	1.5	9.0	38.0	5.8	0.40	0.80	4.0	9.0	1.4	3.1	13.5	30.0	6.5	14.5	540	1200	70	*	*
Liver and glandular meal	65.0	14.0	2.0	4.5	6.0	0.60	1.10	1.2	2.6	18.0	40.0	70.0	154.0	47.0	103.0	4750	10400	90	1300	2860
Malt (barley malt)	13.0	2.0	3.5	70.0	2.5	0.05	0.45	1.7	3.7	1.3	2.8	25.0	55.0	3.6	8.0	400	800	80	*	*
Malt sprouts	26.0	1.3	14.0	44	6.0	0.20	0.70	0.3	0.7	0.7	1.6	20.0	44.0	3.9	8.6	680	1500	70	*	*
Meat meal (55%)	55.0	8.0	2.5	2.0	22.0	8.0	4.00	0.5	1.1	*	1.6	24.0	—	*	*	*	*	65	920	2020

Feed																				
Meat and bone meal (50%)	50.0	8.0	2.5	2.0	29.0	10.0	4.80	0.5	1.1	2.0	4.4	20.0	44.0	1.7	3.7	900	2000	64	900	2000
Meat and bone meal (50%) solvent	50.0	2.0	2.5	2.0	30.0	10.0	4.80	0.5	1.1	2.0	4.4	20.0	44.0	1.7	3.7	800	1760	54	*	*
Meat and bone meal (45%)	45.0	8.0	3.0	1.0	34.0	10.5	5.00	0.5	1.1	1.4	3.1	17.0	37.5	1.1	2.4	800	1760	64	810	1785
Meat meal tankage	60.0	8.0	3.0	1.0	22.0	6.0	3.10	0.2	0.4	1.1	2.4	17.0	38.0	1.1	2.4	980	2160	68	1200	2610
Millet grain	12.0	4.0	9.0	69.0	3.6	0.06	0.30	3.3	7.4	0.8	1.76	26.0	57.0	3.5	7.7	440	880	75	*	*
Millet seed (foxtail)	12.0	4.0	9.0	69.0	4.0	0.05	0.30	*	*	*	*	*	*	*	*	*	*	75	*	*
Milo maize	10.5	2.8	2.5	70.0	1.8	0.04	0.28	1.7	3.9	0.5	1.2	19.0	42.0	5.0	11.0	300	670	82	1500	3300
Milo head chops	9.5	2.5	8.0	64.0	4.0	0.08	0.25	*	*	*	*	*	*	*	*	*	*	68	*	*
Molasses, beet	6.5	—	62.0	—	8.0	0.15	0.02	*	*	*	2.4	19.0	42.0	2.1	4.6	400	875	61	*	*
Molasses, cane (blackstrap)	3.0	—	63.0	—	8.5	0.8	0.06	0.4	0.9	1.5	3.3	15.0	33.0	17.0	38.0	*	*	56	890	1960
Molasses, citrus	7.0	—	52.0	—	6.0	1.3	0.15	*	*	2.8	6.2	12.0	26.0	5.5	12.0	*	*	54	*	*
Molasses, corn sugar	0.5	—	60.0	—	9.0	*	*	*	*	*	*	*	*	*	*	*	*	67	*	*
Molasses, wood	0.5	—	—	—	4.2	1.4	0.05	*	*	*	*	*	*	*	*	*	*	48	*	*
Oats	12.0	4.5	12.0	56.0	3.5	0.1	0.35	2.8	6.2	0.7	1.6	7.2	15.8	5.8	12.9	480	1070	70	1190	2620
Oats, western	10.0	4.5	11.0	58.0	3.5	0.1	0.35	*	*	*	*	*	*	*	*	*	*	72	*	*
Oats, clipped by-product	9.0	2.0	25.0	40.0	11.0	*	*	*	*	*	*	*	*	*	*	*	*	*	*	*
Oat groats (hulled oats)	16.0	5.5	3.0	63.0	2.2	0.07	0.45	3.1	6.8	0.6	1.3	3.7	8.1	6.6	14.6	560	1240	92	1450	3200
Oat meal, feeding	16.0	5.5	3.0	50.0	2.2	0.08	0.40	3.2	7.0	0.9	2.0	4.0	8.8	6.6	14.6	570	1260	91	*	*
Oat hulls	4.0	1.0	32.0	50.0	6.0	0.10	0.15	0.5	1.1	0.8	1.7	4.0	8.8	1.5	3.3	150	330	30	*	*
Oat mill by-product	10.0	3.0	20.0	53.0	6.0	0.10	0.25	*	*	1.4	3.1	*	*	*	*	*	*	35	*	*
Peanut meal and hulls (o.p.)	45.0	5.0	12.0	23.0	5.5	0.15	0.50	3.3	7.3	2.4	5.3	76.0	168.0	22.0	48.0	760	1680	76	1300	2860
Peanut meal and hulls (solvent)	47.0	1.0	13.0	21.0	5.0	0.20	0.60	3.3	7.3	2.4	5.3	77.0	170.0	24.0	53.0	770	1700	77	1130	2480
Peanut hulls	6.0	1.0	61.0	19.0	4.5	0.20	0.05	*	*	*	*	*	*	*	*	*	*	18	*	*
Poultry by-product meal	56.0	12.5	2.5	1.0	15.0	3.50	1.70	*	*	4.5	9.5	18.0	40.0	4.0	8.8	2700	5950	74	1350	2970
Poultry fat	—	96.0	—	—	—	—	—	—	—	—	—	—	—	—	—	—	—	*	3700	8150
Rice grain (brown)	8.0	1.2	2.0	74.0	1.4	0.04	0.22	1.2	2.8	0.27	0.6	16.0	35.2	2.8	6.2	410	906		1100	2420
Rice bran (solvent)	13.0	1.0	14.0	45.0	16.0	0.05	1.4	10.0	22.0	1.2	2.6	135.0	300.0	10.5	23.0	570	1250	55	1000	2200
Rice polishings	12.0	13.0	3.0	51.0	8.0	0.05	1.4	8.8	19.5	0.8	1.8	240.0	530.0	26.0	58.0	590	1300	78	1500	3300
Rice hulls	3.0	0.8	40.0	30.0	19.0	0.08	0.06	*	*	*	*	*	*	*	*	*	*	10	*	*
Rye	11.5	1.5	2.5	70.0	2.0	0.05	0.30	1.7	3.9	0.7	1.6	0.6	1.3	3.1	6.9	*	*	72	*	*
Rye middlings	16.0	3.5	6.5	60.0	3.5	0.05	0.60	3.3	3.3	1.1	2.4	7.5	16.5	3.0	22.0	*	*	76	*	*
Rye distillers grains, dried	22.0	6.0	14.0	47.0	2.8	0.10	0.40	1.8	1.8	1.5	3.3	7.6	16.7	3.0	6.6	800	1760	58	*	*
Whole pressed safflower seed	20.0	6.0	3.8	—	3.8	0.20	0.70	*	*	1.8	4.0	38.0	85.0	1.8	4.0	*	*	57	*	*
Sesame meal	44.0	5.0	6.0	20.0	10.0	2.00	1.20	1.3	2.9	1.7	3.7	*	*	2.9	6.4	680	1500	71	1200	2610

TABLE 27 *(Continued)*

NUTRIENT COMPOSITION OF FEEDSTUFFS

Vitamin values (Thiamine, Riboflavin, Niacin, Pantothenic Acid, Choline) are given in Milligrams per Lb and per Kg.

Ingredient	Protein %	Fat %	Fiber %	NFE %	Ash %	Calcium %	Phosphorus %	Thiamine Lb	Thiamine Kg	Riboflavin Lb	Riboflavin Kg	Niacin Lb	Niacin Kg	Pantothenic Acid Lb	Pantothenic Acid Kg	Choline Lb	Choline Kg	TDN %	Met. Energy Kcal Lb	Met. Energy Kcal Kg
Shrimp meal	45.0	3.0	12.0	1.0	26.0	7.3	1.50	*	*	1.8	4.0	*	*	*	*	2600	5800	43	*	*
Skim milk, dried	33.0	1.0	0.1	51.0	8.0	1.30	1.00	1.6	3.5	9.1	20.0	5.0	11.0	15.0	33.0	630	1390	80	1230	2700
Sorghum distillers' grains	32.0	8.0	12.5	38.0	4.0	0.15	0.55	0.5	1.1	1.4	3.1	26.0	58.0	5.5	12.0	370	820	80	*	*
Sorghum distillers' solubles	26.0	5.0	4.0	48.0	9.0	0.6	1.30	2.0	4.4	7.2	16.0	64.0	141.0	12.0	26.5	*	*	78	*	*
Sorghum gluten feed	24.0	3.5	9.0	46.0	8.0	0.15	0.65	*	*	1.2	2.6	24.0	53.0	8.5	18.7	790	*	83	*	*
Sorghum gluten meal	42.0	4.5	3.5	37.0	1.8	0.05	0.30	*	*	0.7	1.5	15.0	33.0	4.6	10.0	325	715	—	*	*
Soybeans	36.0	17.0	6.0	24.0	5.0	0.25	0.60	5.0	11.0	1.2	2.6	10.0	22.0	7.0	15.4	1300	2860	82	1130	2500
Soybean hulls	12.5	2.0	36.0	33.0	4.8	0.50	0.15	*	*	*	*	*	*	*	*	*	*	48	*	*
Soybean mill feed	13.0	2.0	31.0	35.0	5.5	0.55	0.15	*	*	1.6	3.5	*	*	6.0	13.2	*	*	37	350	770
Soybean meal (expeller)	42.0	5.0	6.0	30.0	6.0	0.25	0.60	1.8	4.0	1.5	3.3	13.5	30.0	6.5	14.3	1200	2650	74	1140	2530
Soybean meal 44% (solvent)	44.0	0.9	6.5	29.0	6.0	0.30	0.65	2.8	6.2	1.4	3.1	12.0	26.4	6.6	14.5	1240	2730	78	1020	2240
Soybean meal (dehulled-solvent)	50.0	0.8	3.0	28.0	6.0	0.25	0.60	1.1	2.4	1.4	3.1	9.5	21.0	6.5	14.3	1250	2760	79	1150	2540
Sun flower seed	16.0	20.0	30.0	21.0	3.0	0.20	0.50	0.2	0.44	*	*	*	*	*	*	*	*	76	1210	2660
Sugar (sucrose)	—	—	—	—	—	—	—	—	—	—	—	—	—	—	—	—	—	95	1670	3680
Tomato pomace, dried	21.0	12.0	29.0	24.0	4.0	0.25	0.50	5.4	11.9	2.8	6.2	*	*	*	*	*	*	56	*	*
Vegetable oils	—	96.0	—	—	—	—	—	—	—	—	—	—	—	—	—	—	—	—	4000	8800
Vinegar grains	18.0	6.0	19.0	46.0	2.5	*	*	*	*	*	*	*	*	*	*	*	*	—	*	*
Whale meal with bone	54.0	9.5	2.0	1.0	26.0	8.2	4.1	0.6	1.3	3.8	8.4	47.0	103.0	1.2	2.6	*	*	80	*	*
Wheat, hard	13.0	1.7	3.0	68.0	1.8	0.05	0.4	2.2	4.9	0.5	1.2	25.0	56.0	5.5	12.0	375	830	79	1500	3300
Wheat, soft	10.0	1.7	2.5	72.0	2.0	0.08	0.30	2.2	4.9	0.5	1.2	26.0	58.0	5.2	11.5	350	770	79	1400	3100
Wheat, durum	13.0	1.5	2.5	71.0	1.7	0.15	0.40	*	*	*	*	*	*	*	*	*	*	77	*	*
Wheat, flaked	12.0	1.5	3.0	67.0	2.0	0.05	0.35	*	*	*	*	*	*	*	*	*	*	77	*	*
Wheat bran	15.0	4.0	10.0	53.0	6.5	0.10	1.10	3.6	7.9	1.4	3.1	75.0	165.0	13.0	29.0	460	1000	66	640	1410
Wheat middlings	17.0	4.5	8.0	55.0	4.5	0.10	0.90	5.8	12.8	0.9	2.0	44.0	97.0	9.0	19.8	460	1000	77	885	1945

Wheat shorts	17.0	4.0	6.0	57.0	4.0	0.08	0.70	7.1	15.6	0.9	2.0	42.0	92.0	8.0	17.6	420	930	72	1200	2640
Wheat mill run	15.0	4.0	9.0	55.0	5.5	0.10	1.00	6.9	15.2	1.1	2.4	51.0	112.0	6.0	13.2	450	990	72	*	*
Wheat germ meal	25.0	8.0	3.0	45.0	4.5	0.05	1.00	12.0	26.0	2.3	5.1	21.0	47.0	5.0	11.0	1350	3000	80	*	*
Wheat feed flour	15.0	2.5	2.0	66.0	2.1	0.05	0.30	2.6	5.7	0.4	0.9	19.0	42.0	6.0	13.2	450	990	72	1390	3060
Wheat red dog	17.0	3.5	2.5	62	2.5	0.05	0.50	8.6	19.0	0.7	1.5	23.0	51.0	6.2	13.6	450	990	74	1250	2750
Wheat screenings	15.0	3.0	7.0	60.0	3.2	0.08	0.35	1.3	2.9	*	*	*	*	*	*	*	*	65	*	*
Wheat chaff	12.0	2.5	19.0	43.0	11.0	0.80	0.20	*	*	*	*	*	*	*	*	*	*	32	*	*
Whey, dried cheese	13.0	1.0	0.1	72.0	8.0	0.90	0.75	1.7	3.8	12.0	26.4	5.0	11.0	21.0	46.5	1100	3300	78	870	1910
Whey product, dried†	18.0	1.0	0.1	59.0	16.0	1.30	1.00	2.0	4.4	20.0	44.0	7.0	15.4	32.0	70.0	1600	3530	*	*	*
Yeast dried grains	19.0	6.0	18.0	45.0	3.0	*	*	*	*	*	*	*	*	*	*	*	*	*	*	*
Yeast culture	12.0	3.5	5.0	66.0	4.5	*	*	*	*	*	*	*	*	*	*	*	*	*	*	*
Yeast, irradiated	47.0	1.1	7.0	31.0	6.0	0.30	1.30	*	*	*	*	*	*	*	*	*	*	72	*	*
Yeast, brewers, dried	45.0	1.1	3.0	37.0	6.5	0.12	1.40	40.0	88.0	15.0	33.0	200.0	440.0	50.0	110.0	1750	3860	73	840	1850
Yeast, torula, dried	47.0	2.5	2.5	32.0	8.0	0.55	1.60	2.8	6.2	20.0	44.0	225.0	500.0	37.0	82.0	1320	2910	70	*	*

† The dried whey product indicated contains 50 to 53 % lactose. Other whey products are available containing less lactose but more protein and vitamins.
* Reliable data not available.
— Indicates no value.

TABLE 28

RATIONS AND SUPPLEMENTS FOR BROOD SOWS

	Ration	Supplement
Ground corn	1,110 lb	—
Ground oats	400	—
Middlings	50	200
17% dehy. alfalfa meal	50	200
50% meat & bone meal	75	300
60% fish meal	25	100
44% soybean meal	250	1,030
Dicalcium phosphate (18%)	10	40
Calcium carbonate	20	80
Salt	8	30
Trace mineral mix	1	4
Hog vitamin premix	5	20
Total batch	2,004 lb	2,004 lb
Calculated Nutrient Content		
Protein, %	16.2	36.2
Fat, %	3.6	2.6
Fiber, %	5.6	7.3
Calcium, %	1.02	4.0
Vitamin A, IU/lb	2,750	10,000
Phosphorus, %	0.62	1.65
Vitamin D2, USPU/lb	375	1,500
Riboflavin, Mg/lb	2.4	7.7
Niacin, Mg/lb	23.3	66.5
Pantothenic acid Mg/lb	8.9	26.0
Choline, Mg/lb	530	1,145
Vitamin B12, Mcg/lb	12.5	51.0
Guaranteed Analysis		
Crude protein, min %	16.0	36.0
Crude fat, min %	3.5	2.5
Crude fiber, max %	6.0	8.0
Calcium (Ca), max %	Not Needed	4.5
Calcium (Ca), min %	Not Needed	3.5
Phosphorus (P), min %	Not Needed	1.6
Iodine (I), min %	Not Needed	0.0006
Salt (NaCl), max %	Not Needed	2.0
Salt (NaCl), min %	Not Needed	1.0

ence between these supplements is that one is based on all natural protein, the other three including a nonprotein nitrogen source in the form of urea.

As we study the composition of these representative feed formulas, we may find it useful to review the following analyses of widely used feedstuffs compiled by P. J. Van Soest of the US Dept. of Agr.

The swine and cattle feed formulas provided here are representative of those being manufactured in the United States today. To present further formulas for rations for other animals and

TABLE 29

BEEF CATTLE SUPPLEMENTS

Ingredients	32% Cattle Supplement All Natural Protein	32% Cattle Supplement with 4% Urea	32% Cattle Supplement with 6% Urea	65% Cattle Supplement with 20% Urea
44% Soybean meal	1320	600	60	50
Wheat middlings	100	750	450	150
Corn gluten feed	100	—	800	—
17% dehy. alfalfa	150	200	200	600
Urea (45%)	—	80	120	400
Dicalcium phosphate	30	30	40	150
Calcium carbonate	100	110	100	60
Salt	50	50	50	80
Trace mineral mix	5	5	5	10
"Dry" molasses product	—	—	—	400
Vitamin A (13.6 million per pound)	2	2	2	4
Vitamin D2 (Fidy 9F)	1	1	1	2
Molasses (wet)	150	180	180	100
Total batch	2008	2008	2008	2006

Calculated Nutrient Content				
Protein, %	32.2	32.2	32.4	64.4
Equivalent protein, %	—	11.24	16.86	56.20
Fat, %	1.05	2.1	2.0	1.04
Fiber, %	7.0	7.5	8.2	11.3
Calcium, %	2.58	2.72	2.73	3.36
Phosphorus. %	0.80	0.81	0.85	1.51
Vitamin A, IU/lb.	13,600	13,600	13,600	27,200
Vitamin D2, IU/lb.	2,000	2,000	2,000	4,000

Guaranteed Analysis				
Crude protein, min %	32.0	32.0	32.0	64.0
This includes no more than—% equivalent crude protein from non-protein nitrogen	—	11.5	17.0	56.5
Crude fat, min %	1.0	2.0	2.0	1.0
Crude fiber, max %	8.0	8.5	9.0	12.5
Calcium (Ca), max %	3.0	3.0	3.0	4.0
Calcium (Ca), min %	2.5	2.5	2.5	3.0
Phosphorus, min %	0.8	0.8	0.8	1.5
Iodine (I), min %	0.0006	0.0006	0.0006	0.001
Salt (NaCl), max %	3.0	3.0	3.0	5.0
Salt (NaCl), min %	2.5	2.5	2.5	4.0

poultry would be of little value, for their composition varies only in relation to the differences in their requirements for various nutrients.

Feed Bag Red Book (Smith 1969, see Bibliography), published annually, is a ready source of a number of other typical feed formulations.

TABLE 30
COMPOSITION OF SOME FEEDSTUFFS

Description	Crude Fiber	Cell Walls	Acid Deterg. Fiber	Lignin	Crude Protein	Dry Matter Digestibility	
						Ruminant	Non-ruminant
			Dry Matter Basis				
Corn (dent) grain	2.2	10.1	2.7	0.6	9.8	86	
Wheat grain	2.3	12.4	3.1	0.7	11.6	85	
Barley grain	7.0	20	6.3	1.0	12.7	80	
Oats grain	12	31	16.4	2.8	13.6	75	
Soybean meal	6.4	12.0	9.0	0.3	44	84	
Sesame meal	6.9	16.8	9.8	2.0	52	77	
Linseed meal	9.0	25	16.4	6.7	38	81	
Cottonseed meal	12.4	27	18.0	6.6	44	19	
Safflower meal	35	58	41	13.7	24	50	
Cooked feathers	1.5	19.5	12.2	—	96	—	70
Raw feathers	1.3	9.0	65.0	—	96	—	<10
Tankage	3.7	25	3.6	—	67	—	68
Rice bran	13	24.1	15.9	4.3	15	61	
Malt sprouts	12.5	45	16	1.1	26	77	
Wheat bran	11	47	12.1	4.0	17.4	73	
Dried beet pulp	21	54	33	2.5	10.0	76	
Alfalfa hay (early)	23.5	40	25	5.3	20	62	50
Alfalfa hay (late)	39	55	40	9.0	14	53	
Orchardgrass (early)	24	52	27	2.7	24	72	38
Orchardgrass (late)	35	70	40	4.7	11	57	20
Timothy	32	65	43	7.0	10	50	
Coastal Bermudagrass	32	76	38	5.6	6.4	46	
Soybean hulls	28	63	45	2.0	15	68	
Soybean straw	43	67	53	12.0	11	44	
Barley straw	38	72	45	5.2	2.6	54	
Wheat straw	42	82	53	7.6	3.2	36	
Corn cobs + husk	32	83	45	5.0	2.3	50	
Rice hulls	43	81	66	14	3.0	<10	

REGULATION OF THE FEED INDUSTRY

The manufactured feed industry is supervised in all cases by state governmental agencies and additionally, in most cases, by the federal Food and Drug Administration. Each state designates a "feed control official." It is unlikely that he will carry this exact title; more likely, he is the state chemist, an untitled staff member of the state department of agriculture, or affiliated with an agricultural experiment station.

It is the feed control official's responsibility to see that the feed products sold in his state conform to the legal minima and maxima for fat, fiber, and protein and also that they contain the ingredients listed on their tags. He also must insure that no unsafe or otherwise prohibited ingredients are used.

A feed tag or label must be affixed to every sack or delivered with every bulk order. Many years ago, the control officers in the various states joined together to form the Association of American Feed Control Officials. This professional society, which also has members in Canada and in the federal government, meets regularly to define and review the various ingredients used in feeds.

Out of their deliberations have come standardized definitions for more than a thousand feed ingredients. Although there is no national feed law in the United States, the control officials from the various states have worked together to develop a "uniform feed bill," which is followed closely by the regulatory bodies of virtually all states.

Violation of proper labeling requirements or failure to live up to the label guarantees can bring heavy fines or even imprisonment. But serious violations are infrequent and the teamwork and cooperation between the control officials and the feed industry is notable.

M. P. Etheredge, Mississippi state feed control official, made this comment (1968) on the current goals of both government and industry: "With the exception of a few physical problems with different sized particles, the feed of today is almost as homogenous as a loaf of bread. Chemists should dedicate themselves to do more analyses by the modern instrumentation procedures; the inspectors should go toward mechanical sampling; and, the manufacturers should daily strive to more nearly reach the goal of complete homogeneity."

C. R. Phillips, director general in the Production and Marketing Branch of the Canada Department of Agriculture, made these observations (1969) on his retirement as president of the Association of American Feed Control Officials: "The cost-price squeeze is hitting our farmers and our feed manufacturers. The relaxed atmosphere of small family operations is disappearing as corporate businesses emerge, both in feed manufacturing and on the farm. We as feed officials are caught in the midst of these changes. We are on a neutral ground between the farm business and the feed manufacturer.

"Can we assist farmers and feed manufacturers by simplifying our rules, by reducing some of the complexity in a world that is becoming increasingly complex and, as a result, improve farm profits?"

The Food and Drug Administration vigorously polices feed manufacturing establishments which use drugs and other medicated additives in their feeds. Specific federal regulations govern the drugs which can be used, the quantities which can be used in different types of rations, the directions which must be given with medicated feeds as to their proper feeding, and the housekeeping practices of the mills. It is FDA practice to inspect every feed mill using medicated additives—which essentially means virtually all plants in the country—at least twice every 12 months.

Certain drugs cannot be used by individual feed manufacturers unless those manufacturers individually have been granted permission for such use by the FDA. A clearance to the producer of the drug for its use in feed is not enough; the individual feed manufacturer also must secure a supplementary clearance.

Drugs and other medications are added to feeds both for therapeutic and growth-promoting purposes. For the latter, antibiotics generally are incorporated into the ration at a low level.

Specialty Rations

Name an animal, bird, or fish and there's a great likelihood that the feed industry is providing one or more dietary items for it. As recently as 30 yr ago, the feeds on the market were concerned mainly with cattle, hogs, sheep, poultry, and work horses and mules. The picture is vastly different today.

There are feeds for tropical fish, trout, pan fish, and catfish. They come in different formulas, forms, and even different flavors and colors.

Rations are manufactured for laboratory mice, rats, and guinea pigs. There are feeds for hamsters and gerbils, for every species of zoo animal and bird, for mink, fox, and chinchilla.

Pleasure horses provide a growing market for formula feeds. Other rations are made for race horses, ponies, and show horses. Refined feeds are made for work horses and even for mules.

These special feed types still are paced in volume by the formulas for dogs, followed closely by those for cats. The cat food market was virtually untapped as recently as 1955. Today it accounts for millions of dollars worth of feed tonnage annually.

Major feed manufacturers employ research and marketing personnel for every type of specialty feed. It is not uncommon for a large feed company to assign some staff members wholly to dog food, others to cat food, laboratory animal rations, and feed for horses.

The feed industry takes an active role in pet and horse shows, fish producers' organizations, and similar activities in the other specialty feed fields. Although the specialty feed market competition is intense, manufacturers generally realize a higher percentage of profit on specialties than they do on their farm feed lines.

The larger feed manufacturers sell under their own brand name a variety of animal health care products, including topical medications, sprays, dusts, brushes, combs, and even leather goods such as harnesses and leashes. The feed companies do not produce these products themselves, instead have them manufactured to order by firms whose primary business is in these fields.

THE FEED INDUSTRY IN THE FUTURE

Allen and Hodges (1968) note: "The feed industry by the end of this century probably will be more consumer-oriented. The industry, through coordinated arrangements with livestock farmers and processors, will move closer to producing what the consumer wants, when and where he wants it, and probably at more uniform price levels."

It appears entirely likely that the future will see feedlots with 100,000 or more cattle in each, million-bird broiler complexes, and vast concrete "pastures" of hogs. Each such animal or poultry production unit probably will be served by its own continuously operating feed mill.

The American public appears more than willing to spend an increasing share of its take-home pay for meat, milk, and eggs. Although the American family will eat less bread and cereal grains directly in the years ahead, its preference for protein foods assures continued strong demand for grains to feed the livestock and poultry which produce the end-products which its affluent tastes desire.

BIBLIOGRAPHY

ALLEN, G. C., and HODGES, E. F. 1968. How feed industry has changed, grown in recent decades. Feed Bag Mag. *6*, 16–19, 25–27, 40.

ANON. 1969. Gen. Cir. E-1. Am. Feed Mfr. Assoc., Chicago.

CLOUGH, M. 1969. Proc. National Agr. Outlook Conf., U.S. Dept. Agr., Washington, D.C.

CRAMPTON, E. W., and HARRIS, L. E. 1969. Applied Animal Nutrition—the Use of Feedstuffs in the Formulation of Livestock Rations. W. H. Freeman, San Francisco.

HEAGY, A. B. (Editor). 1968. Feed Control. Assoc. Am. Feed Control Officials, Lexington, Ky.

HEAGY, A. B. (Editor). 1969. Feed Control. Assoc. Am. Feed Control Officials, Lexington, Ky.

SCHAIBLE, P. J. 1970. Poultry: Feeds and Nutrition. Avi Publishing Co., Westport, Conn.

SCOTT, J. T., JR. 1968. High-lysine corn, its likely future impact. Feed Bag Mag. *1*, 18–19.

SMITH, B. W. 1968. West Germany looking forward to centennial of mixed feed trade. Feed Bag Mag. *1*, No. 3, 42.

SMITH, B. W. (Editor). 1969. Feed Bag Red Book, Editorial Service Co., Milwaukee, Wisc.

SNYDER, J. C., NELSON, L. L., and GUTHRIE, T. L. 1969. Profit Planning and Control, a Computer Oriented System for Feed Industry Management. Am. Feed Mfr. Assoc., Chicago.

STROWD, W. H. 1925. Commercial Feeds. National Miller, Chicago.

WHERRY, L. 1947. Golden Anniversary of Scientific Feeding. Business Press, Milwaukee, Wisc.

CHAPTER 4

Paul R. Witt, Jr.[1]   # Malting

INTRODUCTION

The process of malting consists of the controlled germinating and usually the subsequent drying of a seed. Barley is most generally used, although small quantities of wheat and rye are also processed.

This chapter discusses the purposes of malting and the physical means employed to induce the change in the properties of the kernel which maltsters refer to as modification. While the term "modification" has been associated with the physical properties which characterize a good malt, it definitely includes all those changes in colloidal state and chemical composition which the constituents of barley undergo during malting. These changes are not confined to the germinating period but continue during the early stages of kiln drying and they markedly affect flavor, color and a number of aspects of chemical composition. The conversion of grain to malt is essentially physiological in nature and is a result of the action of enzymes.

In the subsequent discussion, the emphasis will be placed on malting practices in North America and specifically in the United States, unless otherwise noted. Reference will be made to malting barley varieties, mechanical means for preparing the barley for malting, the actual steps of processing, the variations in processing which affect quality and the merchandising of malt as related to process methods.

BARLEY VARIETIES

Generally speaking, commercial malt as it enters channels of distribution consists mostly of the varieties Larker and Dickson.

Malt is blended to satisfy individual specifications and the varieties usually requested in addition to Larker and Dickson are Conquest, a blue aleurone variety of Canadian origin, and the two-row variety Hannchen. Commercial quantities of the two-row varieties Piroline and Firlbecks are grown in western United States while another, Betzes, is grown in Montana and Colorado. While

[1] PAUL R. WITT, JR. is vice president of Northwestern Malt and Grain Co.

two-row varieties are frequently included to serve as extract boosters, many brewers also consider their admixture as a factor contributing flavor and/or foam sustaining characteristics.

Larker and Dickson are mainly grown in the Red River Valley of North Dakota and in southwestern Minnesota and northeastern South Dakota. They are white aleurone barleys with relatively high potential enzymatic activity and modify well during malting. Barley selected for malting is usually within an approximate protein range of 12.0 to 13.5%; however, satisfactory malts have been prepared from grain beyond these limits.

At present, most of the Conquest originates in the prairie provinces of Canada. It is highly enzymatic and lends itself well to modification. While some Hannchen is still planted in the Willamette Valley of California, significant quantities are imported from Canada. Due to cheaper transportation by water, Canadian Hannchen is less costly and sound barley of this type results in malt of satisfactory modification.

The variety Atlas, relatively low in laboratory-determined enzymatic activity, is grown and malted in California and in a small area of Mexico. When used in brewing, it is generally mashed in conjunction with Midwestern malt (Larker and Dickson or a mixture containing Conquest).

A more complete description of the agronomical and morphological aspects of barley varieties is included in *Cereal Science* (Matz 1969). Additional details can also be found in *Classification of Barley Varieties Grown in the United States and Canada in 1945* (Aberg and Weibe 1946) and *Barley Variety Dictionary* (Anon. 1957A).

Requirements governing commercial trading of barley are listed in the "United States Official Grain Standards" (Anon. 1957B). Additional information is available in the "Western Canada Grain Grades" (Anon. 1926).

PREPARATION OF BARLEY FOR MALTING

The process of readying barley for malting can be separated into several steps. The first step is basic and reflects itself in procurement. Maltsters and some of the grain merchandising houses conduct field studies of the barley crop before harvest. Knowledge thus gained provides the buyer with information as to origin of sound and plump grain.

Soundness is an expression which in its strictest sense means viability or germinating capacity, but it has been augmented to include relative freedom from mold or mold damage.

Total protein content is also determined as extensively as sample accrual and laboratory facilities permit. The resulting data can be arranged as a varietal-origin-protein map and the maltster can resolve his storage or binning program with respect to arbitrary ranges within a variety.

As a result of practical experience and from knowledge gained by pilot or experimental malting, the maltster can establish an area of demarcation where perhaps all samples analyzing 12.8% or under will qualify as "low" protein and those above as "high." A statistical mean is not necessarily the most desirable way to set up limits between lots of barley. While the degree of modification of malt is obviously related to the crop year, adaptability of a barley type to the individual physical plant and to the methods of malting is more important when segregating according to protein content.

In years when growing conditions have been affected by considerable local variations in weather, maltsters have sought to achieve malt uniformity by binning and malting selected protein ranges of a definite variety originating in distinct growing areas. For example, studies concerned with a discontinued variety, Kindred, found that mature barley of 12.8 to 13.2% protein originating in the Red River Valley in the general vicinity of Fargo, North Dakota, modified more readily than comparable barley of 12.3 to 12.5% from an area adjacent to Bismarck, North Dakota, where drier conditions are frequently experienced.

In general, it is thought that barleys of lower protein content modify more readily.

Dormancy and "newness" are factors to be considered when readying barleys for malt processing. True dormancy, which is revealed by the lack of response to germination by sound barley, is thought to be related to variety, geographical origin and to drying and/or earliness of harvest. Dormancy has been shown to be related to the accessibility of oxygen to the embryo. The causes of dormancy in seeds have been studied by Crocker (1916), Gracinin (1928), Deuber (1931), Flemion (1934) and Brown (1933) and they have suggested physical and chemical methods of overcoming dormancy. In the malting laboratory, dormancy is generally overcome by steeping or soaking in 0.2 to 0.5% hydrogen peroxide, subjecting to cold (45° to 55°F) storage as dry or as steeped grain and by scarification or physical abrasion.

In practical operations, dormancy has been treated by washing the barley in lime water, rinsing with water and then adding water acidified to pH 2.5 with sulfuric acid. The steeping barley is agitated with compressed air or other mechanical means and the acidified water decanted after 3 to 5 hr.

Custom in the malting industry has demanded that barley "go through the sweat" prior to qualifying as ready for malting. Since good practice demands selection of grain which is dry and since barley is freed from contaminating seeds and foreign material which may still be green (weeds, etc.), very little if any actual sweating is observed in present-day "terminal" barley storage.

Yet, sound barley will still produce undesirable malt if malted too soon after harvest even though it may germinate vigorously and uniformly. Undermodification and hazy laboratory worts frequently characterize malts prepared from immature barley.

Sound and bright barley from the earlier variety Kindred, when malted under comparable conditions at different times after harvest yielded malts whose quality characteristics are described in Table 31.

TABLE 31

EFFECT OF PROGRESSION OF MATURATION OF NEW CROP BARLEY ON MALT QUALITY

Steep Date	Germination %	Extract %	Soluble/ Total Protein %	Wort Clarity
Aug. 21	97	72.5	35.5	Hazy
Sept. 21	99	73.5	37.5	Slightly hazy
Oct. 15	99	74.3	39.2	Clear

In all probability, the "sweat" period customarily referred to is in actuality an "after-ripening." Eckerson (1913) studied microchemically the alterations accompanying the after-ripening of seeds and noted that the initial change in embryo was an increase in acidity. This was correlated with an increased water-holding capacity and an increase in the activity of catalase and peroxidase. Near the end of the after-ripening period there was a sudden increase in acidity and in water absorbing ability and oxidase made its first appearance.

In actual commercial practice many maltsters accept as a rule the deferment of malting "new" crop barley until about October 15.

Cleaning and Grading of Barley

It has been noted that accepted practice demands that each barley variety be stored separately; types within a variety be segregated as to protein level and in some instances as to geographical origin; barley must be dry (under 14% moisture) at time of storage; and it must be cleaned of dust and weed seeds of high moisture content before binning.

The act of grading or kernel sizing prepares the barley for malt processing, the initial step of which is steeping or soaking. Barley is classified as to kernel size or shape primarily to assure uniformity during the malting process.

To demonstrate barley grading, a flow sheet of one possible procedure is outlined in Fig. 38. Barley as harvested, known as "country-run", enters a scalping machine at (A) from the bin, and is fed at the desired rate to the scalping reel (C) which removes sticks, straw, nails and other roughage. Roughage is then carried over at (B), to the rescalping reel (D). Aspiration takes

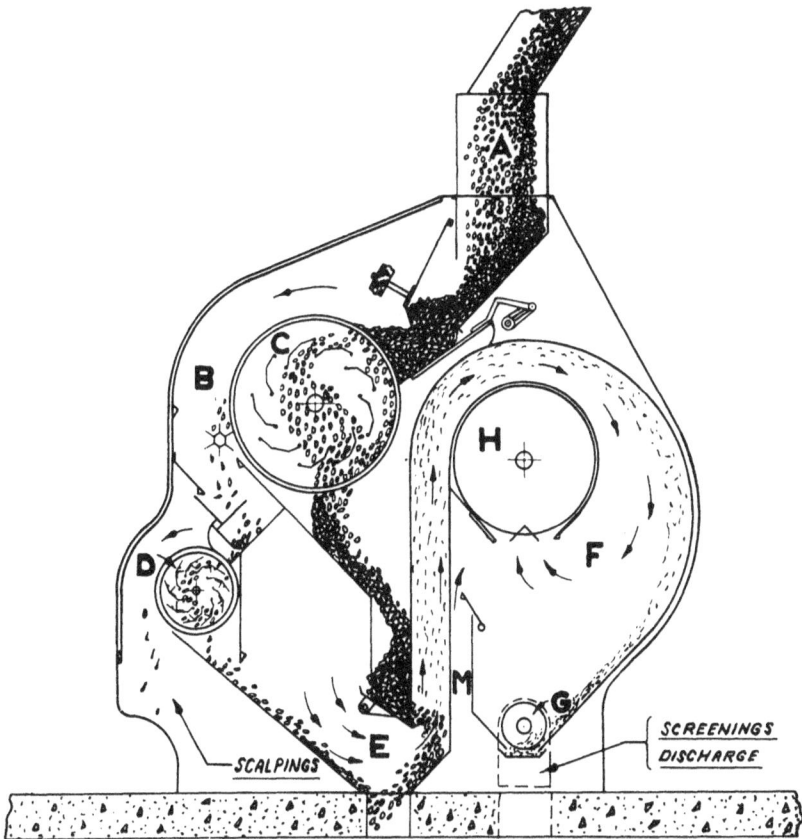

FIG. 38. CLEANING AND GRADING BARLEY PRIOR TO MALTING

place at (E), where the grain leaves the seal gate in an evenly spread stream through which a current of air is drawn. The light screenings are carried up the aspirating leg and the heavier grain drops back into the main grain stream, thus effecting light screenings removal. The air from settling chamber (F) is drawn into the fan through inlet (H). The fan exhausts into a dust collector. Rapid expansion of air in the settling chamber (F) allows screenings to drop to the bottom where they leave the machine by screw conveyor (G). The volume of air passing through the grain at (E) may be controlled by adjusting the bypass valve (M) through which air enters the suction tube directly.

To achieve a more thorough cleaning, grain from the Scalperator may be permitted to pass through a Eureka-type cleaner. "Country-run" grain may also be passed directly to this machine.

The Eureka-type cleaner consists of parallel sieves with a flat plane surface, placed at a slight angle to the horizontal, permitting the grain to flow downward. The sieves are fixed to wooden or steel frames and are oscillated by means of eccentrics on a driving shaft. Series of permanent magnets are often fixed in the path of flow of the grain to the initial or upper level. The uppermost sieve has oval perforations with dimensions of 9/64 to 10/64 in. It permits removal of larger stones, straw ends, ears, etc. The middle screen permits passage of barley and smaller seed, but serves to eliminate some larger grains such as corn. The lower screen usually contains triangular openings so that it retains the barley but sifts out seeds such as flax, mustard, etc.

After the barley leaves the Scalperator or the Eureka-type cleaner, it is directed to machinery designed to separate wheat, oats, seeds and broken barley from the usable barley. An example of a machine used for this purpose is the Carter Disk Separator, which takes advantage of length difference between various materials.

Figure 39 is a view of the mechanics of disk separation. While the picture shows a mixture of wheat and weed seeds, the mixture being separated can be construed to be barley and oats. The barley fits into the pockets of rotating vertical disks and is lifted from the mixture while the oats are too long to fit into the pockets and are rejected.

The principle of operation of Carter Disk is illustrated in Fig. 40 and shows how materials such as oats are removed.

The Hart Uni-Flow Cylinder Separator is another example of a machine designed to remove wheat, oats and broken kernels

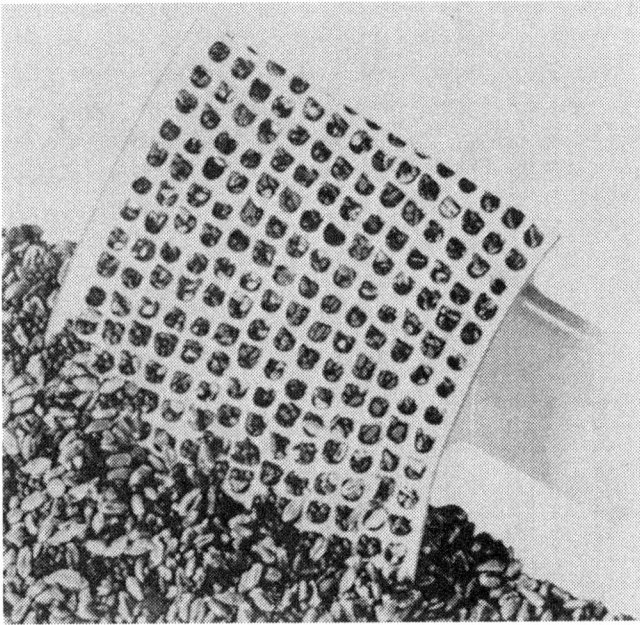

FIG. 39. SECTION OF A CARTER DISK

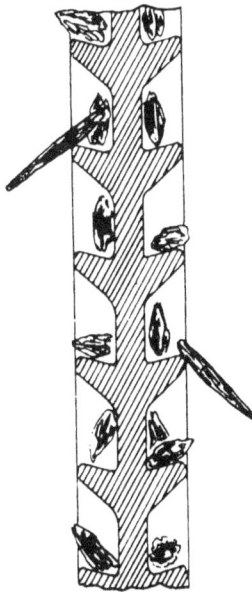

FIG. 40. PRINCIPLE OF
SEPARATION BY THE
CARTER DISK

from barley. Figure 41 illustrates the lifting of material by the Uni-Flow indents (A) and the depositing of it into a conveyor trough (B). The position of the separating edge (C) of the trough is one factor controlling the fineness or coarseness of the separation to be made. The position of the separating edge can be raised for finer separations or lowered for coarser separations.

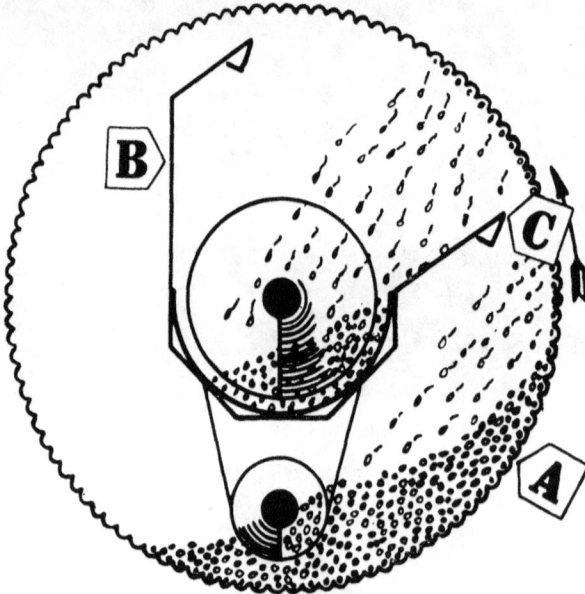

FIG. 41. HART UNI-FLOW CYLINDER SEPARATOR

Material lifted by Uni-Flow cylinder indents (A) is deposited in the conveyor trough (B). The position of the separating edge (C) of the trough is one factor in controlling the fineness or coarseness of the separation to be made. The position of the separating edge can be raised for finer separations; lowered for coarser separations.

After cleaning in the disk- or cylinder-type machine, the barley theoretically contains as the only contaminant barley kernels too small for malting.

To effectively separate barley into sizes insuring uniform malt modification, the effluent barley after wheat and oat removal passes into a slotted flat sieve or cylindrical sieve. Most modern processing plants employ Eureka Ring graders or Hart-Carter Precision graders. As an example the Precision grader provides a desirable method of sizing barley by thickness.

One flow sheet of practical operation is the removal of small corn, weed seeds, etc. by treatment on the Eureka cleaner, and passage of the grain via a Scalperator through the Disk or Uni-Flow machine. The effluent is processed in a Eureka or Precision grader, where the initial slot width may be 7.5/64 in. Barley retained by this screen, if Midwestern in origin, may be discarded as incompatible to malting size. Removal of additional corn and soybeans is also thus effected.

The plump or A grade malting barley is retained on the 6.25/64 in. cylinder slot width. Grain passing through a 6.25/64 in. slot is graded in a cylinder having 5.25/64 in. or 5.5/64 in. slots. The retained product is the smaller size adaptable to malting and is catalogued as "B" grade.

Steeping

The primary purpose of steeping is to introduce water into the barley kernels at a rate and in a manner to permit germination and to produce a well modified and plump malt.

Classical texts indicate that the temperature of the steep water may be raised to a maximum of 68°F to permit a reduction of the steeping time. However, some doubt does exist as to the quality of malts steeped at the higher temperature levels. It is generally true that the degree of water uptake by the barley kernel, as measured in the laboratory, within certain limits varies directly with the temperature of the water. Nevertheless judgments as to the completeness of steeping which are based solely on laboratory determined moistures may result in the transfer of insufficiently steeped barley to the germinating areas.

A warm steep is associated with a rapid water uptake which may not permit proper distribution of moisture within the kernel. Leberle (1956) studied water absorption during steeping. The data appearing in Table 32 indicate the tendency of moisture to distribute unequally within the kernel.

TABLE 32

AREA DISTRIBUTION OF MOISTURE IN STEEPING BARLEY[1]

Kernel Area	Moisture %
Base	47.1
Middle	38.3
Tip	39.1

[1] Leberle (1956).

There is a belief that use of a colder water necessitating a longer exposure to soaking will result in a more uniformly steeped kernel.

Barley is considered to be sufficiently steeped if the moisture content has risen to 42 to 44%. In practical operations, the steep ripeness of the kernel is detected by pinching the two ends between the thumb and forefinger. If the proper absorption has been achieved, the kernel should open readily and, when it is cut, the moisture should have penetrated to the center.

Physiologically, it is thought that sufficient steeping has been achieved when the cell expansion, due to water uptake, has reached its limit. Then the hydrostatic pressure is equivalent to the osmotic pressure of the cell sap and the condition of cell rigidity is spoken of as the "turgidity" of the cell.

Insufficient steeping adversely affects many aspects of malt quality. Primarily affected are wort clarity and the fine-coarse grind extract differences of the finished malt. Attention is being given to wort viscosity. It is felt that understeeping contributes to increased viscosities which may be related to reduced brewhouse yields and to an extended primary beer filtration. The practical maltster frequently observes greater resistance to the "bite", finding it more difficult to chew the malt.

When differences in steeped-out moisture are relatively small, variations in laboratory-measured quality may not be significant. However, observed differences in the plumping tendencies of some malts are noticeable as a result of varied water absorption during steeping. Table 33 lists the results obtained when an "A" grade Kindred from the 1957 crop was steeped at 60°F with appropriate water changes and then transferred as a water slurry ("wet steep") to the compartment at the indicated moisture levels. Prior to sampling of the steeped barley, surface moisture was removed by "venting" with warmer air (68°F and 80% RH) and germination and kilning were conducted comparably in a conventional and accepted manner.

It should again be noted that, with reference to the water absorption-kernel size data, all other facets of the process were held constant. In drying, the green malt temperature did not exceed 130°F until the malt moisture decreased to 12%. The barley was vigorous and germinated 98 to 100%.

That barley plumpness is a marked determinant in the steeping time required is shown by Leberle (1952), who included this factor in his study. The data are presented in Table 34.

When a "wet" steep-out is employed, understeeping may be

TABLE 33

THE EFFECT OF STEEPING TIME AND MOISTURE ABSORPTION ON THE PLUMPING OF MALT
"A" size Kindred, 1957 crop; Water at 60°F

Sieve	Barley Before Steeping %	Steeped[1] 22 Hr %	Steeped[2] 28 Hr %	Steeped[3] 40 Hr %
On $\frac{7}{64}$ in. screen	1.8	38.2	43.4	49.5
On $\frac{6}{64}$ in. screen	71.6	54.5	52.2	48.3
On $\frac{5}{64}$ in. screen	26.0	6.7	4.1	2.0
Through $\frac{5}{64}$ in. screen	0.6	0.6	0.3	0.2

[1] Barley contained 38 to 39 % moisture.
[2] Barley contained 41 to 42 % moisture.
[3] Barley contained 43 to 44 % moisture.

partially corrected after transfer to the compartment or drum. Absorption of water has been found to continue up to an additional 2 to 3% where such a procedure is operative. It is advisable that surface evaporation be held at a minimum during the 8 to 15 hr period following discharge from the steep tank. Water can also be given during this period to facilitate after-steeping.

Attempts to increase the moisture content to any marked degree (over 1 to 2%) after the "chit" or onset of acrospire (plumule) growth has begun, create a serious threat of "overgrown" development (condition of the acrospire or prime bud extending beyond the length of the kernel). This condition is undesirable.

Oversteeping markedly delays the onset of germination, encouraging increased mold and bacterial growth. Undesirable odors are often associated with barley steeped in this manner.

The term "oversteep" connotes excessive water absorption. However, laboratory data indicate that it is related to the appearance of the steeped barley and to its subsequent onset of germination. Such conditions are usually associated with extended periods of submersion.

TABLE 34

THE RELATION OF KERNEL SIZE TO WATER UPTAKE[1]
Original barley moisture, 13.3%

Steep Time, Hr	Kernel size		
	2.2–2.5 Mm	2.5–2.8 Mm	Over 2.8 Mm
	(5.5/64–6.25/64)	(6.25/64–7/64)	(Over 7/64)
		Moisture Content, %	
16	30.9	29.7	29.3
39	37.1	35.8	35.2
62	40.8	39.6	38.9
87	43.1	41.7	41.0

[1] Leberle (1952).

To illustrate, graded mature barley of vigorous germinating capacity was steeped into a 0.03 to 0.04% solution of lime water at 65°F and pH 9.2. After 3 hr, during which the grain was agitated every 15 min, the reaction of the water had fallen to pH 7.2. After draining, the barley was washed with several changes of water at 65°F.

Steeping at 65°F was continued in the manner indicated: (A) permitted to rest with occasional agitation; (B) steeped under continuous aeration; and (C) steeped with continuous fresh water overflow at 65°F.

The water was drained after 9, 20, and 30 hr (after the beginning of steeping), the barley washed with fresh water and steeping continued as outlined in (A) through (C).

Moisture was determined at 26, 32, and 40 hr, after steeping-in, the physical conditions of the barley being noted. The data are presented in Table 35.

As noted in Table 35, only the barley associated with water changes (A) exhibited oversteeped characteristics after 26 hr. Where continuous overflow was employed (C), the barley reflected a desirable degree of steep after 32 hr and produced an excellent finished malt. Even after 44 hr, this barley appeared to be in good condition and produced a good malt. Aeration (B), while compensating in part for the deficiencies in water change, resulted in a product which, after 32 hr, seemed to be oversteeped.

Yet, it is to be noted that the moisture content of the steeped barley resulting from the three procedures is reasonably comparable.

After germination of the "oversteeped" barley has begun, a retardation of rootlet growth is observed. This is often accompanied by a too rapid elongation of the acrospire. The husk of the malt becomes darkened in color.

It is believed that the effects of so-called oversteeping are the results of the competition by molds and bacteria for oxygen together with a saturation of the steep water with respiratory carbon dioxide. In the experiment just described, the water of the unaerated steep (A) had a reaction of pH 5.4 after 26 hr while its aerated counterpart (B) was at pH 7.2. It may be theorized that there occurs a cessation of oxygen uptake by the cell with the substitution of an anaerobic metabolism. Some of the lingering intermediate or end products of these anaerobic processes are toxic and must be dispelled prior to the onset of germination.

Barley variety also affects the selection of length of steeping

TABLE 35

EFFECT OF STEEPING PROCEDURES ON THE CONDITION OF STEEP-OUT BARLEY

Duration of Steep

Sample Identification	26 Hr			32 Hr			44 Hr		
	Moisture %	Physical Condition	Chit %	Moisture %	Physical Condition	Chit %	Moisture %	Physical Condition	Chit %
A	42.1	Soft	5–10	42.9	Soft	5–10	45.5	Soft	..
B	42.8	Firm	20–25	43.4	Beginning to soften	25–30	46.0	Soft	..
C	42.6	Firm	85	43.6	Firm	90	..	Firm	95

time. Data showing the relative water uptake by Montcalm and Kindred (1957 crop year) of comparable kernel size appear in Table 36. The barley was graded to pass through a 6/64 in. screen and to maintain above a 5.5/64 in. screen.

Kernels of lowered germination energy are said by Leberle (1952) to absorb, under comparable conditions of steep, 0.7 to 2.0% more water than their vigorous counterparts. It is a recognized principle in commercial malting that a reduction in steeping time must be made for barley containing kernels of lowered germinating energy.

TABLE 36

WATER UPTAKE OF KINDRED AND MONTCALM BARLEY AFTER STEEPING FOR 26 HOURS AT 54°F

Barley Variety	Moisture %
Kindred	37.8
Montcalm	40.4

It is necessary to supply air to the steeping barley. Aeration protects the barley against the danger of oversteeping. Leberle (1952) postulates that the greater the uptake of water within the barley during steeping and the less maturation (after harvest) the greater the necessity for the application of oxygen through aeration. Leberle has also found that the evils of oversteeping are not always the result of continued submersion, but rather are due to the contact of the respiring barley with acidic material. Removal of the accumulated carbon dioxide by aeration helps remedy this situation.

Aeration also acts as a means of agitation. The latter is desirable as an aid in washing or scrubbing the grain as well as a precaution against the channeling of the overflow water, which is admitted at the bottom of the tank and flows off the surface of the steeping barley.

DeClerck (1952) recommends that aeration by compressed air be carried out 4 or 5 times a day for 15 to 20 min at a time. A volume of 0.23 cu ft per 48-lb bushel delivered at a pressure of 7 psi. is indicated as desirable. Some investigators feel that a limitation should be placed on the duration and frequency of aeration, believing that excessive oxygen encourages mold growth. However, treatment of the grain with permitted antiseptics in the steep tank can control or alleviate the danger of mold contamination.

Water composition, as directly related to the biological process occurring within the steeping kernel, is not of too great importance because of the semipermeable membrane which limits access of the water constituents to the cell interior.

Addition of sodium hydroxide or, more often, lime, to about pH 9.0 to 9.5 is frequently made to the first steeping-in water to assist in the removal of undesirable husk substance.

Warrick *et al.* (1935) studied steep water wastes and observed that approximately 0.28 lb of dissolved solids were removed per bushel; that process usage of 75 gal. of water per bushel occasioned a loss of 0.6% of the weight of barley processed. He analyzed the waste produced by each step of the malting process. Table 37 indicates the degree of waste produced in a malting plant.

TABLE 37

STEEP WATER WASTE FROM THE MALTING PROCESS

	5 Day BOD	Total Organic Nitrogen	Total Solids	Hr of Contact
1st Steep water	960	69	4856	24
2nd Steep water	920	72	2372	12
3rd Steep water	185	4	418	12
4th Steep water	254	7	452	16
Germ drum water	50	12	534	..

Additional physical concepts of steeping are worthy of note. When observed in a dry state, 500 gm of "B" size Kindred barley have been found to occupy 775 ml (equivalent to approximately 1.19 cu ft per bu). This agrees with the figure quoted by DeClerck (1952). However, the same quantity of barley, if allowed to drop into a predetermined volume of water, a practice necessary to achieve proper "liming" and to effect the desired removal of "skimmings," will occupy, after settling, about a 10% greater volume.

After completion of steeping, an approximate 30% increase in barley volume is observed (about 1.55 cu ft per bu). DeClerck states that, to insure sufficient coverage of the steeping grain with water, the total mass should be equivalent to about 1.76 cu ft per 48-lb bu of barley steeped.

Figure 42 portrays a steep tank used in commercial malting. Figure 43 depicts the bottom of a tank. The "screw" valve on the left controls the discharge of "dry" steeped-out barley. The pipe to the right of the cone illustrates the mode of draining off water or admitting fresh water.

Courtesy of Stockland Malting Machinery Co.

FIG. 42. STEEP TANK

Courtesy of Stockland Malting Machinery Co.

FIG. 43. BOTTOM OF A STEEP TANK SHOWING BARLEY OUT-
LET VALVE

GERMINATION

Germination is the physiological process in which the lamella or acrospire and the rootlet of the seed are elaborated thus forming a new plant through the multiplication and enlargement of cells. In the malting process, the conditions of germination which are subject to control are moisture, air, temperature and, to a much lesser extent, the carbon dioxide concentration of the air associated with the germinating grain. In the American system, the steeped barley is transferred to the germinating chambers by pumps as water slurry or by screw conveyors in a partially surface-dry state (see Fig. 44). The bed of grain in the germinating chamber is called a piece.

Fig. 44. Filling the Germinating Chamber with Steeped Barley

Conventional germinating chambers are either permanently fixed compartments or cylindrical drums which can be rotated. The compartment type is a rectangular chamber consisting of 2 parallel longer walls and 2 shorter ends (see Fig. 45). It has a depth of 4.5 to 5.0 ft. However, some installations have malt beds in excess of 6.0 ft. The longer walls are 7 to 8 in. thick and are generally constructed of reinforced concrete which has been troweled to a very smooth surface on the inner side. Mounted on the surface of each of the longer walls are a track and a cog rail which

FIG. 45. GERMINATING CHAMBER

support and assist in controlling the passage of the machine which turns or mixes the malt.

The floor is fitted with perforated trays, permitting passage of air through the bed of germinating grain. The area below the compartment varies in depth from 2 to 7 ft, the figure in a particular case being influenced by the available space and by design requirements for an efficient exhaust of air. The perforated bottoms are usually permanently affixed where depths of 5 to 7 ft exist, since washing with high-pressure water can be accomplished without raising the trays. More shallow sub-compartments require lifting the trays to permit washing.

Aeration of the malting compartment is either updraft or downdraft. It has been reported that some installations permit a combination of the two systems.

Most American installations expose the surface of the germinating grain within the compartments to the atmosphere of a large room. When this arrangement is employed, the updraft system provides the maltster with the greatest degree of control. Each compartment may be appointed with individual heating coils or with gas jets which warm the air directly. Where refrigerated, the water temperature of the individual humidifiers may be controlled. New designs permit selection of desired air velocity by assignment of a single fan to 1 or 2 compartments.

The attempered humidified air is forced through the green malt and is permitted to pass through ducts to the outside. More often, the duct work and air valves are so arranged as to permit recirculation of part or all of the compartment exhaust air. Such an arrangement is generally the case where the water of the humidifiers is refrigerated.

In the downdraft system all compartments are exposed to the same humidified and attempered air. The only controlled variable specific to the individual compartment is the velocity of the air passing through the piece. The compartment room is generally adjusted to 54° to 56°F, the moisture content of the air being brought to near saturation.

The heat of respiration of the germinating grain is utilized in the adjustment or control of the temperature of the piece. While the exhaust system applies a negative pressure common to all compartments, the degree of opening of the draft door in the subcompartment area controls the velocity of air passage through the piece. Air passage, in turn, is regulated to permit the desired temperature buildup to occur.

Within compartment installations, be it an updraft or downdraft system, smaller amounts of air, approximately 2 to 3 cu ft/bu/min, are demanded during the first 3 days of germination. Less resistance is exerted against the passage of air due to the presence of shorter rootlets. Moreover, the germinating grain has as yet not reached the peak of respiration.

After watering and succession to the fourth day of growth, temperature control demands a larger quantity of air, approaching a minimum velocity of 5 to 6 cu ft/bu/min.

The term "continuous" ventilation implies a procedure demanding air passage through the bed of malt from the time of loading to the time of withdrawal for kilning. The referred to process of venting, applied to remove excess surface moisture resulting from a wet steep-out, may be included to initiate the malting period of continuous aeration. An outline of one method will be presented. The steeped barley enters the compartment at 60°F. If, in the instance of a "wet" steeping-out procedure, a colder steep water (50°F) has been employed, the barley may be transferred in slurry with water which has been warmed to 65° to 68°F.

The bed of germinating grain is either vented with warmer air or the normal humidified air to 54° to 56°F is passed through at the rate of 1 to 2 cu ft/bu/min immediately after leveling of the steeped-in barley. After 30 hr, during which the stirring machine has effected 2 or 3 passages, a light watering is applied and the air velocity is increased to 2 to 3 cu ft/bu/min. It is estimated that the quantity of the "pre-water" is slightly in excess of a quarter of a gallon per bushel.

After about 55 hr, a second watering, in extent of 1 to $1\frac{1}{4}$ gallon per bushel is applied and the air velocity is increased to about 7 to 8 cu ft/bu/min. After a warming to about 62°F immediately after watering the piece cools down to about 56°F and this temperature is retained for the remainder of the germinating period.

Larger kerneled barley frequently requires a light third watering. However, such practice is frequently responsible for excessive moisture retention within the green malt at the time of transfer to the kiln.

Malts prepared in a system of continuous aeration are characterized by a lower soluble protein content and by paler colors after boiling. When properly kilned, they exhibit a malty flavor.

In discussions of the method of interrupted ventilation, the term "couching" is frequently found. In earlier texts, the practice of "couching" meant permitting conical heaps of steeped-out barley to rest undisturbed for periods up to 24 hr. The heaped barley would become warm, thus assisting the onset of germination. "Couching" in this manner is not recommended as it promotes a lack of uniformity resulting from the temperature gradient existing from center to surface of the heap.

Today, "couching" means the procedures of leveling a compartment after steeping-in and permitting the temperature within the piece to rise to 64° to 66°F by arresting the air passage. Air is then circulated at the rate of 2 to 3 cu ft/bu/min to maintain a temperature of 60° to 62°F. The draft door of the compartment may be closed during the passage of the turning machine.

After 24 to 36 hr of "couching", aeration is increased to create a temperature of 58° to 60°F. When the rootlet approaches 0.5 cm in length, water is applied at the rate of $1\frac{1}{4}$ gal. per bu. Ventilation is conducted at the rate of 7 to 8 cu ft/bu/min during and after watering. When the surface moisture has disappeared, accompanied by a concurrent drop in temperature to about 54°F, the draft door is closed and the piece permitted to warm to 66° to 68°F.

Cooling as rapidly as possible to 54°F is then effected and the draft door closed to again achieve the upper temperature limit. Each cycle of warming and cooling usually requires 6 to 8 hr.

Continuous aeration at the rate of 6 to 7 cu ft/bu/min is conducted during the last 12 to 15 hr.

Malt prepared by the "interrupted" process, when contrasted with that produced with "continuous" ventilation, exhibits some of these characteristics: (1) darker wort color and darker color after wort boiling; (2) higher alpha-amylase activity and increased soluble protein; (3) increased extract with smaller coarse-fine grind difference; and (4) greater mealiness when crushed and chewed.

It has been stated that the temperature of the ventilating air is a prime factor affecting malt quality.

Hind (1938) presented data for extract and soluble protein composition of malts germinated at three temperature levels within the range of 55° to 72°F. It is assumed that the temperatures have been applied in a manner of continuous ventilation. While the germinating time consisted of nine days, analytical values for a period thought to coincide with a peak in modification are recorded in Table 38.

TABLE 38

EFFECT OF GERMINATING TEMPERATURE ON MALT CHARACTERISTICS[1]

	Temp Range, °F	Days Growth		
		5	6	7
Soluble protein as % total nitrogen	55.4–62.6	38.3	36.8	37.6
	59.0–68.0	35.5	34.2	31.9
	66.2–71.6	32.1	31.2	30.0
Extract, % (dry)	55.4–62.6	79.2	79.4	79.7
	59.0–68.0	78.6	78.2	78.3
	66.2–71.6	78.5	78.2	78.2
Acid (ml 0.10 N NaOH to pH 9.2 for 100 ml wort)	55.4–62.6	16.1	15.3	15.6
	59.0–68.0	14.6	14.1	14.0
	66.2–71.6	14.1	14.1	14.0

[1] Hind (1938).

The previously cited flow sheets of germination employing continuous or interrupted ventilation denote the use of several temperature levels.

Dickson and Shands (1942) studied the effect on quality when germinating 4 barley varieties for 6 days, using a constant moisture level of 45% and temperatures of 53.6°, 60.8°, and 68°F singly and in varied sequence. Generally speaking, a 2-day period of 60.8°F followed by 4 days at 53.6°F produced malts with the greatest recovery and with quality factors which coincide with present-day brewery specifications.

Germination at one fixed moisture level will result in a malt differing in quality from that produced at another level.

Klopper and Kortenhorst are quoted by Kolbach (1955) as steeping barley at 54.5°F for 45 hr and recording a steep-out moisture at 40.2%. Germination was conducted at this level and at additional levels of increased moisture content which were achieved by watering on the first and second days of germination. The control of rootlet growth, diastatic power and soluble protein is noted in Table 39.

TABLE 39

THE EFFECT OF MOISTURE LEVEL WITHIN GERMINATING BARLEY ON THE QUALITY OF THE RESULTANT MALT[1]

	Green Malt Moisture Level %			
	40.2	42.6	43.9	48.2
Rootlets as % dry barley	2.1	3.0	3.4	4.6
Diastatic power[2]	160	180	195	205
Soluble protein as % of total protein	39	41	42	44

[1] Klopper and Kortenhorst as quoted by Kolbach (1955).
[2] Degrees Lintner.

The proper duration of the germinating period is a subject which has reflected a difference of opinion among commercial maltsters. It has been reported that 3-, 4-, and 5-day germinated malts are being prepared from midwestern-type barleys.

First, it may be well to distinguish between the propriety of one processor germinating for 4 days while another utilizes 5. In the four-day process, a "dry" steeping-out, where the onset of germination may have begun in the steep tank, could have been employed. In the instance of five-day malt, the barley may have been discharged into the germinating chamber as soon as the proper steep-out moisture was achieved. Reasonably comparable germinating periods may have existed in both procedures.

To achieve satisfactory modification, plumper kerneled midwestern and two-row barleys often require longer periods of germination. Thus, the length of germinating time may be varied within one process.

DeClerck (1952) and others present curvilinear data for respiration, diastatic and proteolytic activity, noting a peak between the fourth and fifth day of germination. When, in practical operations, germination has been arrested at this period, malts are obtained which yield satisfactory analyses in most areas of quality. However, studies have been conducted which suggest that the arresting of germination at the indicated peak may result in malts lacking in one or more essential qualities.

Meredith and Bendelow (1956) examined additional properties of worts and showed that viscosity was a useful measure of quality factors associated with modification that were not measured in other determinations. He found that the viscosity of Congress wort was a direct reflection of the modification process and of the ease of conversion of some carbohydrate material in the barley into solubles in the wort. He also presented evidence of a close relation between the viscosity of the 158°F worts and the amount of cold-water extract. Low values for cold-water extract are associated with high wort viscosity.

Table 40 shows the increase of cold-water extract of the variety Montcalm when malted 5 and 6 days at moisture levels of 42.0 and 44.0%.

TABLE 40

GERMINATING TIME AND MOISTURE LEVEL AS FACTORS IN COLD WATER EXTRACT DEVELOPMENT[1]

| Moisture, % | Cold Water Extract, % | |
	5-Day Germination	6-Day Germination
42	16.8	19.2
44	18.0	20.4

[1] Meredith and Bendelow (1956).

For purposes of references, Meredith has shown the relationship of cold-water extract and the viscosity of 158°F malt worts prepared from different barley varieties. Values are presented in Table 41.

Another factor which is claimed to effect malt quality is the carbon dioxide content of the air surrounding the grain. Hind

TABLE 41

COLD WATER EXTRACTS AND WORT VISCOSITIES FROM DIFFERENT VARIETAL MALTS[1]

Malt Sample Identification	Cold Water Extract %	Viscosity[2]
A	14.0	2.24
E	16.7	1.85
F	18.1	1.70
L	20.0	1.54

[1] Meredith and Bendelow (1956).
[2] Centipoise.

(1938) cites the study of Hoffman-Bang who found that the greatest percentage of salt-soluble nitrogen could be obtained from malts made at the lowest temperatures and in the presence of carbon dioxide.

Leberle (1952) presents data observed in the interrupted or intermittent aeration of a pneumatic compartment. After the piece had achieved maximum respiration, a 4-hr cessation of ventilation was accompanied by a marked increase of the carbon dioxide content of the air. The data appear in Table 42.

TABLE 42

BUILD-UP OF CARBON DIOXIDE DURING INTERMITTENT VENTILATION[1]

| Age of Piece, Days | Length of Rest, Hr | CO₂ Content of Air | |
		Lower Half %	Upper Half %
2	2	6.0	5.4
2	4	9.6	7.0
3	2	7.0	6.0
5	2	7.5	6.0
6	4	15.6	10.4
7	2	8.7	6.0

[1] Leberle (1952).

Lack of data comparing carbon dioxide buildup during continuous ventilation with that resulting from interrupted ventilation prevents a contrast of quality. Values do exist which contrast malt from the Kropf carbon dioxide system with malt from the pneumatic compartment methods. These are presented in Table 43.

TABLE 43

COMPARISON OF THE QUALITY OF GREEN MALT RESULTING FROM A PNEUMATIC COMPARTMENT AND THE KROPF CARBON DIOXIDE SYSTEM

	Pneumatic	Kropf
Moisture, %	42.0	43.5
Total acids, ml N/1	12.7	13.3
Total formol N, mg/100 gm malt	315	380
Diastase, degrees Lintner	340	380
Invert sugar, %	2.76	3.00
Sucrose, %	5.40	4.72

A brief description of the pneumatic compartment has already been presented. As indicated, the alternate method for conducting germination is within a drum. Rotation of the drum serves as a means of agitation or mixing the green malt. Tempered and humidified air flows to the drum as the result of the negative pressure imposed by withdrawal of the exhaust air. Germinating temperatures within the drum are estimated by the temperature of the exhaust air. DeClerck (1952) provides a more complete description of the Galland drum and its operation.

Most installations require removal of the malt from the compartment with subsequent conveyance to a kiln. For purposes of labor saving, green malt removal machines have been installed. Generally, one man is thus able to transfer malt from the compartment. An example of such a machine is presented in Fig. 46.

FIG. 46. REMOVING GREEN MALT FROM THE GERMINATING CHAMBER

KILNING

Efficient kilning consists of drying the green malt at a maximum rate of water removal within a temperature gradient which permits retention of desired quality factors and which causes other factors of quality to be created.

There are 2 and possibly 3 distinct phases which must be considered in kilning. The operation commences when the moisture level of the green malt is approximately 45%. In this initial stage, enzymatic degradation and breakdown still go forward because the gradual drying accomplished at lower temperatures permits sufficient moisture to be retained in part of the green malt bed. The second phase of kilning, accomplished by heating after the moisture level of the malt has decreased below 10%, may be considered the final drying or curing, during which chemical or physicochemical reactions take place within the constituents of the malt.

When "sulfuring" is practiced, the application of sulfur dioxide may be considered a distinct phase of kilning. "Sulfuring" is generally conducted while the moisture level of the green malt exceeds 40%.

Most of the malt currently manufactured is dried by direct exposure to the heat of furnaces using natural gas or mixtures of propane and butane (Fig. 47). A few installations still employ

FIG. 47. KILNS

fuel oil. However, the use of oil other than for stand-by purposes is to be discouraged because of a possible danger of soot or film deposit on the malt. The combustion of sulfur-containing oils also results in an uncontrolled contact of the product with sulfur dioxide.

Kilns consisting of 1, 2, or 3 levels are generally employed. When the sequence of drying in two or more levels is considered, it is implied that the heated air first passes through the lower bed of malt, which by nature of position and length of exposure, is the driest. Air leaving the lower level is drawn up through the second or middle bed, which is of an intermediate degree of dryness, and finally through the third or upper bed which has accommodated the last or most recent loading of green malt. A fan, or series of fans, situated above the upper level thus exhausts air whose initial drying characteristics are established in the heating chambers below the first level. Exceptions to the outlined manner of heat application which affect the drying properties of the air are presented later in the discussion.

The depth to which a kiln may be loaded is dependent on the draught which is to be passed through the bed of green malt. Generally, an air velocity of 60 to 70 cu ft/bu/min is considered adequate to remove moisture to a sufficiently low level to permit application of the hot air for the duration of the curing period, thus allowing sufficient time for the operation to coincide with the production schedule. DeClerck (1952) presents values for the volumes occupied by malt at degressive stages of moisture during kilning. The data, appearing in Table 44, are calculated on a malt bushel dry basis. The figure showing one bushel (dry basis) at 45% moisture as occupying 2.18 cu ft coincides well with the value of 1 cu ft of green malt weighing about 22 lb.

TABLE 44

EFFECT OF MOISTURE DECREASE DURING KILNING ON THE VOLUME OCCUPIED BY MALT[1]

Moisture Content of Malt %	Volume, Occupied by One Malt Bu, Dry Basis Cu Ft
45.0	2.18
40.0	2.02
30.0	1.69
20.0	1.36
10.0	1.09
5.0	0.98

[1] DeClerck (1952).

The malt moisture to air temperature relationship is extremely important during kilning from the standpoint of malt quality. This relationship is accentuated at the higher moisture levels. Hind (1938) presents data published by Kolbach and Schild in Table 45 which indicates the temperature effect on the increase in soluble nitrogenous compounds.

Practical experience has shown that wort color and flavor parallel, though to a lesser degree, the increase of soluble nitrogen achieved in this manner. Witt (1945) has noted that an increase in diastatic power is also observed.

TABLE 45

INFLUENCE OF TEMPERATURE AND MOISTURE ON THE "SOLUBLE PROTEIN" FORMATION DURING KILNING[1]

Moisture (Green Malt) %	Lowest Temp. of Increase Occurrence °F	Maximum Temp. of Increase Occurrence °F	Maximum Increase Protein[2]
43	72	133–136	18.7
34	79	Above 140	14.3
24	104	144–151	9.3
15	122	151–158	3.7

[1] Kolbach and Schild, as quoted by Hind (1938).

[2] Soluble protein tabulated as the per cent soluble of total protein, assuming a ten per cent total malt protein.

A recent study by Grunewald (1954) compares two malts resulting from drying schedules in which the heat was raised to 140°F after 6 and 10 hr, respectively, following kiln loading. The same air velocities were employed initially. However, increases in temperature were accompanied by decreases in air velocity. The time-temperature diagram and the data reflecting the effect on nitrogen solubility appear in Fig. 48 and Table 46, respectively.

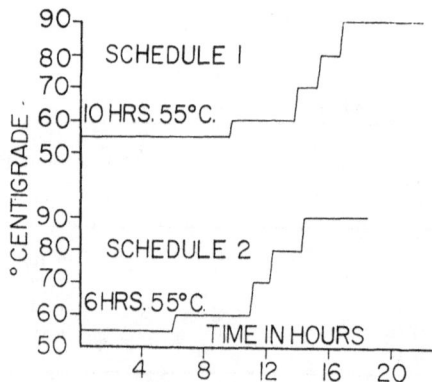

FIG. 48. DRYING SCHEDULE OF TWO MALTS

Temperature was raised to 140°F after 6 and 10 hr, respectively, following kiln loading.

It is to be noted that the lower temperature drying schedule required about 30% more heat and necessitated 2 and 45 min additional kilning time. The investigator indicates that an initial low temperature kilning period was associated with a paler wort color and with a smaller fine-coarse grind extract difference which had only been previously achieved by germinating an additional day.

However, modern kiln design and economical operating procedures may be used to achieve low temperature kilning at higher

TABLE 46

THE FUEL AND LENGTH OF TIME REQUIRED FOR TWO KILNING SCHEDULES STUDIED
AND EFFECTS OF THE TWO METHODS ON PROTEIN BREAKDOWN[1]

	Schedule 1	Schedule 2
Heat requirements (Btu per malt bu)	98,000	75,000
Time requirement (hr on the kiln)	21.5	18.7
Soluble protein increase (% soluble/total)	41.5	44.2

[1] Grünewald (1954).

moisture levels without the sacrifice of air velocity or the expenditure of additional fuel.

A bypass system may be installed in a two level kiln which may be regulated to provide for an admixture of nonheated air with that passing through the malt bed on the lower kiln. In this manner, higher temperatures may be empolyed on the lower kiln without adversely affecting the new malt on the upper kiln.

Examples of kilning methods comparing the use of bypass outside air with a system where all the air employed originates in the heating chambers appears in Table 47.

It is to be noted that, in the instance of the bypass, the air

TABLE 47

HEAT REQUIREMENT FOR DOUBLE KILN EMPLOYING DIRECT HEAT CONTRASTED
WITH ONE USING BYPASS AIR

	Direct Heat		
	Entrance Air Heat Chamber	Exit Air Heat Chamber	Exit Air Malt Bed
Temperature, °F	32	135	73
Relative humidity, %	80	2.3	90
Moisture of air (water per 100 lb dry air)	0.303	0.303	1.573
Heat content (Total Btu per 100 lb dry air)	1096	3528	3528
Water removal (Btu per lb of water)	...	...	1947
Quantity water removed first hour after loading, lb (2,000 bu at rate 60 cu ft/bu/min)	...	...	6788

	Bypass Air Employed (20% Admixture)			
	Entrance Air Heat Chamber	Exit Air Heat Chamber	Entrance Air Upper Kiln	Exit Air Upper Kiln
Temperature, °F	32	135	93	67
Relative humidity, %	80	2.3	24	90
Moisture of air (water per 100 lb dry air)	0.303	0.303	0.720	1.278
Heat content (total Btu per 100 lb dry air)	1096	3528	3040	3040
Water removal total (Btu per lb of water)	...	...	...	1870
Total quantity water removed first hour after loading, lb upper and lower kiln	...	...	...	5492

entering the bed of green malt is 93°F as contrasted with 135°F where no direct outside air is used. In computing the data associated with the bypass air, it is assumed that the malt in the lower kiln has a moisture content of about 10% and that the relative humidity of the air leaving that malt is 16%. It is also assumed that the 20% admixture of outside air reduces the velocity and quantity of air passing through the heating chamber and the lower kiln while permitting passage of the maximum quantity through the upper kiln. A detailed example of the calculation appears later.

The economics of fuel consumption are comparable, each system requiring about 1940 Btu to remove 1 lb of water during the initial drying period. The difference occurs in that 30% less water is removed during the first hour when bypass is used. It is not to be construed that the entire kilning period is lengthened to this degree, as this wide difference exists only with the first 2 or 3 hr of drying. Generally, the entire drying period is only lengthened by a few hours.

Obviously, the factor of increased drying time can be compensated for by augmented fan power. It is reasonable to assume that the cited air velocity of 60 cu ft could be increased to 75 cu ft/bu/min to achieve the desired drying in a comparable length of time.

Control of the humidity of the heated air entering the kiln has been recognized to be of value in achieving malt quality. Pale malts of desirable mellowness and soluble nitrogen have been prepared by drying with an initial air temperature of 120° to 130°F and 15 to 20% RH.

Experimental kilning has been conducted where the humid exhaust air from the germinating chambers has been mixed with outside air prior to entering the heating chambers. A time-temperature diagram of drying employing a mixture of 70% exhaust air and 30% outside air appears in Table 48.

While the example of the use of all outside air was shown to effect an approximate 12% greater moisture removal the first hour, admixture of the malt house exhaust air achieved 10% less heat expenditure per pound of water removal.

Shands et al. (1942) studied some of the changes which occur in malt within the final kilning or curing period. This phase is arbitrarily assumed to begin when the moisture falls below 10 to 12%.

Increased heat is responsible for a marked decrease in diastatic power and a small reduction in extract. The soluble protein remains comparable at the kilning temperature investigated. The data are presented in Table 49.

TABLE 48

HEAT REQUIREMENT FOR DRYING SINGLE LEVEL KILN—DRYING EFFICIENCY OF
GERMINATING COMPARTMENT EXHAUST AIR CONTRASTED WITH THE
CONVENTIONAL USE OF OUTSIDE AIR

	All Outside Air		
	Entrance Air Heat Chamber	Exit Air Heat Chamber	Exit Air Malt Bed
Temperature, °F	32	122	70
Relative humidity, %	80	3.5	90
Moisture of air (lb of water per 100 lb dry air)	0.303	0.303	1.420
Water removal (Btu per lb of water)	...	...	1920
Quantity water removed: lb in first hour after loading (2,000 bu at rate 60 cu ft/bu/min)	...	...	6098
	70% Germinating Compartment Air		
	Entrance Air Heat Chamber	Exit Air Heat Chamber	Exit Air Malt Bed
Temperature, °F	51	122	75.5
Relative humidity, %	96	9	90
Moisture of air (lb of water per 100 lb dry air)	0.762	0.762	1.718
Water removal (Btu per lb of water)	...	...	1754
Quantity water removed: lb in first hour after loading (2,000 bu at rate 60 cu ft/bu/min)	...	...	5347

TABLE 49

THE INFLUENCE OF TEMPERATURE AND TIME OF DRYING ON THE COMPOSITION
OF MALTS MADE FROM ODERBRUCKER AND WISCONSIN BARBLESS
BARLEYS—1939 SERIES

Drying Treatment[1] Temperature in °F	Moisture %	Diastatic Power ° L.	Extract Dry Basis %	Soluble Protein[2]
Oderbrucker variety				
12 hr at 113	6.1	221	72.9	38.9
12 hr at 113; 4 hr at 131; 4 hr at 149	4.5	193	73.0	38.5
12 hr at 113; 4 hr at 131; 6 hr at 149; 2 hr at 167	3.7	165	72.6	38.5
12 hr at 113; 4 hr at 131; 6 hr at 149; 2 hr at 167; 2 hr at 185	3.2	135	72.3	38.5
Wisconsin Barbless variety				
12 hr at 113	7.1	119	74.3	32.5
12 hr at 113; 4 hr at 131; 4 hr at 149	4.9	107	74.6	30.5
12 hr at 113; 4 hr at 131; 6 hr at 149; 2 hr at 167	3.9	85	73.9	30.5
12 hr at 113; 4 hr at 131; 6 hr at 149; 2 hr at 167; 2 hr at 185	3.1	60	73.5	30.0

[1] All samples received 8 hr drying at 77° F plus 4 hr at 95° F, plus additional treatment indicated.

[2] Soluble protein as percent of total protein assuming a total malt protein of 12.5%.

Drying at temperatures about 190°F is frequently associated with a reduction of the soluble protein as a result of the amino acids entering into the reactions responsible for melanoidin formation. When measured analytically, the alpha-amylase activity is reduced but slightly when malt is kilned off below 180° to 185°F.

The activity of limit-dextrinase and of some of the proteolytic and viscosity reducing enzymes, as well as others presently unclassified, is rapidly diminished at temperatures above 140° to 150°F.

NEW PROCEDURES

Kellett (1965) has reviewed procedures of continuous malting, the Popp system and vertical malting.

Innovations in the malting procedure have consisted in plant design or in process application designed to stimulate enzymatic activity or to reduce the loss occasioned by respiration and rootlet growth.

Construction of new malting facilities is planned for the economical utilization of labor, fuel and power. Flexibility with respect to the length of germinating and kilning-times has been incorporated into the design. Rather than requiring a barley variety to adapt itself to a stereotyped procedure, the plan of processing may be tailored to the needs of the raw material.

Continuous Malting

An example of flexibility is the continuous application known as the Domalt System which is operated by the Dominion Malting Company of Canada. Schematically, it is shown in Fig. 49.

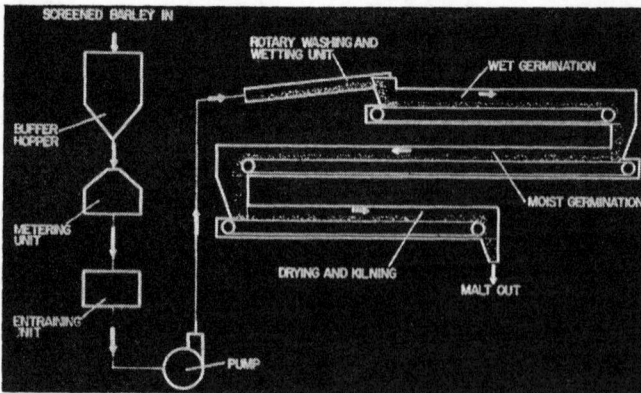

FIG. 49. DOMALT SYSTEM

In this concept, the grain is kept moving slowly, for the desired period of germination, on a flat sectional-screen conveyor belt, through a tunnel-like housing. Successive sections are equipped for individual control of forced-draft circulation of conditioned air upwards through the moving grain bed. The tunnel-section devoted to kilning is separated from the germinating section by a baffle-seal or other suitable arrangement, and is equipped with higher-capacity fans as well with the necessary malt-kiln furnace unit of special design. Ahead of the germinating section, a large diameter slow-speed washing-wetting conveyor is provided to carry out the first steeping stage of the incoming dry barley, by counter-current flow of water and grain. Spray-nozzle assemblies are provided at intervals in the germinating section. A turning apparatus to lift and tumble the grain is provided at intervals within the germinating section. A picture of this device appears in Fig. 50.

As indicated from the above description, steeping is accomplished by sparging rather than by total immersion and germination of the grain commences long before final malt-moisture levels are attained, because of the unlimited access of oxygen to the germ which sparge- or spray-steeping permits (as opposed to the limited oxygen supply available to the germ in conventional steeping).

This rapid onset of germination combined with some acceleration of the total process, permits a reduction in the total time of the malting cycle. It is reported that some vigorous types of Canadian barleys can be processed in as short a time as 80 hr.

While labor requirements are at a minimum, it is said that the power demand is not.

Based on laboratory testing, the quality of the malt from the

FIG. 50. TURNER FOR DOMALT SYSTEM

Domalt system is comparable to conventionally-produced malt. The design lends itself well to experimentation where chemical treatments may be applied to increase enzymatic activity or to result in a reduction in malting loss.

The Graff System

A number of the labor saving features as well as those permitting flexibility in processing characteristic of the Domalt design are claimed for the more stationery drum design of Alan Graff. The inventor describes two types of cylinders which are adapted to improving the malting process.

(1) **The Rotary Malting Drum.**—The drum is designed to accommodate all 3 stages of malting—steeping, germinating and drying—within 1 machine, with no transfers of malt from 1 vessel or chamber to another between stages. Separate steeping facilities to increase overall capacity of the plant may be, of course, provided if desired. With spray- or sparge-steeping at temperatures of 64°F or above, 4 to 6 hr of steeping is sufficient to initiate germination. After emergence of the rootlets has been well advanced by a

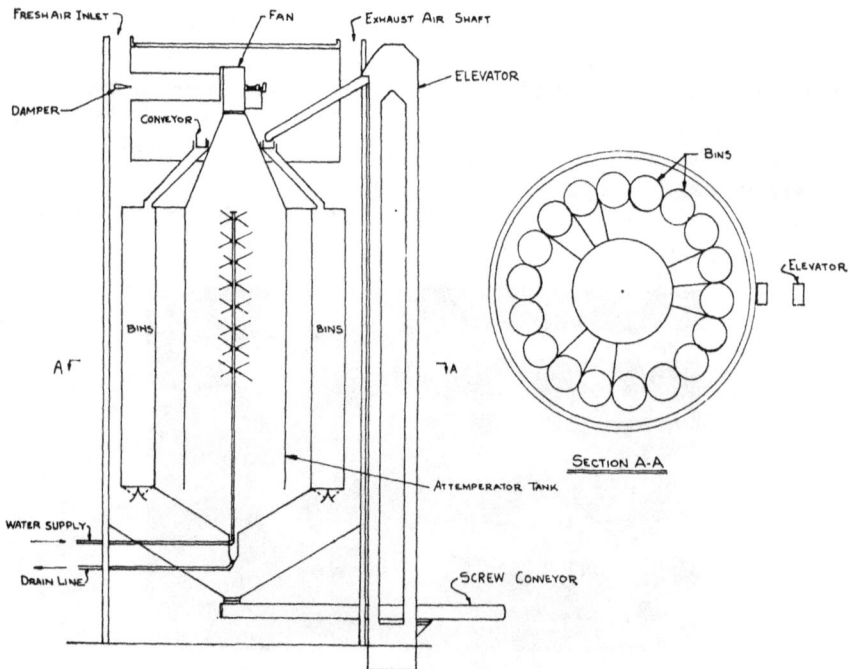

FIG. 51. GRAFF VERTICAL MALTING UNIT

subsequent aeration period, further spray steeping is employed to bring malt moisture to proper levels for modification.

An inner cylinder serves as the air inlet duct and the large fan supplying air for both germinating and drying is mounted to discharge directly into this central cylinder. Air passes from the central cylinder, through the malt bed and out through the perforated screen sections of the outer cylindrical shell.

A unit of this type, installed in Chicago in 1956, had a length of 42 ft and a capacity of 3500 bu of barley.

(2) **The Vertical Malting Unit.**—At the time of this writing, no plant scale installations have been constructed. A sketch drawn from a pilot plant prototype is presented in Fig. 51. The larger detailed drawing illustrates a center pressurized tempering and humidifying area. The air flows upwards through the surrounding bins which hold the germinating barley. Moving or turning the grain is accomplished by conveyance from the bottom of the bin into the top of another.

The recommended dimensions of a single germinating bin consist of a diameter of 3.5 to 4.0 ft and a height of about 40 ft.

While the labor requirement to maintain such an operation is minimal, it is felt that continued surveillance will be necessary.

The Popp System

Named after its inventor, Herman Popp, the system utilizes a novel way of turning the germinating grain (Popp 1965). Steeped barley is loaded to a depth of 7 or 8 ft and the turn is effected by the sudden release of air at high pressure beneath the mass. The load is blown to the top of the vessel, agitating the grain which then falls back to the supporting perforated floor.

The dimensions of a vessel in commercial use is an overall height of 38 ft and a diameter of 18.5 ft. A conical bottom fitted with a door facilitates discharge of the germinated grain. Diagrammatic sketches are shown in Fig. 52 and 53.

Since the equipment may be viewed as a large steeping tank paralleling the smaller counterpart of conventional malting, it is said to lend itself to the RESTEEP system which will be described later.

The advantage of the Popp system is the elimination of mechanical turning devices directly associated with the germinating grain. Moreover, discharge of the green malt is achieved by gravity flow to a screw conveyor.

The unusual depth of the mass would present a disadvantage

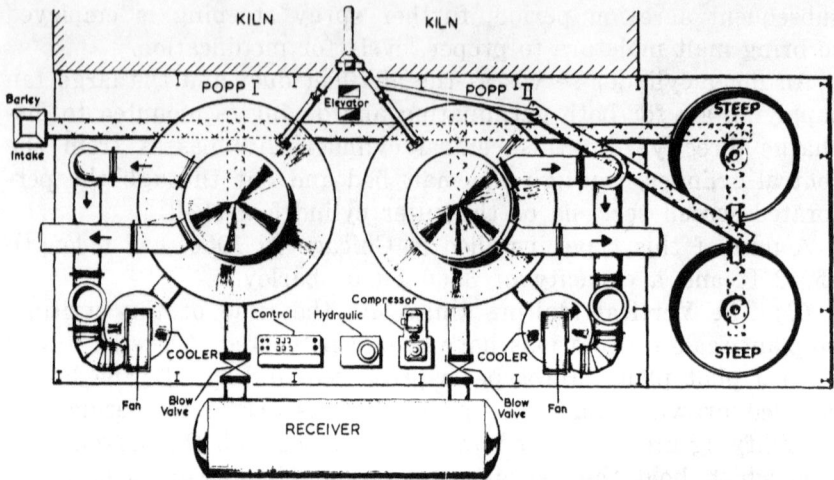

FIG. 52. GENERAL ARRANGEMENT OF THE POPP GERMINATING SYSTEM

for inspection to determine the progress of modification. The danger of power failure, always a hazard in modern malting, would seem to be accentuated in the Popp system.

Vertical Malting

Vertical malt houses, either continual or batch, are found to be usable. One elevation takes the grain to the top where it enters

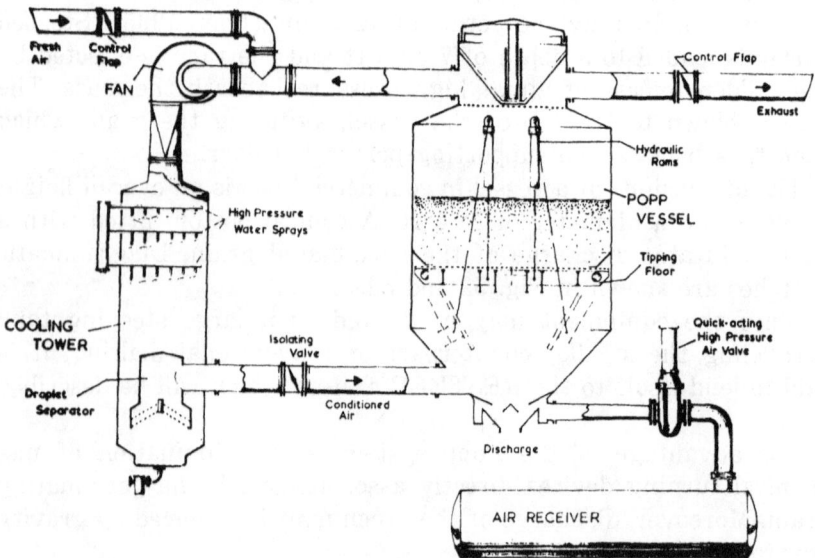

FIG. 53. DIAGRAM OF THE POPP GERMINATING SYSTEM

the process. Gravity flow permits the grain to pass thru the germinating areas and then into the kilning area which is at the bottom of the sequence.

Figure 54 illustrates the Frauenheim gravity system of malting.

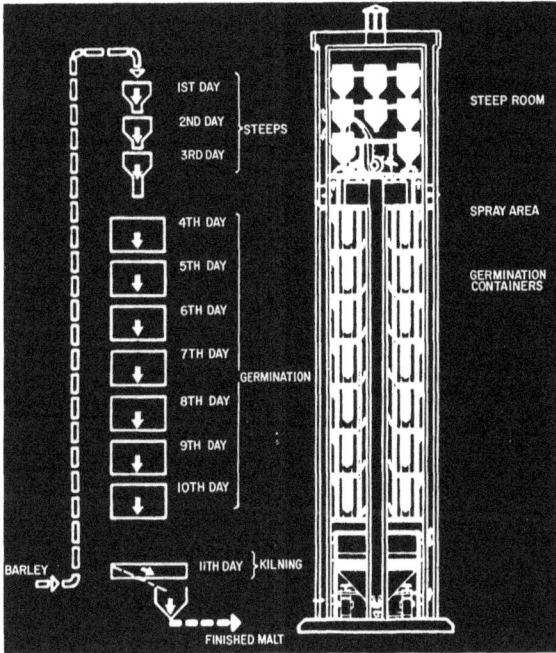

FIG. 54. FRAUENHEIM GRAVITY SYSTEM OF MALTING

The Resteep System

This procedure is one designed to achieve maximum modification with a minimum loss in barley weight due to respiration and rootlet formation (Pollock and Pool 1967).

The design is based on the premise that once the germination or development of the embryo has been initiated, it is unnecessary to maintain active growth for continued modification. This is achieved by "drowning" the kernel soon after it has started to grow.

In general, barley is steeped with maximum aeration to about 40% moisture—a somewhat lower level than is accomplished in conventional practice. A rapid onset of germination is initiated and is maintained for 1 or 2 days. Growth is then arrested by submerging in water. The physical manifestation is one where the "resteep" water becomes slightly acid, contributing to a lightening or brightening of the husk. The rootlets wither and seem to disappear.

After removal of the "resteep" water, the processed grain is permitted to rest under conditions of conventional germination for an additional day or two. It is then kilned in a normal manner. The total malting loss, dry basis, is 2 to 3% compared to 7 to 10% for regular pneumatic malting.

Other than a slightly greater protein solubilization in the wort, malt from the resteep process is comparable in composition or superior to conventionally-prepared malt. An example is presented in Table 50. Comparison is made with a malt treated with 0.06 ppm gibberellic acid (based on relation to barley weight).

TABLE 50

ANALYSES OF PROCTER MALT

| | Conventional | | Resteep |
	Without Gibberellic Acid	With Gibberellic Acid	
Extract, %, fine grind	80.0	80.1	79.8
Diff., %, fine-coarse	2.2	1.6	1.8
Total protein, %, dry	9.06	9.32	9.75
Soluble total protein, %	35.2	40.3	35.9
Wort color (°EBC)	3.5	4.0	3.5
Degree of actual final attenuation, %	61.1	64.2	66.0

Using up to 40% yellow corn grits as an adjunct in infusion (US system) of mashing, satisfactory beer has been prepared with commercial Resteeped malt from the variety Procter.

Several alterations have recently been introduced into the Resteep Process. The necessary period of submergence occasioned by conventional water temperatures has been of a duration to encourage the growth of microorganisms. One procedure is to use the "resteeping" water at 104°F for 1 hr. A rapid cessation of growth without subsequent detrimental effects occurs.

Another approach has been to add traces of formaldehyde to colder "resteep" water requiring a longer period of submergence. Since "resteeping" is thought to be practiced in countries other than the United States, the attitude of the Food and Drug Administration to the use of formaldehyde in malting is not known.

A diagrammatic representation of the equipment is shown in Fig. 55.

The vessel is loaded with barley and then filled with attempered water from an overhead tank. Entry of this water into the air ducts is prevented by a valve. During the ensuing steep, the grain

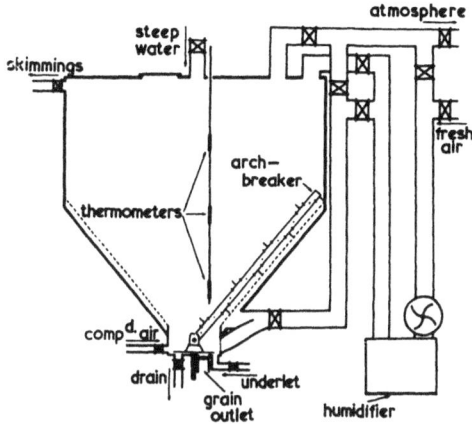

FIG. 55. RESTEEP PROCESS

and water may be roused by compressed air admitted to the underside of the tank. The temperature of the mixture is continuously measured by means of recording thermometers fitted to a rope suspended from the top of the chamber.

After the completion of normal steepings and the period of "resteep", "germination" is controlled by passage of attempered and humidified air through the mass.

Moving of the grain while in the drum as well as its removal is accomplished by an arch-breaker, shown diagrammatically in Fig. 56. This device consists of an arm bearing a series of helically arranged prongs and connected by means of a universal joint to a vertical motor-driven shaft. Turning of this shaft causes a rotation of the arm carrying the prongs which thus dislodge the individual

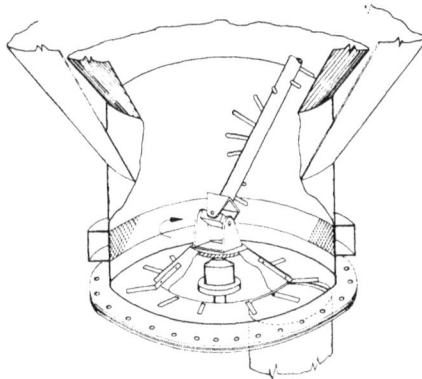

FIG. 56. RESTEEP ARCHBREAKER

grains so that they fall to the bottom of the cone. These are then swept through the outlet by narrow vanes joined to the shaft which drives the arch-breaker.

The modified barley is transferred to a kiln and dried conventionally.

Designed in the manner outlined, a plant of 30 ton capacity was erected in Dublin during 1966 and came into use at the end of that year.

EXAMPLE OF KILN HEATING REQUIREMENT

Admixture of Fresh Air with the Heated Air Exhausted from Lower Kiln as Affects the Drying of the Green Malt on the Upper Kiln

Example A (Control)

No admixture of fresh air; all drying air passes through the heating chambers. This may be construed to be a single level kiln and the malt is a newly-loaded lot with a moisture of 45%.

(1) Fresh air enters heating chambers at 32°F, 80% RH and contains 0.303 lb of water and 1096 Btu per 100 lb dry air.

(2) The air is heated to 135°F, has 2.3% RH and a calorific content of 3568 Btu per 100 lb of dry air.

(3) The air is exhausted from the upper kiln at 73°F, 90% RH and contains 1.573 lb water per 100 lb dry air.

(4) Calculations.

  (a) Heat requirements:

$$\frac{3568 - 1096}{1.573 - 0.303} = 1947 \text{ Btu/lb water removed}$$

  (b) Evaporation rate: 2,000 bu at 45% moisture are ventilated for 1 hr at the rate of 60 cu ft/bu/min under the condition stated. Water evaporated first hour:

$$2000 \times 60 \times 60 \left[\frac{(1.575)}{(13.75 \times 100)} - \frac{(0.303)}{(15.05 \times 100)} \right] = 6788 \text{ lb}$$

Example B

The lower level of a 2-stage kiln contains malt at 10% moisture. The upper kiln is newly loaded, the green malt moisture being 45%. A bypass or opening to the outside air exists between the decks of the lower and upper kiln.

(1) The process is started when the malt in the lower kiln has 10% moisture. The air exhausted from the malt (lower kiln) has

a temperature of 108°F and 16% RH. Contained in 100 lb of dry air are 0.890 lb of water and 3527 Btu.

(2) Admixture of 20% outside air is affected. The air has a temperature of 32°F, 80% RH and contains 0.303 lb of water and 1096 Btu per 100 lb dry air.

(3) The mixture of air entering the bed of malt on this upper kiln has a temperature of 93°F, 21% RH and contains 0.720 lb of water and 3040 Btu per 100 lb dry air.

(4) The air, exhausted from the green malt on the upper kiln, has a temperature of 67°F and 90% RH and contains 1.278 lb of water per 100 lb dry air.

(5) Calculations.

(a) Heat requirements:

$$\frac{[(0.890-0.303)(3527-1096)] + [(1.278-0.720)((0.8\times3527)-1096)]}{1.278}$$
$$= 1870 \text{ Btu}$$

$$\frac{1870}{1.145} = 1633 \text{ Btu/lb water removed}$$

(b) Evaporation rate employed is 60 cu ft per bu per min. Upper and lower kiln each contain 2,000 bu of malt.

$$0.8 \times 7,200,000 \left[\frac{(0.890)}{(14.48 \times 100)} - \frac{(0.303)}{(14.98 \times 100)}\right] = 2375 \text{ lb water}$$

Upper kiln:

$$7,200,000 \left[\frac{(1.278)}{(13.53 \times 100)} - \frac{(0.720)}{(14.07 \times 100)}\right] = 3117 \text{ lb water}$$

Total Water Removal First Hour: $2375 + 3117 = 5492$ lb

Admixture of Germinating Compartment Exhaust air (59°F, 90% RH) with Outside Air (32°F, 80% RH)

Example A (Control)

Air at 32°F, 80% RH heated to 122°F. All drying air passes through the heating chambers. The calculations parallel Example A, 4 of No. 1.

(4a) Heat requirements:

$$\frac{3214 - 1096}{1.420 - 0.303} = 1896 \text{ Btu/lb water removed.}$$

(b) Evaporation rate (2,000 bu at 60 cu ft per bu per min)

$$7{,}200{,}000 \left[\frac{(1.420)}{(13.4 \times 100)} - \frac{(0.303)}{(14.60 \times 100)} \right] = 6127 \text{ lb evaporated first hr}$$

Example B

Seventy percent exhaust air (59°F, 90% RH) mixed with 30% outside air (32°F, 80% RH). All drying air passes through the heating chambers. This example, as well as "Example A", above, may be construed to treat a single level kiln and the malt is a newly-loaded lot with a moisture of 45%.

(1) Germinating exhaust air at 59°F and 90% RH contains 0.9594 lb of water and 2439 Btu per 100 lb dry air.

Outside air at 32°F and 80% RH contains 0.3025 lb of water and 1096 Btu per 100 lb dry air.

(2) Seventy percent germinating exhaust air plus 30% outside air, when entering heating chambers, has a temperature of 51°F and 9% RH and contains 0.762 lb of water and 2036 Btu per 100 lb dry air.

(3) When exhausted from the bed of malt, the air has a temperature of 75.5°F and 90% RH and contains 1.718 lb of water per 100 lb dry air.

(4) Calculations:

(a) Heat requirements:

$$\frac{3713 - 2036}{1.718 - 0.762} = 1754 \text{ Btu per lb water removed}$$

(b) Evaporation rate (60 cu ft per bu per min; 2000 bu of malt)

$$7{,}200{,}000 \left[\frac{(1.718)}{(13.7 \times 100)} - \frac{(0.762)}{(14.9 \times 100)} \right]$$
$$= 5347 \text{ lb water evaporated first hr}$$

SULFUR DIOXIDE IN KILNING

A large quantity of the malt finding its way into commercial channels is sulfured. Sulfur dioxide, originating from the burning of lump sulfur or the evaporation of liquefied gas, is applied to the green malt immediately after kiln loading. When burned, pans containing the sulfur are placed adjacent to the heating chambers and the gas is drawn up into the malt as dispersed in the heated air.

When considered quantitatively, the gas from the burning sulfur appears to be more effective than that evaporated from the liquid phase, probably due to the presence of trioxides in the former. Gas from the liquid phase has value as its rate of application can be more readily controlled.

Among the primary objectives of the application of sulfur dioxide are the bleaching of the malt and the destruction of possible mold and bacteria whose presence during initial drying may be detrimental to quality.

Use of sulfur dioxide is associated with an increase of soluble protein and extract. Witt and Adamic (1957) studied the effect of kiln sulfur dioxide on the malt proteolytic activity exerted during mashing. The degree of sulfuring was considered to be relatively heavy and was equivalent to the passage of 2250 cu ft of air containing 0.08% sulfur dioxide through 1 bu of malt.

The increase in total soluble protein as affected by sulfuring approximated 30%. It was concluded that:

(1) Part of the increase was independent of mashing and existed as preformed "proteins", probably enzymatically formed during the initial kilning phase.

(2) Depression of the mash acidity was associated with the acceleration of mash proteolytic activity.

(3) Kiln sulfuring was responsible for an actual net increase of proteolytic activity which, in itself, was independent of the effect of mash pH value.

Use of kiln sulfur dioxide is known to produce malts having a residual sulfur dioxide content. However, laboratory beers, when compared with those from unsulfured malts, do not reflect the increases of sulfur dioxide observed in the sulfured malts (Witt and Ohle 1950).

BIBLIOGRAPHY

ABERG, E., and WEIBE, G. A. 1946. Classification of barley varieties grown in the United States and Canada in 1945. US Dept. Agr. Tech. Bull. *907*.

ANON. 1926. Western Canada Grain Grades. Dawson-Richardson Publications, Winnipeg, Man.

ANON. 1957A. Barley Variety Dictionary. Malting Barley Improvement Assoc., Milwaukee, Wisc.

ANON. 1957B. Official Grain Standards of the United States. US Dept. Agr. Serv. Regulatory Announcements *AMS-177*.

BROWN, A. H. 1933. Effects of sulfuric acid delinting on cotton seeds. Botan. Gaz. *94*, 755–770.

CROCKER, W. 1916. Mechanics of dormancy in seeds. Am. J. Botany *3*, 99–121.

DECLERCK, J. 1952. A Textbook of Brewing. Versuchs und Lehranstalt fur Brauerei, Berlin.

DEUBER, C. G. 1931. Chemical treatment to shorten the rest period of tree seeds. Science *73*, 320–321.

DICKSON, A. D., and SHANDS, H. L. 1942. The influence of the drying procedure on malt composition. Cereal Chem. *19*, 411–419.

ECKERSON, S. H. 1913. A physiological and chemical study of after-ripening. Botan. Gaz. *55*, 286–299.

FLEMION, F. 1934. Physiological and chemical studies of after-ripening of Rhodotypos kerriodes seeds. Contribs. Boyce Thompson Inst. *6*, 91–102.

GRACININ, M. 1928. Orthophosphoric acid as a stimulant to germinating energy and an activator of the germinating capacity of seeds. Biochem. Z. *195*, 457–468.

GRAFF, A. R. The Graff Rotary Malting System. Unpublished.

GRUNEWALD, J. 1954. Effect of kilning conditions on malt quality. Brauwelt *92*, 1–16.

HIND, H. L. 1938. Brewing: Science and Practice. John Wiley & Sons, New York.

KELLETT, O. S. 1965. International aspects of malt house design. Master Brewers Assoc. Am.—Tech. Quart. *2*, No. 1, 69–78.

KOLBACH, P. 1955. Protein degradation during malting and its control. Wiss. Beil. *6*, 71–76.

LEBERLE, H. 1952. Brewing of Beer. F. Enke, Stuttgart. (German)

MATZ, S. A. 1969. Cereal Science. Avi Publishing Co., Westport, Conn.

MEREDITH, W. O. S., and BENDELOW, V. M. 1956. Additional criteria of malting quality in varietal studies. Proc. Am. Soc. Brewing Chemists *1956*, 77–82.

POLLOCK, J. R. A., and POOL, A. A. 1967. A review of resteeping and multiple steeping methods of malting. Master Brewers Assoc. Am. Tech. Quart. *4*, No. 4, 217–226.

POPP, H. 1965. International aspects of malt house design. Master Brewers Assoc. Am.—Tech. Quart. *2*, No. 1, 71–72.

SHANDS, H. L., DICKSON, A. D., and DICKSON, J. G. 1942. The effect of temperature change during malting on four barley varieties. Cereal Chem. *19*, 471–480.

WARRICK, L. F., RUF, H. W., and NICHOLS, M. S. 1935. Malt house waste treatment studies in Wisconsin. Sewage Works J. *7*, 564–574.

WITT, P. R., JR. 1945. Effect of kilning on the amylolytic activity of barley malts. Cereal Chem. *4*, 341–349.

WITT, P. R., JR., and ADAMIC, E. 1957. The effect of kiln sulfur dioxide on proteolytic activity during mashing. Proc. Am. Soc. Brewing Chemists *1957*, 37–45.

WITT, P. R., JR., and OHLE, R. L. 1950. The effect of boiling on the color and on the indicator time test of laboratory wort. Proc. Am. Soc. Brewing Chemists *1950*, 37–43.

CHAPTER 5

Donald W. Ohlmeyer[1]

Samuel A. Matz

Brewing

INTRODUCTION

The fermentation of cereal grains to produce beer is as old as history itself. The basic processes used today are fundamentally the same as those used in ancient times, but the application of scientific methods has resulted in a better understanding of the reactions taking place and has thus enabled closer control of the various steps so that a product having optimum flavor, color and stability can be obtained with greater reliability.

In this chapter, a brief general description of the brewing process will be given first, then the details of the various steps will be dealt with in later sections.

The main ingredient of beer is malt. Malt is usually prepared from barley which has been soaked in water, drained, and sprouted until the acrospire (plumule) has grown to an average length equal to three-fourths of the kernel length. The green (moist) malt is then dried to halt growth (see Chapter 4). Malting develops the amylolytic enzymes which change starch to fermentable sugars, the source of the alcohol in beer. Malt also contains proteins, protein breakdown products, and a host of other chemical entities which are sources of flavor and act as nutrients for the yeast during fermentation. In addition, there are substances such as cellulose which are more or less inert during the brewing process. An extensive discussion of the malting process is given in Chapter 4.

Corn and rice are used as adjuncts in present-day brewing, to obtain the paler, snappier, less filling beer preferred by today's consumers. These adjuncts are heated to boiling in the presence of a small amount of malt to furnish amylolytic enzymes. The heating gelatinizes the starch, permitting more rapid attack by the enzymes, which then liquefy the gelatinized starch without converting it to sugar. The heating is usually in stages, the temperature being held for some time at specified temperatures to give the enzymes time to act.

The initial process in brewing is mashing. The dried malt is milled to crush the husks so that the starches can dissolve more

---

[1] President, Ohlmeyer's Kloh, Inc., Chicago, Ill.

easily. The milled malt and water are put into the mash tub where a series of controlled time-temperature treatments changes favor the action of enzymes on the proteins and starch, the latter being converted to fermentable sugars. The boiled adjunct from the cooker is added during the mashing, and thoroughly mixed in. The mash mixture, now called wort, passes to the lauter tub, which has a false bottom. The grain husks collect on the false bottom and form a filter bed, through which the sugar-containing liquid, called wort, is strained.

A conventional brewhouse contains as minimum basic equipment, a mash cooker and mash tub, a lauter tub or mash filter, and a wort kettle.

The clear liquid wort is collected in the grant and then goes to the kettle where it is boiled for about 2.5 hr, with addition of hops and (usually) corn sugar (dextrose). During boiling some of the hydrolyzed protein is coagulated, any remaining enzymes are inactivated, and the liquid is sterilized. When the boiling is finished the hot wort is strained in the hop strainer to remove the leaves and stems of the hops, and then collected in the hot wort tank. After the temperature of the wort is reduced to a suitable level in the cooler and the cooler pan, yeast is mixed in and the mixture pumped to the settlers where it remains for 10 or 12 hr. From the settlers, the wort is pumped to the fermentation tanks where it remains until fermentation is complete and the sugars have been converted to ethyl alcohol and carbon dioxide. Storage, addition of chillproof, and carbonation follow (see Fig. 57).

There are several different types of beer, with variations in flavor, color, carbonation, and alcoholic content which result from differences in the kinds and proportions of ingredients and in the processing conditions. Most lager beer-types, such as Pilsener, Munich, Dortmund, and Vienna, are produced by bottom fermentation, that is, with a yeast which sinks to the bottom of the tank during fermentation. Stout, porter, and ale are top-fermented beers.

Where the product is not otherwise identified in the following discussion, the process is that used for brewing American lager beer. Most American lager beers do not closely resemble the usual European types.

The term "lager beer" simply means that the beer has been aged after fermentation. Pilsener beer is a pale beer having a high hop rate and a very stable head. Munich, or Munchener beer is darker and sweeter than Pilsener and has a more pronounced malt flavor,

[BREWING]

[FERMENTATION and FINISHING]

[MALTING]

MALT Cleaning

MALT Milling

Weighing

PURE YEAST CULTURE

Cooled Wort

Yeast

STARTING TANK — Addition of Yeast Starting Fermentation Separation of Precipitated Solids

CO_2

FERMENTER

Yeast to Drier Plant

GRAIN ELEVATORS — Accumulation and Storage of Barley Cleaning and Grading

COOKER — Liquefaction

MASH TUB — Conversion

STORAGE TANK — Ageing and Mellowing

MALT HOUSES — Steeping Germination Kilning Cleaning

LAUTER TUB

MASH FILTER — Extraction

Spent Grain Driers

FILTERS — Prefiltration and Carbonation

BREW KETTLE — Hop Extraction Concentration Coagulation

PREFILTRATION and STORAGE — Chill Proofing and Bonding CO_2

GRAIN ELEV. — Storage of Finished Malt Blending

HOP STRAINER — Separation of Hops

FILTERS — Polish Filtration and Carbonation

SETTLING TANK — Separation of Trub

To Bottle and Can Filters

To Racking Room (Draught Beer)

Cooling of Wort — Wort Coolers

To Starters

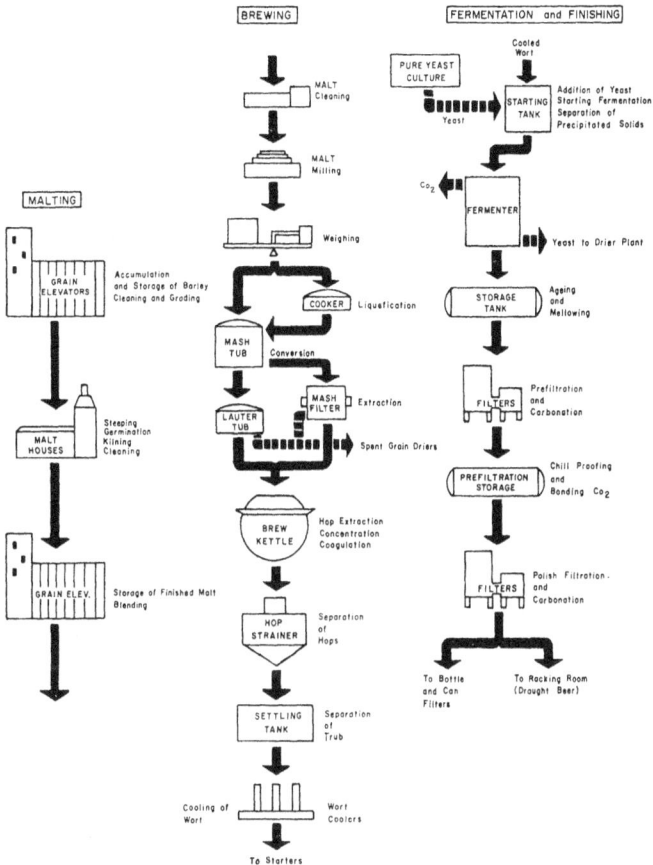

Courtesy of Jos. Schlitz Brewing Co.

FIG. 57. MALTING AND BREWING FLOW CHART

due not only to use of a darker malt but also to a lower hop rate. Bock beer is a traditional spring time beer, being brewed in the winter. It is darker in color and usually sweeter and fuller bodied than other types. Caramelized or roasted malt is used in its preparation.

RAW MATERIALS

The selection of the proper raw materials is vital to the brewing industry, as it is impossible to brew a superior product from inferior starting materials. Some pertinent information regarding raw materials will be briefly discussed.

Barley

The traditional grain for brewing is barley. Over thousands of years, varieties suitable for malting and brewing have been selected until the present day barleys have certain characteristics favorable to the conventional brewing process which would be difficult to reproduce in other grains. Nonetheless, new methods of brewing as well as new and cheaper sources of enzymes and carbohydrates have opened the way for other grains to partially or completely replace barley malt. Tradition remains a powerful force in restraining the switch to other beer ingredients.

Two factors probably have been the most important in promoting the use of barley as the primary grain for brewing: (1) it can be grown in a great variety of climatic conditions, and thus is found all over the world, and (2) its straw-like husk makes it easier to process. The husks are not removed by threshing, and therefore protect the embryo during malting. They also assist in the clarification of wort by forming an efficient filter bed.

The two main types of barley are the 2-rowed and the 6-rowed varieties. The two-rowed barleys, *Hordeum distichum*, are considered in Europe to be the best for brewing. They are generally sown in the early spring. In the United States, these barleys are cultivated in the Pacific Northwest. Six-rowed barleys, *Hordeum hexastichum*, known as winter barleys because they are sown in the fall, are the most commercially important barleys grown in North America, and are extensively used for brewing. Barley grown in the Red River Valley is considered best for this purpose. Present practice is to select a pure line and strain of barley, having known malting characteristics.

The preparation of barley malt is described in a preceding chapter. The malt used for brewing commonly has a diastatic power of 120° to 140° Lintner, and alpha-amylase activity of 25 to 40 units. Distillers' malt is stronger, 190° to 250°L and alpha-amylase strength of 50 to 70 units.

For unusual flavor or color effects, special types of malt are available. These are generally barley malts which have been subjected to modified kilning procedures. Higher than normal temperatures lead to dark "caramel" malts which are valued for their coloring effect in bock beer and other dark varieties. Caramel or roasted malts are available in different color gradations. Crystal malt is kilned or roasted for 30 to 45 min at 140° to 167°F in the absence of air, then the cover is removed and the mass is heated at around 300°F for 1 to 2 hr in a normal draft.

Adjuncts

Barley malt is the traditional source of fermentable sugars for wort, and in some countries still constitutes the only source. There are, however, economical substitutes which are being used more and more. These include unmalted grains such as barley and rice which are added with the malt and saccharified by the malt enzymes, sugar, and syrups or dried syrups made by reacting grain or purified starch with bacterial or fungal enzymes. The sugar and syrups may be added to the cooled wort. Corn syrup of high dextrose equivalent is a typical example of the class.

Adjuncts containing unmodified starch, such as broken rice or corn grits must be precooked separately before they are added to the mash. The mix for precooking consists of a relatively small amount of finely ground malt, the ground adjunct, and water. In the cereal cooker this mixture is brought to about 158°F[1] and held for about 30 min, then boiled for 10 min.

Rice, corn, unmalted barley, sorghum, and some of the other adjuncts give a paler beer, and, frequently, a milder flavored beer.

Rice is used in the form of grits, which are the broken grains obtained in the course of dehulling and polishing. Lipid content should not exceed 0.5 to 0.7% and free fatty acid content should be low to avoid rancid flavors in the beer. Pre-boiling of corn can be avoided by using flakes, obtained by passing degermed ground meal between heated cylinders to gelatinize the starch. The fat content of the meal should not exceed 1%, and the fat must not be rancid, as otherwise free fatty acids would pass into the beer.

Corn syrups and corn syrup solids may be added to the kettle to increase the amount of fermentable sugars. Commercial preparations of hydrolyzed corn starch contain variable amounts of dextrins, which are not objectionable as they give the beer a mellower flavor. The "dextrose" used should have a neutral reaction in aqueous solution, and should contain no iron or starch.

In most cases, even where substantial amounts of adjuncts are used, barley malt is needed for enzyme activity, but several patents have been issued covering the addition of nonmalt enzymes to adjuncts. For example, a recent patent (Bavisotto 1967) claims that malt: adjunct ratios as low as 40:60 can be used if 0.1 to 1.0% (malt basis) of proteolytic enzyme preparation (papain, bromelin, ficin, etc.) is added to the mash.

---

[1] In brewing technology, temperatures are expressed as degrees Reaumur (°R). Zero on the Reaumur scale is equal to zero on the Centigrade scale. One degree Reaumur is equal to 1.25° Centigrade, or 2.250° Fahrenheit.

Hops

Brewers' hops are the dried fruit (cones) of the perennial plant *Humulus lupulus*. Male and female inflorescences are borne on separate plants and only the female ones are used in brewing. The flowers contain small lupulin glands which secrete the bitter resins and essential oils. Hops are picked in late August or early September, and must be dried as rapidly as possible to a moisture content of 12 to 13% to prevent oxidation and polymerization of the essential oils and bitter acids. However, the temperature should not rise above 122°F during the drying.

The hops grown in Czechoslovakia are considered by some to be the finest. In the United States, hops are grown mostly in Washington, Oregon, California, Idaho, and New York. American hops are characterized by a rather fruity flavor. The average chemical composition of kiln-dried hops, as given by DeClerck (1958) is shown in Table 51.

TABLE 51
COMPOSITION OF DRIED HOPS

	%		%
Moisture	12.5	Ether extract (mostly resins)	18.3
Ash	7.5	Tannin	3.0
Cellulose (gross)	13.3	Nitrogenous matter	17.5
Essential oils	0.4	Nonnitrogenous extract	27.5

From the standpoint of brewing, the important constituents of hops are the bitter resins and the essential oils. The bitterness of hops is due to resinous materials. Hop oils contribute both flavor and aroma. Two resins are known: humulone (α-acid, α-resin) and lupulone (β-acid, β-resin). It has been shown by Rigby and Bethune (1952, 1953, and 1955) that humulone is a mixture of three closely related substances, called humulone, cohumulone and adhumulone. The structure of these compounds has been elucidated by Howard and Tatchell (1954). Lupulone also is a mixture of at least three related compounds called lupulone, colupulone, and adlupulone. Structures for the first two compounds have been proposed by Riedl (1954), Howard and Pollock (1954), and others. These resins, which are responsible for the bittering and antiseptic powers of hops, seem to be peculiar to this plant. During storage the resins are progressively oxidized and polymerized.

The composition of hop oil is very complex, as shown in Table 52. About 70 to 80% of the oil consists of terpene-like hydrocarbons, but they are thought to contribute less to the aroma than the oxygenated compounds which are present. The terpenes consist chiefly of myrcene and humulene (α-caryophyllene), with smaller amounts of other terpenes. These oils are also subject to resinification and polymerization when the hops are stored. A large proportion of the resin and oil content is lost during brewing.

Another important component of hops is tannin. This has been found by Vancraenenbroeck and Lontie (1955) to consist of a mixture of at least three different groups of substances, including flavanols and leucoanthocyanins. Harris (1956) has detected a number of individual compounds by means of paper chromatography using different solvents. Hop tannin is converted to reddish-brown phlobaphene on oxidation, a behavior which is characteristic of catechol tannins, and is precipitated with barley albumin during the boiling in the kettle.

Hops for brewing are selected chiefly by their external appearance and place of origin. Chemical analyses for moisture content and resins are useful.

There is a trend toward the use of hop extracts in order to reduce the loss of resins and oils and for greater convenience and better

TABLE 52

COMPOUNDS FOUND IN THE OIL STEAM-DISTILLED FROM "BULLION" HOPS
(CHROMATOGRAPHIC ANALYSIS[1])

Carbonyls	Free Alcohols	Carboxylic Acids
Hexanal	Butyl	Butyric
2-Hexenal	Isobutyl	Isobutyric
Heptanal	Amyl	Isovaleric
2-Heptenal	Isoamyl	Caproic
Heptanone	Hexyl	Isocaproic
Octanal	Octyl	Methylethylacetic
2-Octenal	Nonyl	Enanthic
2-Octanone	Decyl	Isoenanthic
Nonanal	Geraniol	Caprylic
2-Nonenal	Linalol	Isocaprylic
2-Nonanone	Nerol	Pelargonic
Decanal	Terpineol	Isopelargonic
Decanone	Undecyl	Capric
Citral I	Dodecyl	Isocapric
Citral II	Nerolidol	Ceranic I
Undecanal		Geranic II
Undecanone		Undecanoic
Dodecanal		9-Methyl decanoic
Dodecanone		Lauric
Tetradecanal		Tridecanoic
		Isotridecanoic
		Myristic
		Isomyristic
		Pentadecanoic
		Isopentadecanoic
		Hexadecanoic

[1] Jahnsen (1963).

standardization of the finished product. These extracts can be added to the brewers' copper or to the fermented beer (Howard 1967). Six gallons of refined hop extract will replace a 200 lb bale of dried hop blossoms.

Hydrogenation or sodium borohydride reduction of the hop extract is said to reduce or eliminate the formation of odorous mercaptan-like compounds under the influence of light. The hydrogenation reduces the carbonyl group of the isohexenoyl side chain of an isohumulone to a secondary alcohol group.

Water

Water is of the utmost importance in brewing, as its salt composition has a pronounced effect on the character and flavor of the beer brewed with it. The direct effect of the salts on the flavor of the beer is slight, for the reason that their amount is small in comparison to the total mineral content of the beer. However, different salts affect enzymatic and other reactions, as well as the solubility of the malt proteins, in different ways, thus exerting a powerful indirect effect on the taste. The salts are in the form of ions in a very dilute solution, so that it is inaccurate to speak of specific salts being present. The individual ions and their quantitative relationships to each other are the determining factors in the suitability of the water for brewing.

TABLE 53

COMPOSITION OF VARIOUS BREWING WATERS
(Mg/1000 Ml)

	Pilsen	Dortmund	Munich	Burton
Total solids after evaporation	51	1,110	284	1,790
Calcium (CaO)	9.8	367	106	520
Magnesium (MgO)	1.2	38	30	145
Sulfates (SO_3)	4.3	240	7.5	756
Chlorides (Cl)	5.0	107	2.0	34
Nitrates (N_2O_5)	Trace	Trace	Trace	22

In past times, the quality of the water supply of a brewery was a critical factor in its success (see Table 53). It was literally impossible to produce a satisfactory product with certain water supplies. At the present state of the art, corrective measures can be applied with little difficulty, and any potable water can be adapted to be suitable for any desired type of beer.

For example, it may be necessary to either lower or raise the sulfate and chloride levels. The carbonate concentration is an important factor (Nissen 1965).

Pilsener beers require a very soft water, and dark beers a medium hard to hard carbonate water. For lightly hopped beers, a minimum residual carbonate hardness of 3 to 5° is needed in order to avoid a relatively flat or bland taste in the finished beer. Very soft water can be corrected with up to 0.15 gm of gypsum or calcium chloride per liter (Narziss 1966).

Water low in alkalinity and rather high in calcium sulfate hardness is needed so that the mash pH will be low enough (4.6 to 5.7) for optimum enzyme activity. If the water is too soft, addition of calcium sulfate (gypsum) and sodium chloride may be required. The addition of hardening salts is called "burtonizing." A high wort pH leads to uneven saccharification during mashing, difficulties in the separation of the wort from the spent grain (resulting in a low yield of extract), incomplete coagulation of the proteins during boiling, sharper bitter flavor from the hops, and a biologically unstable beer susceptible to infection by lactic organisms.

The acidifying effect of calcium from the water supply (or added gypsum) is due to the reaction of calcium ions with secondary phosphate ions derived from the malt, as indicated by the equation

$$3\ Ca^{++} + 2HPO_4^{--} \rightarrow Ca_3(PO_4)_2 + 2H^+$$

The triorthophosphate salt of calcium has a very low solubility, but the reaction does not go to completion.

Nitrites or nitrates, the latter being partially reduced to nitrite during fermentation, may interfere with yeast activity even if they are present at very low concentrations. Iron contributes an off-flavor and may discolor the beer by interfering with the clarification process. More than 1 ppm iron also has a deleterious effect on enzymes in the mashing step since it can combine with these proteins as well as other barley components.

Bonnet (1955) states that magnesium concentrations in excess of 30 mg/liter impart a reddish tint and acridness to beer. Chlorides are said to mellow and sweeten or soften beer, especially at levels of 20 to 60 mg/liter. Higher concentrations are to be avoided. Copper in trace amounts has been blamed for accentuating the "gushing" phenomenon (Hoak 1956). Fluoride seems to stimulate yeast metabolism at concentrations of 1 to 5 mg/liter (Webber and Taylor 1954), whereas inhibition has been observed at levels higher than 10 mg/liter. Silica may increase the turbidity if more than 50 ppm are present, although this effect is relatively unimportant (Pozen 1940).

Bonnet (1955) and others have compiled lists of suggested limits for brewing water, and these are summarized in Table 54.

TABLE 54

SPECIFICATION LIMITS FOR BREWING WATER[1]

Color	0–10 ppm
Turbidity	0–10 ppm
Taste and odor	None-low
Total dissolved solids (300 for any one substance)	500–1500 ppm
pH	6.5–7.0
Alkalinity, as $CaCO_3$	75–80 (light beer)
	80–150 (dark beer)
Iron	0.1–1.0 ppm
Manganese	0.1 ppm
Iron + Manganese	0.1 ppm
Carbonate	50–68 ppm
Nitrite	0
Nitrate	10 ppm
Chloride	60–100 ppm
Silica	50 ppm
Hydrogen sulfide	0.2 ppm
Fluoride	1.0 ppm
Calcium	100–200 ppm (light beer)
	200–500 ppm (dark beer)
Calcium sulfate	100–500 ppm
Magnesium	30 ppm

[1] Bonnet (1955).

Flavor materials present in the raw water may be carried through the entire brewing process and appear in the finished beer. There are many types of these contaminants, but phenolic substances, and particularly chlorophenols, can be very troublesome, leading to medicinal or other unpleasant tastes and odors. Charcoal treatment to adsorb these compounds is the most reliable means of correction.

Yeast

American lager beer is generally fermented with a pure strain of *Saccharomyces carlsbergensis*. Strains of *S. cerevisiae* may be used for ales and porter. A bottom yeast is generally used for beer manufacture while top yeast is used for ale. The cells of top yeast gather in clumps and are carried to the top of the fermenting liquid while the cells of bottom yeast stay suspended in the bulk of the liquid or drift toward the bottom.

Probably not more than 20% of the yeast produced in batch fermentation is required for subsequent batches. The excess is disposed of through various commercial channels, e.g. to pharmaceutical and food flavoring manufacturers.

Most brewers prefer to isolate their stock culture from a stock brewery yeast culture and often derive it from a single cell culture. Strandskov (1965) described a typical procedure in which a pure mass culture of *S. carlsbergensis* is used. Yeast reaching the fermenter was actively budding. Microbial infection was found in this procedure to be restricted to *Flavobacterium proteus*, which maintained a fairly constant relationship to yeast count in the fermenters of about 1.5% and did not adversely affect flavor.

The vigor of a yeast strain in producing ethanol and carbon dioxide, and its generation of the minor constituents which affect beer flavor, are the most important characteristics to the brewmaster. The property of flocculence is also of brewing importance. The flocculating ability of yeast determines its capacity of forming aggregates of cells which fall out of suspension in bottom fermentations or are carried to the surface in top fermentations. The more highly flocculating yeasts are removed from suspension relatively soon, and the degree of attenuation (extent to which the sugars are converted to alcohol) tends to be reduced. Thus, highly flocculative yeasts give rise to sweeter, poorly attenuated beers while "powdery" yeasts lead to dry, well attenuated beers. On the other hand, the powdery yeast are more difficult to remove from the beer and tend to be carried over into the storage tanks and even into the finished container.

MASHING

The malt, which is received at the brewery as substantially whole kernels, must be reduced in size to facilitate extraction and enzyme action. Current techniques generally require grinding the malt (see Fig. 58) into a dry or wet grist, retaining the husks in fairly

Courtesy of Stroh Brewing Co.

FIG. 58. MALT MILLS

large pieces while obtaining a maximum amount of finely ground barley flour. The husks will serve as the principal effective component of the filter bed in the lautering process.

There are several variations of the method by which the ground malt is dispersed in water. In general, these variations have little effect on the finished product and are chosen to give maximum efficiency.

The object of mashing the malt and adjuncts is to bring into solution the relatively small quantities of substances which are soluble without enzyme action, and to change insoluble substances into soluble ones through enzyme action. The enzymatic reactions have already been initiated during malting, but are speeded up and controlled by regulating the temperature during mashing. For example, the ratio of starch breakdown during malting to that during mashing is about 1 to 10. The chemical structure of the starch, proteins, gums, pectins, etc. is changed in such a way that the finished wort contains 10 to 14% total extract and no starch or dextrin giving a color with iodine. The pH should be 5.2 to 5.5, and must be regulated during mashing for optimum conversion of starch to sugars and to prevent turbidity caused by complex nitrogenous substances resulting from incomplete breakdown of proteins. Beers brewed from worts that are too alkaline show increased color and lack of stability.

The pH is controlled by adding lactic acid, mineral acids or calcium sulfate (gypsum) to the mash. The calcium sulfate converts phosphates in the malt to acid phosphates. Malt contains about 1% phosphates (calculated as PO_4). Twenty percent of this is soluble phosphate, of which about $\frac{1}{2}$ is present as alkaline phosphate and the other $\frac{1}{2}$ as acid phosphate. The alkaline phosphate must be changed to acid phosphate in order to provide buffer action for regulating the pH. The calcium sulfate reacts with the alkaline phosphates according to the equation

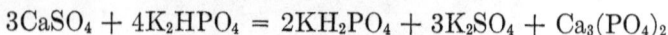

$$3CaSO_4 + 4K_2HPO_4 = 2KH_2PO_4 + 3K_2SO_4 + Ca_3(PO_4)_2$$

Another function of the calcium sulfate is to prevent the conversion of acid phosphates to alkaline phosphates by the calcium carbonate present in the water, according to the equation

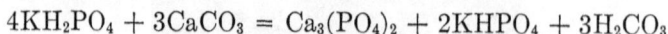

$$4KH_2PO_4 + 3CaCO_3 = Ca_3(PO_4)_2 + 2KHPO_4 + 3H_2CO_3$$

The pH of the mash immediately after mixing with soft water is about 5.8 and the titratable acidity is about 0.05%. The acidity at once begins to increase due to liberation of acid phosphates, lactic acid and other acids, such as amino acids, from the malt. The mash may be held at a temperature of about 95°F for 30 to 60 min to promote acidification. If held too long at this temperature, butyric fermentation may result. This period is known as the "lactic acid rest" and may be omitted if an outside acidifying agent, such as lactic acid, is used.

Subsequently the temperature is increased by stages, by means of steam and by adding adjunct mash from the cooker at appropriate times. "Rests" at specified temperature levels give time for specific enzymatic reactions. In more modern practice, these temperature steps are consolidated.

The optimum degradation of proteins occurs between 113° to 140°F, while the greatest conversion of starches to fermentable sugars takes place between 140° to 149°F. The group of enzymes converting starch into dextrins and maltose is called "diastase." The separate enzymes known to be present in this group are α-amylase and β-amylase. It was suggested by Waldschmidt-Leitz and Mayer (1935) that a third enzyme, called amylophosphatase, was the enzyme responsible for the liquefaction of starch during mashing, by splitting inorganic phosphates from starch without saccharification. The present view (DeClerck 1957) is that this liquefaction is due to the action of α-amylase. During mashing liquefaction is pronounced at 122° to 176°F, with an optimum range of 158° to 167°F. Enzymes of fungal and bacterial origin, which are sometimes used to assist liquefaction of unmalted cereals, have high liquefying power and are destroyed at temperatures of about 190°–200°F.

In order that the starch can be liquefied in a reasonable length of time, it must be gelatinized. The temperature of gelatinization varies with the types of starch and the size of the granules. For example wheat starch gels at about 140°F, while rice starch gels at 176° to 185°F. Gelatinization is accompanied by a sharp rise in viscosity, but this decreases again on liquefaction.

Starch is hydrolyzed by both α-and β-amylases. Optimum conditions for β-amylase are 122° to 145°F and pH 4.7—5.3, and it is destroyed at temperatures over 172°F. It is already quite developed in ungerminated barley, but suffers during the kilning of the malt because of the high temperature. It splits off maltose directly from the starch molecule, leaving behind a compound of the amylodextrin type. While sugar formation from up to 60% of the original starch proceeds rapidly, the iodine test continues to be positive for a long time, and with pure β-amylase complete conversion may never be attained. During mashing the length of time the mash remains at temperature favoring β-amylase activity (140° to 149°F) determines the amount of maltose formed, and thus the alcohol content of the finished beer.

The α-amylase has its optimum activity at 158°F and pH 5.7. It is more resistant to high temperatures but less resistant to low pH than β-amylase. It is practically inactive in ungerminated barley. Hydrolysis of starch by α-amylase leads chiefly to the formation of dextrins of the type giving a negative iodine reaction. Thus a high mashing temperature (above the optimum for β-amylase) will result in wort having high dextrin content and low fermentable sugar content, yielding low alcoholic beer.

Beers low in alcohol because of high dextrin content are not so palatable as beers which have a higher alcohol content. In order to keep the amount of fermentable sugar low the temperature of the mash must be raised quickly from 122° to 158°F to suppress β-amylase activity, and this rapid rise prevents formation of the intermediate protein degradation products which are necessary as yeast nutrients and for adding body to the beer.

Starch degradation products other than maltose and dextrins are present in wort, evidently due to the action of amyloglucosidase in malt. Enzymes other than the diastase group seem to be present in malt. Maltase is present in detectable amounts, but is largely denatured by heat during the kilning. The application of paper chromatography in brewing, introduced by Harris et al. (1951) and developed and simplified by McFarlane and Held (1953), has shown the presence of many different carbohydrates. These have been classified as higher dextrins, maltotetraose, maltotriose, maltose, sucrose, glucose, isomaltose, panose, and fructose.

In addition to starch hydrolysis, an important function of mashing is to bring about changes in the proteins present in malt. American 6 row barleys contain 12 to 14% protein, of which roughly $\frac{2}{3}$ is permanently insoluble. The remaining $\frac{1}{3}$, or about 3 to 5% of dry substance, plays an important part in brewing. The proteins in brewing adjuncts such as grits, flakes etc, are inert as far as mashing is concerned. The boiling of the adjuncts in the cooker seems to coagulate the portion of these proteins which might otherwise be soluble.

During mashing part of the insoluble malt proteins are converted to soluble ones, but a small fraction of the dissolved proteins are subsequently coagulated when the wort is boiled in the kettle. The balance of the dissolved protein material is unaffected by boiling, and is therefore known as permanently soluble protein.

Several different proteolytic enzymes are present in malt, but their exact number and nature are not known. They break down the soluble, complex proteins into simpler compounds of medium molecular weight, known as albumoses, peptones, and polypeptides. These products have a very favorable influence on the palatability and foam of beer. It is essential that all of the original high molecular weight soluble protein be degraded, as otherwise it will cause turbidity and haziness in the beer. The proteinases have an optimum temperature range of 122° to 140°F and an optimum pH range of 4.3 to 5.7. Further degradation of the medium molecular weight products to peptides and amino acids is brought about by a group of enzymes known as peptidases. These low molecular weight products are important as nutrients for the yeast. However, an excess of these compounds may cause sluggishness or degeneration of the yeast. They also tend to increase the surface tension and decrease the foam stability of beer. High mashing temperatures favor the formation of peptones while lower temperatures favor the formation of peptides and amino acids.

There are two traditional types of mashing, decoction and infusion. Infusion mashing as employed, for example, in Britain for top fermentations, requires that conversion and extraction of wort be carried out in a single stage and at a single temperature in the mash tun. The mixed grist and water are allowed to stand in the mash tun and the temperature only gradually raised without boiling. Decoction mashing is also important, e.g., in central Europe for producing bottom fermentation beers. In this method, a part of the mash is withdrawn, boiled, and then returned to the mash tun to raise the temperature of the whole mash. There are also other systems adopted to meet specific needs. Proteolytic activity and saccharification are both favored by the decoction method (Rennie 1967). British malts are generally highly modified to compensate for the enzymatically less favorable conditions of infusion mashing.

The temperatures and time intervals in a typical US mashing process are as follows:

The *albumin* or *protein rest* is at 113°F for 30 to 60 min. During this period the proteinases are very active, forming low molecular weight nitrogen compounds, such as amino acids, amides, and peptides, from the proteins. Beta-amylase, which splits maltose off from starch, is only moderately active. Acidification and formation of buffering compounds occur.

The *sugar rest* is at 133° to 144°F for 5 to 30 min, or it may be entirely omitted. This temperature is near the optimum temperature for β-amylase, which is thus very active. Alpha-amylase, which forms dextrins, is moderately active. Proteinases are very active, but at this temperature produce more medium molecular nitrogen compounds than amino acids. For beers of restricted alcohol content, this temperature range should be passed over as quickly as possible.

The *dextrinizing rest* is at 136° to 162°F for 15 to 45 min. The dextrin-forming α-amylase is very active, while maltose production is much reduced. The proteinase which forms medium-molecular weight compounds is somewhat active, while that forming smaller molecules is only slightly active.

Conversion is at 162° to 167°F for 10 to 20 min, depending on the malt and previous mashing procedure. This is the optimum temperature for quick and complete conversion of starch to dextrin by α-amylase. The conversion may be carried out at 154° to 162°F, if desired. The temperature should not be higher than 167°F, or the diastase (amylases) may be destroyed.

The mashing-off temperature is usually 167° to 176°F, and is held for 5 to 10 min. At this temperature most of the enzymes are inactivated, although α-amylase retains slight activity. The mashing-off temperature may be varied somewhat, depending on the desired alcohol content of the finished beer. For medium to high alcoholic beers a low mashing-off temperature may be chosen, because maltose formation during sparging (subsequent washing of mash residue to remove wort) is not objectionable. A higher mashing-off temperature, 172° to 174°F, is used for low to medium alcoholic beer, to keep maltose formation within the desired limits. High mashing-off temperatures make the wort less viscous and facilitate running off (wort separation), but too high a temperature may lead to belated but incomplete action on previously unconverted starch particles, causing turbid, slow-running worts and starch turbidity in the cellar. With western malts it is advisable to keep the mashing-off temperature at 165° to 167°F.

The mashing schedule is shown in graph form in Fig. 59.

CLARIFICATION OF THE WORT

After the mashing process is completed, the water-soluble materials must be separated from the spent grain. This is accomplished by filtering the liquid through a bed of the spent grain sup-

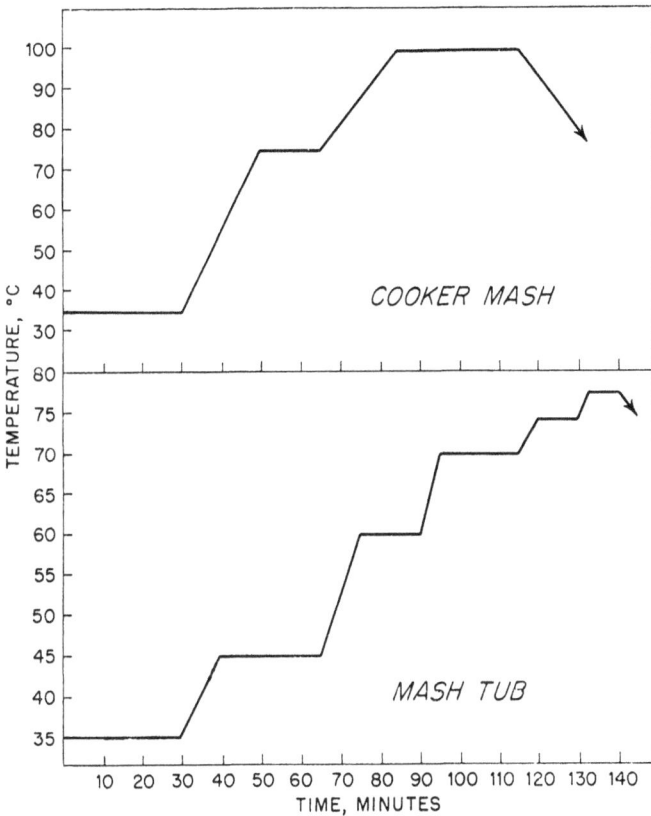

FIG. 59. MASHING SCHEDULE FOR COOKER AND MASH TUB

ported by a false bottom made of perforated plates. The filtration may take place in the mash tub itself, in a separate vessel known as a lauter tub (see Fig. 60), or in a filter press.

Filtration in the mash tub is rather inefficient, and is not common practice. Most frequently the filtration is carried out in the lauter tub. The mash is pumped into the tub and allowed to stand for about ½ hr, so that the spent grain can settle and form a bed on the false bottom. Then, taps leading from the bottom of the tub are opened and the wort is allowed to flow into the grant, a vessel in which the wort can be inspected and from which it can be transferred either to the kettle or back to the top of the lauter tub. The wort coming through into the grant during the first 10 to 15 min is usually opalescent, and is pumped back into the lauter tub to be refiltered. As soon as the filtered wort is clear, it is transferred to the kettle. This fraction is known as the first wort.

Courtesy of Falstaff Brewing Corp.

FIG. 60. THE GRANT, SHOWING ALSO THE LAUTER TUB AND THE TAPS

The process is continued until the runnings again appear opalescent and the rate of flow decreases, indicating that almost all of the liquid has run through. At this point the taps are closed. However, the filter bed still contains a good deal of extract which must be recovered. This is done by spraying hot water onto the filter bed, raking it (see Fig. 61) to mix it thoroughly with the water (mashing over), allowing the grain to settle again, and drawing off the liquid as before. This process is called sparging. It is repeated until the last wort contains about 1% but not less than 0.75% extract. During sparging, the filtered liquid is pumped back as before until it comes through clear.

The sparging conditions must be controlled to ensure recovery of as much extract as possible without also extracting deleterious substances, such as silica and tannins, from the husks. In order to maintain optimum conditions the sparging water is treated to prevent alkalinity, and the temperature of the sparge liquor should not be higher than 175°F. It should, however, not be much below this temperature, or the viscosity of the filter bed will increase to an extent such that the flow of the liquid is retarded. Figure 60 shows a view of the grant and the lower part of a lauter tub.

In some large breweries, filtration in the lauter tub is considered to be too slow for efficient operation, and a plate and frame mash filter is used instead. These filters are constructed similarly to the filter presses used in sugar refineries, but high pressure is not used. The filter consists of hollow frames alternating with grooved plates covered with filter cloth. The wort passes through the filter cloth into the grooves and then through faucets into a collector

Courtesy of Stroh Brewing Co.

FIG. 61. INTERIOR OF LAUTER TUB SHOWING RAKING
MECHANISM

similar to a grant. Sparging is carried out by passing the brewing water either from the pressure side or the inlet side of the filter press.

WORT BOILING

After the wort is separated from the spent grain, it is boiled for about 1.5 to 2 hr (see Fig. 62). The boiling of wort is carried out to sterilize and stabilize it, as well as to extract the constituents from hops which give beer its typical taste and aroma.

During boiling, most of the microorganisms are killed, some of the nitrogenous substances are coagulated, hop components are extracted, and some hop and malt components are volatilized. The slightly acid pH of the wort contributes to the sterilization, and makes it unnecessary to raise the temperature above the normal

FIG. 62. BREW KETTLE

boiling point. The enzymes are also destroyed at this temperature, thus preventing further breakdown of starch and proteins. Dextrose is sometimes added to the kettle to bring the fermentable sugars to the proper level. The most striking phenomenon of the process is the hot break, or separation of a precipitate formed mainly of proteinaceous materials and malt and hop tannin.

Colloidal proteins are coagulated by the heat aided by tannins extracted from malt husks and hops. Care must be taken to coagulate as much of the unstable protein as possible. One important factor is the pH, which should be near the isoelectric point of the proteins. This is generally considered to be about 5.2, and the pH should reach this value near the end of the boiling period (the pH drops by about 0.2 to 0.3 pH unit during boiling). Thus the pH of the wort should be 5.4 or 5.5 at the start of the boil. Vigorous boiling also aids in the coagulation of the proteins, as albumin particles, already denatured by heat, tend to accumulate on the surface of the wort around the steam and air bubbles, and agglomerate more easily because of their proximity to each other. Excessively vigorous boiling in an open kettle, however, may have detrimental effects, owing to conversion of coagulated proteins to soluble forms through oxidation.

Hops are added chiefly to furnish two kinds of substances: essential oils and bitter resins. The resins are responsible for the typical bitter flavor of beer. They are insoluble in cold wort, but somewhat soluble in hot wort at its normal pH. As they are only sparingly soluble in water, their solubility in wort is attributed to their adsorption on the surface of stable colloids. Some of the resins are adsorbed on colloidal particles which are later precipitated, and thus are thrown out of solution again. This loss is greater at lower pH, so that the pH of the wort should not be allowed to fall below 5.2. During boiling the resins are partially hydrolyzed, oxidized, and polymerized.

The essential oils of the hops are very volatile, and evaporate during boiling. Therefore, if a good aroma is desired, more hops must be added near the end of the boiling period. Occasionally hop oil is added to the finished beer instead.

Hopping is a phase of brewing in which detailed scientific knowledge is rather scanty (at least up to the last few years), and the best procedures for attaining the desired flavor and aroma remain an important part of the "art" of brewing.

Another change occurring during boiling is a darkening of the color, caused by a caramelization of sugar, formation of melanoidins and oxidation of tannin to phlobaphene. These changes, which are accompanied by development of a malty flavor in the beer, are accentuated by high pH and excessive aeration, while they may be prevented or greatly reduced by adding sulfur dioxide prior to boiling. The reducing power of wort is increased during boiling by the formation of so-called reductones, which are tautomeric compounds. Aeration of the wort during boiling lowers its reducing power.

Boiling conditions have a considerable effect on the quality of the finished beer. The head retention and shelf-life are affected. Yeast behavior is dependent to some extent on the hop components and nitrogenous substances remaining in the wort after boiling. According to Hudson (1966), beers from worts mashed at 145°F do not differ in their shelf-life whether boiled for 30 min or 2 hr, but, if they have been mashed at 145° or 155°F, the stability is greatly increased and the head retention decreased, by boiling for the longer time.

The spent hops and most of the hot break are removed by the hop strainer, in the Whirlpool.

WORT COOLING

The hot wort must be cooled before fermentation, as high temperature will inactivate the yeast. The wort is cooled to 57° to

61°F, when top-fermenting yeast is to be used and to 43° to 46°F for bottom-fermenting yeast.

During cooling two precautions must be observed: (1) care must be taken not to infect the wort with microorganisms, as the temperature range between 104° to 68°F is most favorable for their growth, and (2) enough oxygen must be dissolved in the wort to shorten the lag phase of fermentation. In addition, during the cooling the protein coagulated during the boiling, the "hot break" should be entirely eliminated and the turbid matter appearing during cooling, the "cold break," should be at least partly precipitated so that it does not remain colloidally suspended.

Wort cooling was originally carried out in large open vessels called coolships. Heat was dissipated to the surroundings without mechanical assistance. This was a slow, inefficient process and very conducive to contamination by microorganisms from the atmosphere. Later, refrigeration was applied to smaller open vessels to increase the rate of cooling.

Modern brewery practice is to solve these problems by using closed coolers and pumping measured amounts of sterile air into the wort in the coolers. A common but older design is that of concentric pipes, the cooling liquid flowing through the central ones and the wort and air through the outer ones. Plate coolers are also very common. Aeration in closed systems can be accomplished by forcing sterile air through porous filter tubes, forming minute bubbles in the cooled or cooling wort. Too much air must be avoided, as an excess can result in oxidation of part of the coagulated protein, preventing efficient precipitation. The proteins precipitate partly by agglomeration of oppositely charged particles, such as the positively charged derived albumins and the negatively charged nucleo-proteins. If these particles are approximately in balance, precipitation will be nearly complete. Oxidation may change the charge on some of the particles, thus destroying the balance.

The wort enters the cooler at a temperature near 205°–210°F, and is cooled to the fermentation starting temperature which, depending upon the type of beer, may vary from 41° to 64°F (Moller and Christensen 1968). The rate of wort cooling has very little influence on the properties of the finished beer.

Cold break formation takes place below 140°F. The consistency and amount of this precipitate will vary with the character of the malt, the type and amount of adjuncts, the mashing procedure, the runoff, the hopping rate, and the extent of wort boiling (Roessler 1968A).

The "pneumatic" removal of cold break can be carried out in large brewing units, usually with a pitching vessel, or it can be achieved through centrifugation or filtration through kieselguhr (Narziss and Kieninger 1968).

To minimize bacterial growth, the coolers are designed to cool the wort rapidly to below the optimum growth range. In the later stage of fermentation, the danger of infection is much less, as the disappearance of sugars and appearance of alcohol due to yeast action provides a less favorable medium. After fermentation, the temperature of the beer is kept low throughout the later processing, so that it remains essentially sterile, and it is finally pasteurized.

Courtesy of Falstaff Brewing Corp.

FIG. 63. PITCHING YEAST IN A STERILE STORAGE AREA

FERMENTATION

Lager beer is produced with bottom fermenting yeast. For this the wort is cooled to 43° to 50°F and pumped into a collecting tub or hot wort tank. It remains here for about 4 hr to give the turbidity (hot and cold break) time to settle out, and is then transferred to the pitching tub, leaving the sediment behind. Here it is pitched (mixed) with the required amount of yeast, usually ¾ to 1 lb per barrel of wort, and the foam formed during mixing is skimmed off. After about 24 to 48 hr the "kraesen" appear. These are heads of foam caused by evolution of gases, which carry hop resins and nitrogenous material to the surface with them. As soon as kraesen appear the wort is pumped to the fermentation tank, which can be either open or closed. Usually closed fermentors are used so that the carbon dioxide produced can be collected and injected back into the beer at a later stage in the processing.

The major change occurring during fermentation is the conversion of fermentable sugars to alcohol and carbon dioxide. The reactions can be summarized by the equation

$$C_6H_{12}O_6 \xrightarrow{\text{yeast}} 2CO_2 + 2C_2H_5OH + 22 \text{ calories}$$

It can be seen from this equation that heat is generated in fermentation. Therefore the temperature will rise, but it should not be allowed to exceed 46° to 48°F. Temperature control is accomplished by coils in the tank, with cold liquid flowing through them. These coils are called attemperators.

The progress of the fermentation is followed by determining the amount of sugar present with a saccharometer. The saccharometer readings begin to fall almost immediately after the wort is pitched with yeast, and fall more rapidly as fermentation progresses. The fermentation becomes very active 15 to 18 hr after pitching and reaches its height 70 to 80 hr after pitching, with maintenance of this level for 48 to 72 hr. It then decreases in intensity, and fermentation is usually complete in 7 to 9 days. The rate of drop of the saccharometer readings is 0.2 to 0.5% per day in the early stages, reaches 1.0 to 1.5% per day in the latter stages, and then falls slowly to about 0.10% in the final stages.

During the active period the foam rises in height and becomes darker in color owing to hop resins, proteins, dead yeast cells, etc. When the rate of fermentation begins to decrease the head of foam starts to collapse because not enough gas is being produced to support it. With the decrease in fermentation less heat is generated and the rate of attemperation is moderated, but the temperature should not be allowed to rise above 59°F.

Fermentation is considered to be complete when the desired degree of sugar conversion (called attenuation) has taken place. If secondary fermentation during storage is desired, some sugar must remain; otherwise fermentation is continued to the stage of complete attenuation. When fermentation is complete the beer is cooled slowly to 39°F and pumped off carefully, in a manner to avoid disturbance of the sediment at the bottom of the tank.

The sediment consists of three layers. The bottom layer is a sludge composed of materials settling out of suspension in the wort. The middle layer is the yeast crop, which is saved for later fermentations. The top layer is mostly hop resins. Yeast not required for brewery operation is filtered, dried and sold for use as stock feed and for human consumption. Figure 59 shows a special sterile storage area for keeping yeast which is to be used for pitching.

More recent practice is to collect the yeast without prior separation and transfer it through sanitized yeast lines to the yeast room. Here the yeast is cleaned and washed. Disinfectant procedures, mainly relying on lowering of pH, are used to establish bacteriostatic conditions. Use of the antibiotics tyrothricin, penicillin, and polymyxin has also been suggested. If the yeast is to be stored longer than five days, it is compressed (dewatered) for storage (Roessler 1968B).

The total amount of yeast collected is about 4 to 5 times the amount needed for pitching.

In addition to the large amounts of carbon dioxide and ethanol which are produced in the fermenting vessel, other compounds are generated, but in much lower quantities. Some of these originate from side reactions of the yeast, while others are produced by infecting microorganisms. Volatile acids, which are initially low in the wort, rise sharply during fermentation, but the rate drops rather quickly. A second and more gradual rise occurs during the latter part of the fermentation process. Beers which are entirely free from infection during processing are invariably low in volatile acids, containing only about 30 to 50 ppm (Owades and Dono 1965).

The growth of the yeast is a phenomenon separate from fermentation, but in brewing the two go hand in hand. The yeast nutrients are furnished by the wort. The essential classes of substances which must be present for yeast nutrition are mineral salts, sources of carbon, nitrogen and energy sources, and growth factors. The necessary mineral salts are normally present in the wort, coming partially from the malt and partially from the brewery water. A small portion (about 2%) of the fermentable sugar in the wort is utilized by the yeast as a source of carbon and also of energy, by fermentative breakdown within the cell. The nitrogen requirements are satisfied by the amino acids and peptides present—more complex protein fragments and proteins themselves cannot be metabolized. Yeast can synthesize many of the growth factors which it requires, the specific capacity varying from one strain of yeast to another. The factors which must be present in the growth medium appear to be present in normal wort, stemming from the malt, which is rich in the vitamin B complex and other growth factors.

The growth of yeast during fermentation can be divided into five well-defined phases: (1) a latent phase, lasting a few hours, during which occurs no significant visible change in the cells; (2) a lag phase, with slow growth, lasting only a short time and dependent to some extent on the concentration of available oxygen; (3) a logarithmic phase, with rapid increase in growth, this phase continues for about 4 hr at 64°F but only 8 min under ideal conditions at 86°F; (4) phase of negative acceleration, during which the rate of reproduction falls slowly; and (5) stationary phase, in which the rate of multiplication is approximately equal to the rate at which the cells die and autolyse.

Several changes take place in the wort during fermentation, in addition to the attenuation. The pH falls, due to formation of carbon dioxide and organic acids, mostly lactic acid. The normal pH after bottom fermentation is 4.2–4.4, or occasionally 4.0 or less, depending on the quantity of buffers present. The nitrogen content also falls by about ⅓, due chiefly to the assimilation of amino acids and peptides by the yeast, and partly to precipitation of complex proteins caused by the lowered pH in the presence of alcohol, and concentration at the surface of the carbon dioxide bubbles and the yeast cells. At the end of fermentation the beer contains about 0.3% dissolved carbon dioxide, and practically no dissolved oxygen. Hop resins are largely eliminated by the fall in pH, adsorption on the yeast cells, and coagulation in the head. The bitter material in the head must be skimmed off carefully to avoid too sharp a taste in the finished beer. Finally, the color of the wort becomes paler during fermentation, partly because of eliminating of colored substances in the scum at the surface, partly because of the reducing action of yeast on oxidized tannin, and partly by the normal fall in color intensity accompanying the drop in pH.

It is highly desirable for the yeast to flocculate near the end of the fermentation. The better the flocculation, the more readily the yeast is removed and the fewer the number of cells which are carried over. Different strains of yeast can vary markedly in their flocculation characteristics, and these can also be affected by wort conditions. *Saccharomyces carlsbergensis* forms no yeast cap, and the cells settle readily.

After fermentation is complete, as shown by a specific gravity of 1.009 to 1.014, the beer is gradually cooled to the lagering temperature of about 27° to 30°F, and during a further holding period of 2 or 3 days additional suspended solids will deposit. Finally, the beer is pumped into the lager cellar.

Infection, or the introduction of undesirable microorganisms into the wort or beer, may have several sources. The wort, the atmosphere, the yeast, or the equipment are common avenues for the introduction of infection. A number of organisms may appear in the wort, but *Aerobacter aerogenes* is the most common. The bacteria appearing in the fermentation area are the classic brewery organisms, *Lactobacillus pastorianus*, *Pediococcus cerevisiae*, and *Flavobacterium proteus*. In the post-blending area, various lactobacilli species, and in the filling and bottling shop, *Saccharomyces diastaticus* are common organisms (Greenspan 1965).

STORAGE

When the beer, at about 39°F is pumped out of the fermentation tank, it passes through a cooler where it is quickly chilled to 32°F and then sent through a diatomaceous earth filter. Diatomaceous earth is preferable to filter mass at this point, as the latter has a tendency to become clogged. The brilliant filtered beer is then pumped into pressure tanks in a storage cellar of the type shown in Fig. 64. Here it is stored at 32°F and is carbonated with carbon dioxide previously collected from the fermentation tanks. In rare cases, the beer is not chilled but is stored at 39° to 43°F, in which case secondary fermentation takes place. At 32°F no further fermentation occurs, but the beer undergoes an important period of aging. During this time yeast cells and other insoluble nitrogenous materials settle out, and the flavor of the beer is improved and mellowed as a result of complex chemical changes involving aldehydes, esters, and higher alcohols.

The storage period may be as long as 6 months, or, in the case of English draft beer, as short as 1 or 2 days. American lager beer is usually held for 3 to 12 weeks. Other types of beer are held in "ruh" storage for 5 to 7 days at 32°F.

During storage, the brewer strives to achieve a balance between the beneficial changes taking place, such as esterification, and undesirable changes related to autolysis, oxidation, etc. (Harrison 1966). The net result is generally a more pleasant aroma, a more mature, less yeasty flavor, and greater resistance to haze development during distribution and marketing.

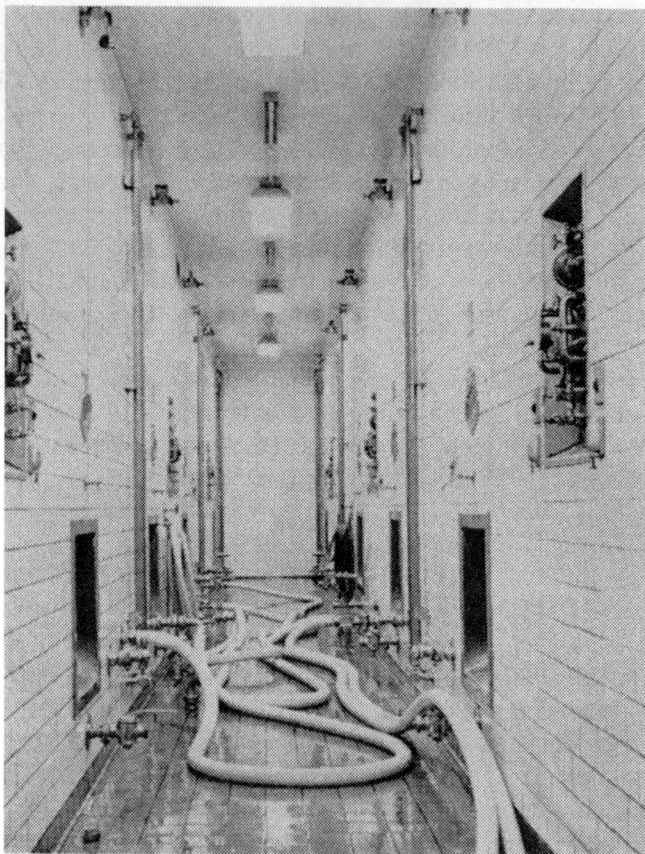

Courtesy of Falstaff Brewing Corp.

FIG. 64. CELLAR STORAGE TANKS BEHIND WHITE TILE WALLS

In continuous aging processes, the beer is transferred from one tank to another, with recarbonation at each transfer. The purpose of these operations is to flush out any dissolved air and also to permit slow and intimate contact of the carbon dioxide with the beer, so that it is not only mechanically held in solution but also adsorbed by the colloidal particles present. The beer is kept under 5 to 6 psi counter pressure during storage and transfer. A period of about two weeks is considered sufficient for adequate carbonation.

American beers are slightly more carbonated than European products. About 2.8 volumes of carbon dioxide is a common level. Fermentation gas is often collected and liquefied for later use in product carbonation. Carbon dioxide may be added through large porous stones in the lagering tanks, or it may be injected into the beer stream as it flows to the packaging equipment.

Natural carbonation or krausening is accomplished by adding from 10 to 20% of vigorously fermenting wort to the fully fermented beer in the lager tank. The tanks are closed after this addition and gas allowed to build up to the selected pressure.

Some time during storage the beer is usually chillproofed. When beer is chilled to low temperature, a colloidal haze appears, but it disappears again when the beer is warmed slightly. The nature of chill haze has been a subject of controversy for many years. Sandegren (1947) and St. Johnston (1948) considered it to be composed of β-globulin degradation products and tannins. Its composition has been clarified by recent work of Biserte and Scriban (1953). Protein (about 25%) and tannins have been shown to be present, the protein fraction being an extremely complex mixture of degraded globulin-like bodies, hordein and glutelin derivatives. Chillproofing is a treatment to minimize the appearance of chill haze. Chillproofs include materials such as tannic acid added to precipitate the proteins, materials to adsorb them (such as certain types of bentonites), and proteolytic enzymes to hydrolyze them to more soluble fractions.

After its sojourn in the storage cellar, the beer is finished, and is ready for a final filtration followed by filling into casks or bottles.

A TYPICAL BREW

Although ingredients used and their proportions vary according to the type of beer and individual preferences, the following "recipe" for a typical brew will give an idea of the amounts of materials needed.

Cooker Mash

The cooker is filled with 3500 lb of grits (milled corn), 675 lb of malt, 70 bbl of water, 2.1 qt of lactic acid and 3 lb of calcium sulfate. The mixture is brought to boiling.

Mash Tub

The tub is filled with 10,000 lb of malt, 100 bbl of water, 1.58 qt of lactic acid and 2.0 lb of calcium sulfate. This is heated according to the mashing schedule, and at the end of the heating period the cooker mash is added.

Lauter Tub

The above mixture is first filtered through the spent grain bed without additions. For sparging, 3.2 qt of lactic acid and 5.0 lb of calcium sulfate are added to the sparge water and 1.4 qt of lactic acid and 2.0 lb of calcium sulfate during the mashover (raking of the grain bed to mix it with the sparge water).

Kettle

The clarified wort and spargings are filled into the kettle as they arrive from the grant. The filling takes about 4 hr. The mixture is then boiled for 2.5 hr with the addition of 16.0 lb of salt, and 4.25 lb of $K_2S_2O_5$. Hops are added in 2 portions as follows: 50 lb at 30 min, and 50 lb at 20 min before the end of the boiling period. This mixture is added to the kettle over a period of 20 min, before the end.

Two hours before the end of boiling, mixing is started, in a separate vessel, of 1000 lb of dextrose and 5 to 6 bbl of water. The temperature should not be allowed to rise above 162°F at any time during the mixing. This mixture is added to the kettle over a period of 20 min, starting 1 hr before the end of the boiling period.

Cellar Additions

During storage 3 lb of $K_2S_2O_5$ per 850 bbl, 1 lb of chillproof per 75 bbl and 5 lb of gum arabic per 100 bbl of beer are added. (A standard US barrel of beer is 31 gal.)

Filtration, Racking, and Bottling

After aging the beer possesses its final flavor and aroma, but it is still slightly hazy. It is therefore given a final filtration through a pulp mass composed of cellulose fibers mixed with about 1% asbestos, through filter sheets, or through diatomaceous earth. The type of filter usually used is shown in Fig. 65. This filtration is not a simple sieving process, as most of the particles present at this time are of colloidal size. Their elimination is due rather to adsorption on the filter mass. Thus not only particles which can be optically detected as a haze, but also other substances having a high surface tension will be absorbed. On this account the final filtration removes some proteins, hop resins, coloring matter, higher alcohols and esters.

The beer is then racked into casks or filled into bottles or cans, and is ready for marketing. The filtration, racking, and bottling

Courtesy of Stroh Brewing Co.

FIG. 65. FILTER ROOM

should be carried out in a manner to avoid loss of carbon dioxide, infection, and oxidation. To prevent loss of carbon dioxide, the beer is kept cold and racked or bottled against a counter pressure of carbon dioxide or air. Carbon dioxide is preferable, as a counter pressure of air may result in the dissolving of some air in the beer, leading to gradual oxidation. The danger of infection of the beer is minimized by sterilizing the filters and machinery and keeping them strictly clean.

Beer to be packed in bottles or cans must be pasteurized or filtered through membranes which will remove virtually all microbial contamination. Pasteurization changes the flavor somewhat, so that draft beer tastes different from bottled beer. The availability of membranes for filter sterilization has made it possible to bottle draft-style beer. Beer should be given a rough filtration to remove coarse material, and then treated with pectolytic and hemicellulytic enzymes before membrane filtration.

RECENT DEVELOPMENTS

The Pablo continuous lautering system (Harsanyi 1968) continuously separates the mash into beer wort and spent grains in a series of centrifugations. The wort extractor consists of a round, horizontally aligned tank that houses a conically shaped sieve drum. The mash is introduced into the narrow end of the drum, while the wide end discharges the extracted grains. Wort passes through the perforated walls of the drum and is collected in appropriate vessels. The grains remaining in the drum are further extracted *in situ* by spraying with dilute wort under pressure.

Continuous fermentation is an obvious and inevitable means for improving the uniformity of beer and increasing the efficiency of brewing. In 1957, the Morton Coutts and Labatt systems were patented. These consist essentially of three stirred tanks in series, with a recycle of yeast from the third vessel back to the first or propagating vessel. These systems are known to be commercially successful, but the extent to which they have displaced conventional fermentation equipment is not known.

Tower fermentation appears to be reaching a practical stage of development. In this process, the wort is passed through a bed of yeast held within a vertically aligned closed fermentation vessel. The highly anaerobic conditions in the tower favor high alcohol yields and minimize the risk of infection while the flow pattern tends to purge the fermenter of nonviable cells and any other nonflocculent organisms. Royston (1966) described the APV tower fermenter which can run continuously for 6 months and produce a completely acceptable product 20 times faster than batch fermentation methods. Another tower fermentation system, the Lefrancois plant, has been successfully used for the production of yeast food, and Cauwe (1968) describes an adaptation suitable for continuous brewing. Cost of such an installation would be about a third of that required for a conventional plant, yields are greater, (i.e., losses are smaller), and the beer is ready for racking ten hours after brewing. In the Lafrancois tower, the wort is maintained in the form of a perpetually circulating foam, the foam being created by blowing carbon dioxide into the base of the fermenter.

Atkinson (1968) described the construction of a pressure precoat filter suitable for single-stage filtration of wort or beer and comparing favorably in results with the usual multi-stage filtration.

Closed, instead of open, fermentation vessels are becoming more common. They allow the use of in-place cleaning (Curtis 1968).

ANALYTICAL CONTROLS IN BREWING

Before the development of suitable analytical methods, brewing was entirely an art, and empirical methods for controlling quality were handed down from generation to generation. While many phases of brewing are still largely an art, and a thorough practical knowledge of the factors to be considered at every stage is indispensable, the brewer now has at his command a host of scientific analytical methods to aid in controlling quality and to detect a possible source of trouble before it becomes serious. A pilot plant is included among the test facilities of most modern breweries (see Fig. 66).

Chemical and physical analyses are generally performed on the raw materials used, on the intermediate products obtained during brewing and storage, and on the finished product. A complete

Courtesy of Stroh Brewing Co.

FIG. 66. BREWERY PILOT PLANT

listing of all the tests used is beyond to scope of this chapter, and only a few representative analyses will be given to indicate the nature of the examinations which may be carried out. Directions for carrying out most analyses in general use are given in *Methods of Analysis of the American Society of Brewing Chemists* (Anon. 1964), hereafter referred to as ASBC Methods.

Typical reports for malt, hops, brewing syrups, and cereal adjuncts are given in Tables 55, 56, 57, and 58. The values in these tables were determined and reported in accordance with the ASBC Methods, with the exception of soluble protein (Table 55). Details concerning this determination will be found in the experimental section.

TABLE 55
MALT—PHYSICAL CHARACTERISTICS

Bushel weight (lb)		39¼
Growth		
Length of acrospire (fraction of kernel length)	0–¼	1%
	¼–½	0%
	½–¾	3%
	¾–1	92%
	Overgrown	4%
Mealiness		
Mealy		95%
Half-glassy		4%
Glassy		1%

CHEMICAL ANALYSIS

Moisture, %	4.7
Extract in the finely ground malt (plato), % as is	74.20
Extract in the finely ground malt (plato), % dry basis	77.90
Conversion in minutes	5 min
Odor of mash	aromatic
Degree of clarity	clear
Speed of filtration	normal
Color of laboratory wort, Lovibond, ½ in. cell, Ser. 52	1.4–1.6
Diastatic power—degree Lintner, as is	124
maltose equivalent, as is	496
Total protein, as is, %	10.33
Total protein, dry basis, %	10.83
Soluble protein, as is, %	3.93
Soluble protein, dry basis, %	4.12
Ratio soluble/total protein	38.10

TABLE 56
ANALYSIS OF HOPS—PHYSICAL EXAMINATION

	Yakima	California Seedless
Cones		
Color	light green	light green
Luster	dull	dull
Size	medium	small-medium
Condition	fair	good—some wind whipped
Lupulin		
Amount	plentiful	plentiful
Color	yellow	yellow
Condition	sticky	sticky
Aroma	aromatic	aromatic
Leaves and stems (%)	0.6	0.2
Seeds (%)	4.9	0.1

CHEMICAL EXAMINATION

	%	%
Moisture	6.7	7.2
Resins		
Alpha resins	4.8	3.1
Beta resins	8.2	7.4
Gamma resins	1.4	1.6
Soft resins	13.0	10.5
Total resins	14.4	12.1
Brewing value $\left(\text{Alpha resins} + \dfrac{\text{Beta resins}}{4}\right)$	6.8	5.0

TABLE 57

ANALYSIS OF BREWING SYRUPS—SUGAR

Physical Characteristics
Samples A and B

Consistency, as is....viscous—lumps	Odor, as is.....none
Color, as is.........colorless	Taste, as is....sweet

CHEMICAL ANALYSIS

	Sample A	Sample B
Clarity of 10% solution	clear	clear
Color of 10% solution	colorless	colorless
Specific gravity of 10% solution	1.03200	1.03310
Extract, as is, %	83.0	86.0
Degrees Baumé	43.9	45.3
Moisture, %	17.0	14.0
Fermentable extract, as is, %	54.0	72.2
dry basis, %	65.0	84.0
Dextrose equivalent	no figures available	86.0
Iodine reaction	negative	negative
pH Value	5.4	5.4
Protein, %	0.1	.01
Ash, %	0.1	.01
Diastatic power	100	. . .
Dextrose, %	no figures available	86.0
Sucrose, %	no figures available	. . .

TABLE 58

ANALYSIS OF CEREAL ADJUNCTS

Physical Characteristics

Color, as is	buff
Odor, as is	clean and normal
Husks, germs, etc	under 1.0%
Mold	none
Weevils, etc	none

Chemical Analysis

Moisture	not more than 11.0%
Oil	less than 1.0%
Extract, as is	79.0%
Extract, dry basis	89.0%
Time of conversion	15 min
Ash	2.0%
Protein	8.0%

A complete analysis of a typical American beer is given in Table 59. Many of these tests are rarely performed.

These analyses may appear to be long and unnecessarily detailed. While they are numerous when considered individually, certain groups of tests serve to point out the most subtle off-taste characteristics, such as sweetness, bitterness and oxidation. Other groups will point out deviations in characteristics such as foam stability and shelf-life. For example, the group of analyses associated with foam stability includes surface tension, surface activity, foam sigma, foam density, and protein and sugar distribution. Not all of the tests are conducted on an individual beer sample.

TABLE 59

ANALYSIS OF A TYPICAL BEER

Specific gravity[1]	1.01210
Beer Balling, %[1]	3.093
Saccharometer, %[1]	3.10
Alcohol by weight, %[1]	3.63
Alcohol by volume, %[1]	4.60
Real extract, %[1]	4.73
Extract of original wort (2A + E), %	11.99
Original balling, %[1]	11.80
Reducing sugars, %[1]	1.160
Degree wort sugar, %[1]	71.30
Degree attenuation, %[1] (real degree of fermentation)	60.00
pH[1]	4.35
Color,° L.[1]	2.94
Air, cc.[2]	1.20
Nitrogen, cc.[2]	1.01
Oxygen, cc.[2]	0.19
Oxygen/air, %[2]	15.80
CO_2 % by wt.[1]	0.460
Acidity, %[1]	0.135
Erythro-dextrins (iodine reaction)[1]	0
Amylo-dextrins (iodine reaction)[1]	0
Dextrins, %[1]	2.73
Iron, ppm[1]	0.175
Indicator-Time test, sec.[3]	290.
Surface tension, dynes[4]	46.00
Surface activity[4]	0.367
$CaSO_4$, ppm[5]	256.00
NaCl, ppm[5]	153.00
Foam sigma	109.00
Foam density	20.40
Tannins, ppm[6]	55.40
Viscosity, cp[4]	1.057
SO_2, ppm[7]	13.20
Ash, %[1]	0.148
Diacetyl, ppm[1]	0.210
Copper, ppm[8]	0.245
Fractional carbohydrates, %[9]	
Glucose	0.001
Fructose	Trace
Sucrose	Trace
Maltose	0.10
Maltotriose	0.20
Maltotetraose	0.45
Higher saccharides	3.04
Total saccharides	3.83
Fractional proteins, %[4]	
Total protein	0.299
High molecular	0.0710
Medium molecular	0.100
Low molecular	0.0951
Non-protein N	0.0290
High/total	23.30
Medium/total	33.20
Low/total	31.60
Non-protein/total	9.6
Calories[4]	168.30

[1] Anon. 1964.
[2] Roberts *et al.* (1947).
[3] Gray and Stone (1939).

[4] Any standard method.
[5] Stone *et al.* (1951).
[6] Stone and Gray (1948).

[7] Monier-Williams (1940).
[8] Stone (1942).
[9] McFarlane *et al.* (1954).

In the case of bitterness, the pertinent analyses are sulfur dioxide, iron, copper, tannins, carbon dioxide, air, and melanoidins. These substances are normally present in beer, but in very small amounts, well below the organoleptic threshold (the level at which a difference in flavor can be detected by tasting the beer). The approximate organoleptic threshold for each of these substances has been established, and is used to evaluate the analytical results. For example, if the amount of iron is 0.35 ppm or below, no off-flavor can be detected, but a value of 0.70 ppm will result in a distinct flavor. If 0.35 ppm are removed, the flavor will again be normal. Another example is sulfur dioxide, which is normally present in the amount of 3.5 to 8 ppm. This is below the organoleptic threshold of most beer drinkers, but if the amount is increased to 16 ppm astringency will be noticeable to almost everyone.

With the threshold values established, it is easy to tell which substance or combination of substances is responsible for any off-flavor occurring, and the cause of the trouble can be corrected. Before accurate analytical methods were developed, correction of the trouble was necessarily based on trial and error, and the length of time needed to find the source of the deviation was largely a matter of luck. Even more important than correction of an existing off-flavor is the ability to detect any "creeping trouble" before it actually results in detectable change of flavor. For example, if the sulfur dioxide values for successive batches of beer showed a steady increase from 3.5 to 7 ppm steps could be taken to find the reason for this increase and correct it before the organoleptic threshold was reached.

One of the most important factors affecting appearance, taste, foam stability and shelf-life of beer is the oxidation of proteins in beer. It has already been pointed out that the proteins can be divided into groups according to their molecular weight, and that this factor must be taken into account during mashing. St. Johnston (1948) has classified the principal protein fractions of wort and beer as follows:

(1) **Protein T.**—The chill-sensitive globulin-tannin complex. Isoelectric point 6.4, completely soluble at 140°F with solubility diminishing below this temperature.

(2) **Protein C.**—The coagulable albumin. Isoelectric point 6.0, readily coagulated by heat.

(3) **Protein O.**—The oxyprotein or nucleoprotein. Isoelectric point 3.9.

(4) Fraction D.—Soluble fraction. Probably proteoses associated with dextrins.

At the pH of beer, about 4.3, the nucleoproteins will carry a negative charge and albumins a positive charge, so that mutual precipitation is possible.

A similar classification of protein fractions is that proposed by Lundin, as quoted by Pawlowski and Doemens (1932). He designates the fractions as A (high molecular weight), B (medium molecular weight) and C (low molecular weight).

The equilibrium between these protein fractions can easily be upset by oxidation. According to Sandegren (1947) in his work on barley protein in brewing, proteins having sulfhydryl groups can be oxidized to higher molecular weight compounds by the following mechanism

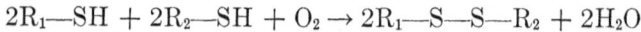

$$2R_1\!-\!SH + 2R_2\!-\!SH + O_2 \rightarrow 2R_1\!-\!S\!-\!S\!-\!R_2 + 2H_2O$$

In addition, albumins can be oxidized to nucleoproteins with attendant change of the charge from positive to negative. The negatively charged oxidation products can then combine with unchanged positively charged proteins, thus increasing the molecular weight and causing mutual precipitation.

The net result of oxidation is to shift the equilibrium between the different fractions so carefully set up during brewing. The distribution of protein fractions at various stages during brewing, and four months after bottling, is shown in Table 60. Examining the values in the table, it can be seen that the amount of protein in all fractions decreases during fermentation, accompanied by an increase in the ratio high/total and a decrease in the ratio low/total. This can be ascribed to the use of proteins, especially amino

TABLE 60

PROTEIN DISTRIBUTION IN BEER AT VARIOUS STAGES

	Kettle Sample %	After Fermentation %	After Storage %	After Pasteurization %	After 4 Months %
High molecular	0.1537	0.1452	0.0716	0.0860	0.1250
Medium molecular	0.1328	0.0959	0.0811	0.0858	0.1051
Low molecular	0.2066	0.1109	0.1965	0.1796	0.1180
Nonprotein	0.031	0.027	0.300	0.028	0.029
Total protein	0.5241	0.3790	0.3792	0.3792	0.3800
High/total	29.31	38.31	18.88	22.66	33.00
Medium/total	25.34	25.30	21.38	22.67	27.65
Low/total	39.40	29.25	51.84	47.34	31.05
Non-protein/total	5.95	6.14	7.90	7.33	8.30

acids and other low molecular weight components, by the yeast. It will be noted that the total protein remains essentially constant once fermentation has been completed, but that the distribution in the different fractions changes markedly. The decrease in the high molecular fraction and in its ratio to the total during storage is a result of chillproofing. The chillproofing agents in common use today are essentially proteolytic enzymes, added for the specific purpose of hydrolyzing the high molecular weight proteins, which are associated with tannins in complex compounds and tend to precipitate at low temperatures, to lower molecular weight compounds which do not precipitate. The progressive increase in the high molecular compounds at the expense of the low molecular fraction after bottling is due to oxidation and the mechanisms described above. It may be mentioned that these changes are accelerated by pasteurization, oxidation being favored by the high temperature and by the dissolving of air from the bottle head space during heating, resulting in more intimate contact of the oxygen with the beer. As oxidation continues, it has a noticeable effect on the appearance and flavor of the beer. The brilliance decreases progressively and finally a definite haze, known as oxidation haze, is visible. This haze is the result of formation of high molecular weight protein complexes having colloidal dimensions. It is of a different nature from chill haze, as it does not disappear on warming. With the advent of oxidation haze the beer has a definite "oxidation flavor" noticeable to the average consumer.

The distribution of protein fractions at the time the beer is bottled can be used to estimate its shelf-life. Thus, if the proportion of high molecular components is higher than usual, the time required to reach the level of visible haze will be less than when this proportion is low at the start. At present, the only test in general use to give an indication of the probable shelf-life of beer is the ITT (indicator time test) developed by Gray and Stone (1939) of the Wallerstein Laboratories. This test measures the reducing power of beer, which increases as the proteins are oxidized. However, it is influenced by other factors in addition to the state of oxidation of the proteins. While it is quite reliable in following the progress of oxidation in a single beer, its use to compare one beer with another sometimes leads to false conclusions. For example, the presence of hydrogen peroxide in beer has long been known to have a detrimental effect on its stability and the presence of as little as 0.005% has been known to cause haziness. The peroxide is formed from the reaction of molecular oxygen and auto-oxidizable substances such as catechol, ascorbic acid, isoascorbic acid, and electron transfer systems, notably reduced flavin nucleotides. In the presence of peroxide, the ITT may show a zero or low value, even in the presence of haze and oxidation taste.

In view of the deleterious effects of oxidation, many attempts have been made to prolong the shelf-life of beer by adding antioxidants. Gray and Stone (1939) have recommended the addition of ascorbic acid, and this has found rather wide application in the industry. However, according to Barton-Wright (1956), its use alone may be dangerous because in removing oxygen it is oxidized to dehydroascorbic acid, which is an oxidizing agent for glutelins. He considers reductones prepared from sugars to be more efficient.

Another type of antioxidant has been suggested by Ohlmeyer (1957). This is the glucose oxidase-catalase system, which removes oxygen by an enzymatic reaction. It will be noted in Table 59 that a typical beer contains 0.001% glucose. This amount, though very small, is enough to bind several times the amount of oxygen normally present. In this system the glucose oxidase catalyzes the oxidation of glucose to gluconic acid and hydrogen peroxide. The catalase then breaks the hydrogen peroxide down to water and oxygen, but only half of the original amount of oxygen taken up is released. The overall reaction is expressed by the equation

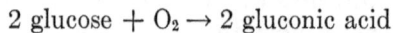

$$2 \text{ glucose} + O_2 \rightarrow 2 \text{ gluconic acid}$$

The addition of proteolytic enzymes to beer, as is done in chillproofing, will also lengthen their shelf-life and delay the appearance of oxidation haze as well as decrease chill haze. However, these enzymes must be used with discretion, as too great a shift of the equilibrium toward low molecular fractions at the expense of the high and medium fractions will lead to foam instability and a poor head on the beer. This is caused by the changes in surface tension. All other things being equal, a higher and more stable foam is obtained with a low surface tension than with a high one. As the molecular weight of the proteins decreases, the surface tension of the beer increases and the foam is adversely affected. These relationships are illustrated in Fig. 67. In this figure the best compromise between optimum haze resistance and optimum foam stability is indicated by the shaded area. Thus, judicious use of proteolytic enzymes will result in an increase in the proportion of proteins in this optimum range without undue shifting of the equilibrium toward low molecular weight fractions.

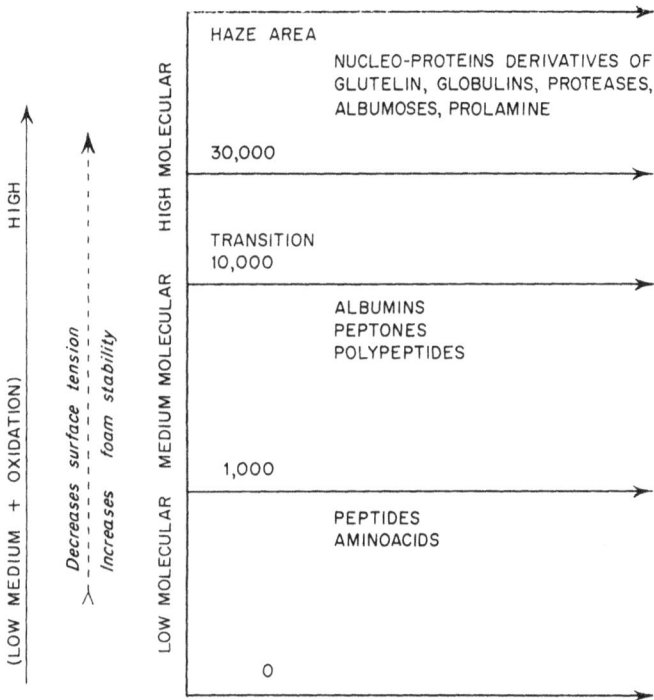

FIG. 67. EFFECT OF PROTEIN DISTRIBUTION ON FOAM STABILIZATION

LABORATORY METHODS OF EXAMINING BEER

Methods for which literature references are not given will be briefly described below.

(A) *Surface tension* was determined with a du Nuoy tensiometer.

(B) *Surface activity* is calculated from the formula

$$\frac{72.75 - \text{surface tension of beer (dynes) at } 68°F}{72.75}$$

The amount 72.75 represents the surface tension of water (dynes) at 68°F.

(C) *Protein fractions* were determined by adaptations from the work of Luudin (as quoted by Pawlowski and Doemens 1932).

Total Soluble Protein

Place a porcelain chip and about 0.1 gm CuO-selenium metal powder mixture into an 800 ml Kjeldahl flask. Pipette 25 ml of decarbonated beer at 68°F into the flask, add 28 to 30 ml of the sulfuric acid digestion mixture. Digest until 30 min after all charring has disappeared. Cool, carefully add 200 ml of distilled water. Transfer flask to distilling apparatus. Place 25 ml of O.1N sulfuric acid in a 250 ml Erlenmeyer flask, add 2 drops methyl red indicator. Place on receiving end of distilling apparatus. Make sure tip of distilling apparatus is in the O.1N acid solution. Cool condensers. Pour about 150 ml of 30% NaOH slowly down the side of the flask so as to form a layer. Add a few pieces of zinc. Attach tightly to the condenser. Shake flask, light Bunsen burner and distill until 100 to 150 ml have been collected. Titrate with O.1N sodium hydroxide until all the red color has disappeared. Record number of milliliters of O.1N sodium hydroxide used.

Calculations.—

$$(\text{ml } H_2SO_4 \times \text{Normality}) - (\text{ml NaOH} \times \text{Normality})$$
$$\times \, 0.056 = \text{Nitrogen}/100 \text{ ml}$$
$$\text{Nitrogen} \times 6.25 = \text{Protein}/100 \text{ ml}$$

Permanently Soluble and Coagulable Protein

Weigh 100 gm of decarbonated beer (at 68°F) into a tared 500 ml flat bottomed, round flask. Reflux for 3 hr. Cool. Restore to original weight with distilled water. Filter; do *not* wash. Pipette 25 ml of filtrate into Kjeldahl flash. Proceed as with Total Soluble Protein. (Distill into 15 ml O.1N sulfuric acid.)

Calculations.—

$$(\text{ml } H_2SO_4 \times \text{Normality}) - (\text{ml NaOH} \times \text{Normality})$$
$$\times \, 0.056 = \text{Nitrogen}/100 \text{ ml}$$
$$\text{Nitrogen} \times 6.25 = \text{Protein}/100 \text{ ml}$$

Formol Proteins: (Low Molecular Fraction)

Use 25 ml of beer.

(1) Adjust potentiometer at pH 4.0 with buffer solution.

(2) Adjust 25 to 30 ml formaldehyde to pH of 9.0 with O.1 N sodium hydroxide.[1]

To start with, add about 1 ml of the sodium hydroxide solution immediately, adding the remainder dropwise. Allow about 1 to 2 min of stirring to adjust pH.

---

[1] This solution should be used only on the day it is made up. It should not be made as a stock solution.

(3) Put sample of beer in place. Adjust pH to approximately 5 to 6.

(4) Begin stirring sample. Add NaOH solution until pH is at 6.8.

(5) Pipette 10 ml of formaldehyde, add to solution stirring constantly.

(6) Titrate with sodium hydroxide solution, noting initial reading on burette, until pH of 9.0 is reached. Note final reading.

Calculations.—

$$\text{ml NaOH} \times \text{Normality} \times 0.056 = \text{Nitrogen/100 ml}$$
$$\text{Nitrogen} \times 6.25 = \text{Protein/100 ml}$$

Phosphomolybdate Precipitable Proteins (Medium Molecular Fraction)

Into a wide mouth 200 ml volumetric flask place

100 ml beer

80 ml distilled water

10 ml 50% sodium molybdate

Put in water bath at 68°F for 1 hr. Make up to mark. Add 10 ml 50% sulfuric acid; mix by inverting. Allow to stand overnight. Filter. Use 50 ml portions of filtrate for proteins. Proceed as for Total Soluble Proteins. (Distill into 15 ml of 0.1N sulfuric acid.)

Calculations.—

$$[(\text{ml } H_2SO_4 \times \text{Normality}) - (\text{ml NaOH} \times \text{Normality})$$
$$- \text{correction for blank}] \times 0.0588 = \text{nonprecipitated Nitrogen}$$

Total Soluble Nitrogen-nonprecipitated Nitrogen = Phosphomolybdate precipitable nitrogen/100 ml

$$\text{Nitrogen} \times 6.25 = \text{Protein/100 ml}$$

Tannin Precipitable Proteins (High Molecular Fraction)

Into a wide mouth 200 ml volumetric flask place

100 ml beer

90 ml distilled water

4 ml 50% sulfuric acid

Put in water bath for 1 hr at 68°F. Make up to mark. Add 10 ml of 16% tannin; mix by inverting. Chill in an ice bath for at least 1 hr. Allow to stand overnight in refrigerator. Filter. Use 50 ml of filtrate. Proceed as with Total Soluble Protein. Distill into 15 ml 0.1N sulfuric acid.

Calculations.—

[(ml H_2SO_4 × Normality) − (ml NaOH × Normality)
       − correction for blank] × 0.0588 = nonprecipitated Nitrogen
Total Soluble Nitrogen-nonprecipated Nitrogen
                       = Tannin precipitable Nitrogen/100 ml
            Nitrogen × 6.25 = Protein/100 ml

Nonprotein Nitrogen

Phosphomolybdate − Tannin = Nonprotein Nitrogen

Blanks.—

Phosphomolybdate   Tannin

Phosphomolybdate	Tannin
10 gm dextrose	10 gm dextrose
100 ml water	100 ml water
10 ml sodium molybdate	4 ml 50% H_2SO_4
Make to mark	Make to mark
Precipitate with 10 ml 50% H_2SO_4	Precipitate with 10 ml tannin

       Proceed as under Phosphomolybdate and Tannin.

DISTILLED PRODUCTS

Ethanol for whiskey is obtained by fermenting grain mashes. Industrial alcohol is generally made from cheaper sources of carbohydrates, or synthetically. From a flavor standpoint, the fermentation conditions are not as critical for whiskey as for beer since the distillation step eliminates most of the minor flavoring ingredients which develop in the mash.

The corn, rye, or other grain is ground, mixed with water, and cooked under pressure to thoroughly gelatinize the starch and sterilize the mash. After partial cooling, barley malt or other enzyme sources are added to hydrolyze the starch to fermentable sugars, principally maltose. The mash is then pumped to fermenters, the temperature adjusted, and yeast added. If sour mash whiskey is to be made, the material is inoculated with some of the residue from previously fermented mash.

Under the conditions usually employed, the fermentable sugars are completely converted in 48 to 72 hr. Alcohol is then distilled from the mash without further separation of the components. The continuous column still is generally used. It can be seen that the fermenting procedure is somewhat simpler and less critical than beer manufacture.

Alcohol for whiskey is distilled at slightly under 160 proof. About 2.5 gal. of absolute alcohol or 5 gal. of 100 proof alcohol can be obtained from a bushel of corn, slightly less from a bushel of rye. The alcohol is adjusted to the desired proof with plain or distilled water and placed in charred barrels for aging.

The flavor of whiskey depends upon the grain used, the fermentation conditions (including inoculum), method of distillation, proof of collection, and method and conditions of aging.

BIBLIOGRAPHY

ACRAMAN, A. R. 1966. Processing brewers' yeasts. Process Biochem. *1*, No. 6, 313–317.

AINSWORTH, G. 1968. Money down the drain? Process Biochem. *3*, No. 12, 11–14, 20.

ANON. 1964. Methods of Analysis, 8th Edition. American Society of Brewing Chemists, Milwaukee, Wisc.

ATKINSON, A. A. 1968. Single-cycle filtration in brewing. Process Biochem. *3*, No. 4, 38–40.

BARTON-WRIGHT, E. D. 1956. Antioxidants for beer. Bull. assoc. anciens etud. brass. univ. Louvain *52*, 24.

BAVISOTTO, V. S. 1967. Process for producing brewery wort with enzymes. US Pat. 3,353,960. Nov. 21.

BISERTE, G., and SCRIBAN, R. 1953. Study of the proteins and polypeptides of barley. European Brewery Conv. Proc., 4th Congr., Nice, 48–66.

BONNET, J. 1955. Water in the agricultural industries, Bull. Centre Belge étude et document. eaux (Liege) No. 28, 70–80.

BROWN, C. M., and WRIGHT, R. L. 1968. Production of brewers' yeast. Process Biochem. *3*, No. 12, 21–22, 28.

CAUWE, Y. 1968. Continuous fermentation. Intern. Brewer Distiller *2*, No. 2, 54–58, 60–62, 64–65.

CLARK, D. F. 1967. Separation techniques brewing. Process Biochem. *2*, No. 9, 41–42.

CURTIS, N. S. 1968. Brewing advances. Process Biochem. *3*, No. 4, 17–19.

DE CLERCK, J. 1958. A Textbook of Brewing. Chapman and Hall, London.

GRAY, P. P., and STONE, I. 1939. Oxidation in beers. I. A simplified method for measurement. Wallerstein Labs. Commun. *2*, 5–16. 1940, The measurement and study of beer foam. Ibid. *3*, 159–171.

GREENSPAN, R. P. 1965. Some aspects of microbiological control in a brewery. Am. Soc. Brewing Chemists Proc. *1965*, 57–59.

GRIFFIN, O. T. 1966. Biochemistry of mashing. Process Biochem. *1*, No. 4, 241–243.

HARRIS, G. 1956. Chromatographic separation of hop and malt tannins. J. Inst. Brewing *62*, 390–406.

HARRIS, G. 1965. The polyphenol composition of non-biological hazes of beer. J. Inst. Brewing *71*, 292–298.

HARRIS, G., BARTON-WRIGHT, E. C., and CURTIS, N. S. 1951. Carbohydrate composition of wort and some aspects of the biochemistry of fermentation. J. Inst. Brewing *57*, 264–280.

HARRISON, J. G. 1966. Beer storage. Process Biochem. *1*, No. 2, 113–118.

HARRISON, J. G. 1967. Beer stability. Intern. Brewer Distiller *1967*, July/Aug., 23–31.

HARSANYI, E. 1968. The Pablo continuous lautering system. Brewers Dig. *43*, No. 7, 46–49, 90.

HEPNER, I. L. 1967. A tale of two breweries. Process Biochem. *2*, No. 12, 41–42, 46.

HOAK, R. D. 1956. Industrial water requirements. Public Works *87*, No. 11, 160.

HOWARD, G. A. 1967. Brewing behaviour of hops. Process Biochem. *2*, No. 2, 31–34.

HOWARD, G. A., and POLLOCK, J. R. A. 1954. Natural occurrence of lupulone analogs. Chem. & Ind. (London), 991.

HOWARD, G. A., and TATCHELL, A. R. 1954. The structure of cohumulone. J. Chem. Soc., 2400–2405. The synthesis of dl-cohumulone. Chem. & Ind. (London), 514.

HUDSON, J. R. 1966. Wort boiling. Brewers' Guild J. *1966*, 435–441.

HUDSON, J. R. 1967. Beer and beer processing. Brewers' Guardian *96*, No. 5, 56–58, 60–63.

IMRIE, F. K. 1968. Malting, mashing, and wort substitutes. Process Biochem. *3*, No. 4, 21–22, 28.

JAHNSEN, V. J. 1963. Composition of hop oil. J. Inst. Brew. *69*, 460–466.

LEACH, A. A. 1967A. Collagen chemistry in relation to isinglass and isinglass finings—a review. J. Inst. Brew. *73*, 8–16.

LEACH, A. A. 1967B. Finings and fining problems. Brewers' Guardian *96*, No. 11, 35–43.

LENSE, K. 1964. Katechismus der Brauerei-Praxis, 13th Edition. Verlag Hans Carl, Nurnberg, Germany.

LEWIS, M. J. 1968. American lager beer. Process Biochem. *3*, No. 8, 47–48, 62.

MCFARLANE, W. D., and HELD, H. R. 1953. Quantitative chromatography of wort and beer carbohydrates. Am. Soc. Brewing Chemists Proc. *1953*, 67–78.

MCFARLANE, W. D., HELD, H. R., and BLINOFF, G. 1954. A simplified method for determining the total fermentable sugars in wort. Am. Soc. Brewing Chemists Proc. *1954*, 121–127.

MOELLER, W. M. 1968. Priming and conditioning agents for beer. Brewers Dig. *43*, No. 9, 84, 86, 88, 92, 126.

MOLLER, N. C., and CHRISTENSEN, B. K. 1968. Establishment of optimal conditions of wort cooling. Intern. Brewer Distiller *2*, No. 4, 49–50, 52–57.

MONIER-WILLIAMS, G. W. 1940. Official and Tentative Methods of Analysis, 5th Edition. Assoc. Offic. Agr. Chemists, Washington, D.C.

NARZISS, L. 1966. Developments in the malting and brewing industries in Germany. J. Inst. Brew. *72*, 13–24.

NARZISS, L., and KIENINGER, H. 1968. Progress in wort treatment from brewhouse to fermentation cellar. Intern. Brewer Distiller 2, No. 2, 40–42, 44–46, 48–50, 52–53.

NISSEN, B. H. 1965. Water for the brewery. Tech. Quart. Master Brewers Assoc. Am. 2, 169–172.

OHLMEYER, D. W. 1957. Use of glucose oxidase to stabilize beer. Food Technol. 11, 503–507.

OWADES, J. L., and DONO, J. M. 1965. The determination of volatile acids in beer. Am. Soc. Brewing Chemists Proc. 1965, 157–160.

PAWLOWSKI, P., and DOEMENS, A. 1932. Die Brautechnischem Utersuchungsmethoden. Lehr- and Versuchsanstalt fur Brauer, Munich.

PIERCE, J. S. 1966. Amino-acids in malting and brewing. Process Biochem. 1, No. 8, 412–416.

POZEN, M. A. 1940. Water in the brewery. Modern Brewery Age 23, No. 3, 67.

RAINBOW, C. 1966. Flocculation of brewers yeast. Process Biochem. 1, No. 9, 489–492.

RENNIE, H. 1967. Infusion and decoction mashing. Process Biochem. 2, No. 7, 9–11.

RIEDL, W. 1954. Synthesis of several analogs with modified acyl groups. Ann. 585, 38–42.

RIGBY, F. L., and BETHUNE, L. J. 1952. Countercurrent distribution of hop constituents. Am. Soc. Brewing Chemists Proc. 93–105. Cohumulone, a new hop constituent. J. Am. Chem. Soc. 74, 6118–6119.

RIGBY, F. L., and BETHUNE, L. J. 1953. The isolation and properties of cohumulone. Am. Soc. Brewing Chemists Proc. 1953, 119–129.

RIGBY, F. L., and BETHUNE, L. J. 1955. Rapid methods for the determination of total hop bitter substances (iso-compounds) in beer. J. Inst. Brewing 61, 325–332. Components of the lead-precipitable fraction of Humulus lupulus, adhumulone. J. Am. Chem. Soc. 77, 2828–2830.

ROBERTS, M., LAUFER, S., and STEWART, E. D. 1947. Determination of air dissolved in storage beer. Am. Soc. Brewing Chemists Proc. 1947, 87–92. Determination of air and oxygen in storage and finished beer. Ibid. 1947, 92–100.

ROESSLER, J. G. 1968A. Yeast management in the brewery. I. Wort production and treatment. Brewers Dig. 43, No. 7, 38, 40, 42, 44.

ROESSLER, J. G. 1968B. Yeast management in the brewery. II. Yeast handling and treatment. Brewers Dig. 43, No. 9, 94, 96, 98, 102, 115.

ROYSTON, M. G. 1966. Tower fermentation of beer. Process Biochem. 1, No. 4, 215–221.

ROYSTON, M. G. 1968. Economics of lautering. Process Biochem. 3, No. 6, 16–21.

ST. JOHNSTON, J. H. 1948. The separation of the protein constituents of beer. J. Inst. Brewing 54, 305–320.

SANDEGREN, E. 1947. On the importance of proteins in brewing. Brewers Dig. 22, No. 8, 47–52.

SEGEL, E., GLENISTER, P. R., and KOEPPL, K. G. 1967. Beer foam. Tech. Quart. Master Brewers Assoc. Am. 1967, No. 4, 104–113.

STONE, I. 1942. Determination of traces of copper in wort, beer, and yeast. Ind. Eng. Chem., Anal. Ed. 14, 479–481.

STONE, I., and GRAY, P. P. 1948. Silica and tannin in worts and beers. Am. Soc. Brewing Chemists Proc. *1948*, 76–94.

STONE, I., GRAY, P. P., and KENIGSBERG, M. 1951. Flame photometry—sodium, potassium, and calcium in brewing materials. Am. Soc. Brewing Chemists Proc. *1951*, 8–20.

STRANDSKOV, F. B. 1965. Yeast handling in a brewery in the United States. Wallerstein Lab. Commun. *28*, No. 95, 29–33.

VANCRAENENBROECK, R., and LONTIE, R. 1955. Study of the tannins and polyphenols of hops. Bull. assoc. anciens etud. brass. univ. Louvain *51*, 1–14.

VERMEYLEN, J. 1962. Treatise on the Manufacture of Malt and Beer, Vols. I and II. Assoc. Royal des anciens Eleves de L'Institut Superieur des Fermentations, Gand, Belgium.

WALDSCHMIDT-LEITZ, E., and MAYER, K. 1935. Amylophosphatase from barley. Z. Physiol. Chem. *236*, 168–180.

WEBBER, H. F. P., and TAYLOR, L. 1954. Effect of fluorine on fermentation. Brewers Digest *29*, No. 12, 59–61.

WEISSLER, H. E. 1967. Brewery products. *In* Encyclopedia of Industrial Chemical Analysis, Vol. 7, F. D. Snell, and L. S. Ettre (Editors). John Wiley & Sons, New York.

WEST, D. B., LAUTENBACH, A. F., and BECKER, K. 1952. Studies on diacetyl in beer. Am. Soc. Brewing Chemists Proc. *1952*, 81–88.

WEST, D. B., LAUTENBACH, A. F., and BRUMSTED, D. D. 1963. Phenolic characteristics in brewing. Am. Soc. Brewing Chemists Proc. *1963*, 194–199.

CHAPTER 6

Samuel A. Matz

Manufacture of Breakfast Cereals

INTRODUCTION

History and Present Status of the Industry

Breakfast cereals may be conveniently divided into two major categories: (1) those cereals, such as oatmeal, which require cooking before they are served, and (2) fully cooked, ready-to-eat cereals such as corn flakes. The former class is probably about as old as civilization, since it is very likely that gruels and porridges made from crushed grains were among the first cereal foods of mankind. Prepared breakfast foods have a short and interesting history which has been well described by Carson (1957).

The original motivation for the development of precooked breakfast foods seems to have been the desire of some vegetarians to create additional variety for their diets. The Seventh Day Adventist Church, many members of which avoid the consumption of animal foods, was closely tied up with the early history of prepared breakfast foods. The close association of certain factions in this group with the Battle Creek Sanitarium and the cereal experimentation which they inspired at this institution gave the city of Battle Creek a head start in the breakfast cereal industry. Today this city still houses the largest producers.

The first ready-to-eat breakfast cereal was probably "Granula," developed by Dr. James C. Jackson about 1863 at Dansville, New York. This health food was made by baking a coarse whole meal dough in thin sheets until it was hard and brittle, breaking and grinding the cake into small chunks, baking the chunks again, and grinding the resultant material into small granules.

The first breakfast food made by Dr. J. H. Kellogg of Battle Creek was named, by an interesting coincidence, "Granola." Kellogg made biscuits about ½ in. thick from a dough composed of wheat meal, oatmeal, and corn meal, and baked them until they were desiccated and beginning to turn brown. The hard biscuits were then ground and packaged. It was C. W. Post, the founder of Post Cereals, who first recognized that convenience and flavor were more forceful and more generally applicable selling points than were the healthfulness and vegetable origin of prepared

221

breakfast cereals. Post became the first great merchandiser of these foods as a result of his grasp of this concept.

Since these early ventures in Battle Creek and elsewhere, prepared breakfast cereals have become recognized by persons in all walks of life as economical, convenient, and flavorful foods suitable for daily consumption by all age groups. The original granules and flakes have multiplied into an array of forms, colors, and flavors which staggers the imagination. Additives and protein combinations have been developed to give the finished product a nutritional adequacy which is equalled by only a few other foods. As a consequence of these changes, breakfast cereals have held or increased their levels of per capita consumption in recent years in spite of the overall decline in cereal food use. Their performance in this respect is superior to that of all other groups in this category with the exception of macaroni products.

The introduction of breakfast cereal factories into foreign countries has proceeded rather rapidly in the last two decades. These foods have had immediate acceptance in all areas where a good product has been offered. The future of the foreign market seems to be very promising.

THE TECHNOLOGY OF BREAKFAST CEREAL MANUFACTURE

At the beginning, breakfast cereal manufacturing was strictly an art. However, the larger producers ultimately came to the conclusion that only by the application of scientific principles to their operations could they keep pace with their competitors in new product development and product control. The relative dearth of publications in the field indicates a considerable desire for secrecy rather than a lack of experimentation. Pressure to secure competitive advantages has increased and forced a faster advance in recent years.

In the manufacture of uncooked breakfast cereals, there are two rather general processing steps. One of these is the reduction of particle size and the other is the elimination from the product of some of the fibrous substances found in the whole grain. The effect of these practices is to reduce cooking time and to improve the texture and perhaps the digestibility of the food. There is usually no attempt to alter materially the natural flavor of the grain by hydrolyzing its starches or caramelizing its sugars although it is true that the heat treatment applied to oatmeal changes the flavor somewhat. Recent advances in technology have resulted in cooking times being decreased to the point where addition of boiling water to the packaged food will give a fully prepared product.

Ready-to-eat breakfast cereals, although they are extremely diverse in appearance, composition, and flavor, have at least two unifying processing principles. One of these is the creation of a crisp texture by drying the cooked product with its content of gelatinized starch to a moisture content of about 3 to 5%, and the other is a flavor change which results from the dextrinization, gelatinization, and caramelization of the cereal starches and their degradation products.

In subsequent sections of this chapter, an attempt will be made to give the salient points of the most important processing techniques used in the breakfast cereal industry.

Processing of Uncooked Breakfast Cereals

Wheat Cereals.—The wheat cereal having the largest consumption is farina. This product is nothing more than the wheat middlings which are more fully described in the chapter on Milling. Middlings are chunks of endosperm free of bran and germ. When reduced in size, middlings become flour. In the manufacture of farina, it is necessary to use hard wheat as a raw material since soft wheat yields a product which becomes excessively pasty upon cooking. About 30% of the wheat can be secured as farina by good milling techniques.

Particle size is thought to be a critical factor influencing consumer acceptance. The federal specification for farina requires that: 100% of the product pass through a US Standard No. 20 woven-wire-cloth sieve; not more than 10.0% pass through a No. 45 sieve; and not more than 30.0% pass through a No. 100 sieve. Vitamin and mineral enrichment is usually applied to farina. Vitamins are usually added in the dry state.

Disodium phosphate has been used (at about the 0.25% level) to increase the speed at which farina cooks. An "instant" farina has come upon the market. This farina is ready-to-eat after about 1 min of boiling time. Apparently it is farina which has been treated with proteolytic enzymes or in some other way. It is believed that bromelin was originally used in this process, but other enzymes may have been substituted in current production. The function of the proteolytic enzymes is said to be the opening up of microscopic pathways for water penetration in the granule.

Farina flavored with malt or with cocoa is marketed. Generally, these products are simple mixtures of the flavoring ingredient with the middlings.

Whole wheat meal, cracked wheat, flaked wheat, and farina with bran and germ are sold to a rather limited extent. The stability of these foods is limited by the tendency of the bran and germ oils to become rancid unless the product is specially processed.

Oat Cereals.—Gunderson and Brownlee (1938) gave a complete description of rolled oat processing (see Fig. 68). Methods have not changed much since that time.

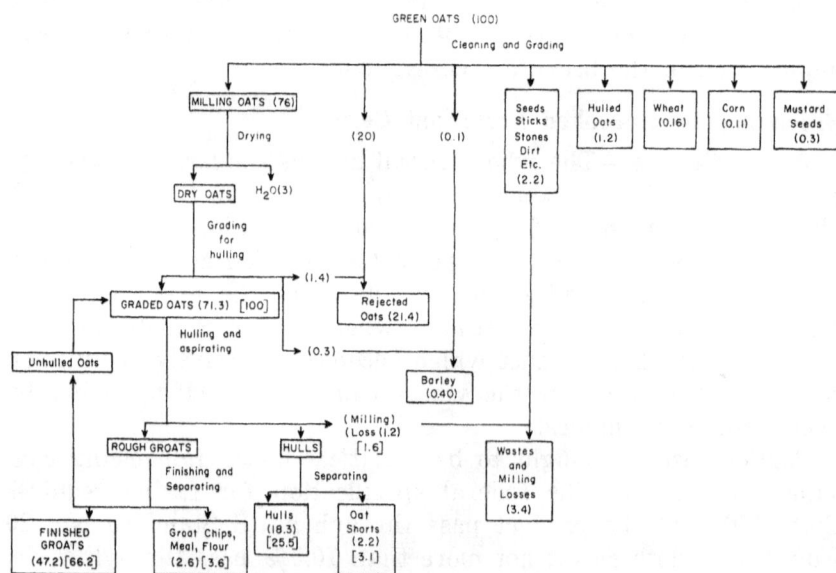

FIG. 68. FLOW DIAGRAM OF THE MILLING OF OATS

Figures in *parentheses* are gross yields, based on green oats. Figures in *brackets* are net yields, based on graded oats going to the hullers.

Most oats for human consumption are marketed as rolled oats. Some demand also exists for ground oats and steel-cut groats. The initial step in the preparation of rolled oats is the roasting process. Thoroughly cleaned grains are heated to 212°F for 1 hr. This roasting procedure reduces the moisture content to about 6% and probably partially dextrinizes the starch. In addition, the hulls become fragile rather than tough, and are more easily removed during the subsequent processing.

After cooling, the roasted grains are separated according to size. The hulls are then removed by passing the kernels between two large circular milling stones mounted horizontally with the grinding surfaces separated by a short distance. The upper, or rotating, member of this pair of stones has a very slightly convex milling surface, while the lower stone, which is stationary, has a flat surface. Oats enter at the center of the upper stone, and are carried to the periphery by centrifugal motion. The distance between the rolls is everywhere more than the width of a dehulled kernel, but is less than the length of the kernels. As the grains travel outward, the hulls are removed by the tearing and abrading action which they undergo. The grain with its hull removed is called a groat. The mixture of hulls, groats, and broken grains from the mill is screened and subjected to other procedures to separate the components. In some plants, hulling is achieved by impact methods, using devices such as the Entoleter.

The next step is flaking. Whole groats may be flaked, or they may be first cut into pieces by rotary granulators. The smaller the piece size, and the thinner the flake, the more quickly the product can be cooked. For example, the so-called quick oats, which are flakes made from a particle about $\frac{1}{3}$ to $\frac{1}{4}$ the size of the whole groat, cook in about 5 min, while flaked whole groats, ("regular oats") are thoroughly cooked after 10 to 15 min boiling. On the other hand, the quicker cooking oats do not stand up as well under prolonged heating such as they might encounter on the steam table of a cafeteria. Regular oats will maintain a satisfactory texture for about 3 hr on the steam table, while the quick oats become undesirable after about 1 hr under the same conditions. By making a thicker flake from the whole groat, oats withstanding 6 hr of heating can be obtained. Steel-cut oats (not flaked) are even more resistant to overcooking.

Corn Cereals.—Corn meal in the form of mush and boiled hominy grits are often used as hot breakfast cereals. The preparation of these products is described in the chapter on Milling.

Rice Cereals.—Whole milled rice is occasionally cooked as a breakfast cereal. Full details on this product are given in the chapter on Rice Processing. A product recently introduced consists of milled rice ground into particles about the size of those which make up farina. Because of the particle size, the product is instant cooking; that is, it does not require additional heating after the addition of boiling water to the rice. A short period of standing after the addition of water is necessary, of course.

FLAKES

General Considerations

Flaking is a relatively simple process, consisting in its most elemental form of cooking fragments of cereal grains (or sometimes whole grains), flattening the soft particles between rollers, and toasting the resultant flake at high temperatures. Apparently the first commercial production of such a food occurred around the turn of the century when J. H. Kellogg and W. H. Kellogg made whole wheat flakes in a barn behind the Battle Creek Sanitarium. Many complications have been introduced into the process since that time in attempts to improve the flavor and the efficiency of operations, and to increase the uniformity of flake size and appearance which is so desirable to the manufacturer and perhaps even to the consumer.

In the basic processing steps, the raw material undergoes the following changes: (1) the starch is gelatinized and probably slightly hydrolyzed; (2) the particle undergoes a browning reaction due probably to interaction of proteins and sugars; (3) enzymatic reactions are stopped, rendering the final product more stable; (4) dextrinization and caramelization of the sugars occur as a result of the high temperatures in the roasting oven; and (5) the flake becomes crisp as a result of the reduction of its moisture content to a very low level.

Flakes owe their popularity with consumers to their crisp but friable texture, to their sweet but rather bland flavor, and to the ease with which a portion may be readied for consumption.

In the subsequent paragraphs of this section, a fairly straightforward corn flake operaton will be described first, followed by a discussion of a wheat flake production method which is considerably more complex. Bran flakes, 40% bran flakes, and rice flakes are also marketed and are produced by quite similar methods.

Corn Flake Production

Hybrid yellow corn is usually used, although white corn provides an equally satisfactory raw material. The corn is broken (milled) so as to yield a No. 4 to No. 5 grit, free of germ and bran. These large pieces represent about one-half of a corn kernel, and they retain their identity throughout the processing, each particle eventually emerging as a corn flake. The hulls resulting from the milling operation are used in animal feeds, and the germs are pressed in expellers to yield the corn oil of commerce and a cake suitable as a feed ingredient. Into a cylindrical pressure cooker are placed about 1700 lb of the grits and 36 gal. of a flavoring syrup consisting of sugar, malt (nondiastatic), salt, and water. Occasionally niacin is also added at this point. During the cooking period the charge accumulates additional water from the steam introduced into the retort, rising to about 33% moisture.

Cooking is done in the slowly rotating retort at 15 to 23 (typically 18) psi steam pressure for 1 to 2 hr. Different lots of corn may vary considerably in the duration of the cooking time required. The end point can be judged by examining a small sample of the charge which is blown out through a gate valve for this purpose. A uniform translucency in the kernels indicates an adequate cook. At this time, the pressure is reduced to the atmospheric level, the retort is opened, and the contents are dumped out onto a moving belt.

FIG. 69. DUMPING COOKED FLAKE INGREDIENTS FROM RETORT

After the lumps from the cooker are broken down to individual particles by a revolving reel, they are distributed to a set of driers. The latter devices are essentially large tubes or tanks extending vertically for several stories. The wet kernels enter the top and are dried by a countercurrent of hot (150°F) air as they travel to the bottom. Another type of drying set-up consists of horizontal rotating cylinders having numerous steam-heated pipes passing longitudinally through them. Louver driers may also be used.

The dried particles now contain 19 to 23% moisture, but this water is unevenly distributed, so the material is transferred to tempering bins for several hours (as many as 24) so that the moisture may equilibrate. After tempering, the hard, dark brown grits are ready for flaking.

The flaking rolls are steel cylinders weighing over a ton each, and revolving at a speed of about 180 to 200 rpm (see Fig. 70). Hydraulic controls maintain a pressure of over 40 tons at the point of contact of the rolls. The rolls are cooled by internal circulation of water. The cooked dried grits are pressed into thin flakes as they pass through the rolls. The product is still rather

FIG. 70. A SET OF FLAKING ROLLS

flexible at this time, lacking the desired crispness and preferred flavor of the finished corn flake.

From the rolls, the flakes pass directly to the rotating toasting ovens (see Fig. 71), which are usually gas fired. The moist flake is tumbled through the perforated drums and passes within a few inches of the gas flames. Treatment may be 50 sec at 575°F, or 2 to 3 min at 550°F. In addition to being thoroughly dehydrated by the process, the flakes are toasted and blistered. They emerge from the oven with less than 3% moisture.

FIG. 71. OVEN FOR TOASTING FLAKES

From the ovens, the flakes are carried by belts to expansion bins where they are permitted to cool to room temperature. On the way, the product is cooled by circulating air and is usually treated with a spray of a solution of thiamine and perhaps other B vitamins.

The Manufacture of Wheat Flakes

Plump kernels of wheat are used as the raw material for wheat flakes. After cleaning and classifying according to size, the wheat is tempered with added moisture in steel bins of small diameter at approximately 80°F for 24 hr. The wheat may be transferred one or more times during this period if such a procedure is necessary in order to keep the temperature within reasonable limits.

After tempering, the wheat is steamed at atmospheric pressure until it reaches about 203°F and 21% moisture.

The steamed wheat is "bumped" between smooth steel rollers set considerably farther apart than are flaking rolls. This treatment flattens the grain slightly and ruptures the bran coat in several places making the kernel more permeable to the moisture added during the cooking step. The flattened kernels are transferred to the pressure cookers, which are similar to those used for corn flakes, and the other ingredients are added. These ingredients include sugars, salt, malt, and sometimes a coloring substance such as caramel.

The retort contents are cooked at 20 psi steam pressure for 90 min while the vessel rotates slowly. After cooking, the grains are soft, translucent, and brown. They contain about 45 to 50% moisture. The starch has, of course, been completely gelatinized. Rotation of the opened retort dumps the contents onto a moving belt which transfers the cooked mass to a chute leading to a "Wiggler."

The Wiggler consists of a horizontal perforated disc, through which warm air is blown in an upward direction, and a rotating arm carrying vertically-oriented inflexible fingers around its upper surface. The clumps of slightly adherent grain are dropped onto the center of the perforated disc and are broken up and the individual grains moved to the outer edge of the disc by the moving fingers.

The individual grains fall from the edge of the disc and are transferred pneumatically to a horizontal rotating cylinder fitted with internal louvers. In this drier, air at 250° to 300°F is passed over the grain, reducing it to 28 to 31% moisture. Holding bins are used to store the material until it can be transferred to the presses.

At this point, the grains are still intact and are rather tough and chewy in texture. Subsequent processing is designed to secure the required crispness. First, the wheat pieces travel through a drier. This could be a Proctor and Schwarz drier composed of 3 sections, the first at 280°F, the second at 290°F, and the third unheated. Rate of movement of the material is adjusted to yield a product of about 21% moisture. A spray of B-complex vitamins is applied at this stage.

Screw conveyors or drag chain conveyors transport the partially dried pellets to the flaking rolls. Just before falling into the flaking rolls, the pellets are heated to about 180° to 190°F, and they become plasticized. The large steel flaking rolls are practically identical with those used for making corn flakes. The pressure applied to the pellets increases their diameter several times and decreases their thickness proportionately.

When they leave the rolls, the flakes contain 10 to 15% moisture and are still slightly flexible. To obtain the desired crispness, they are toasted and dehydrated to less than 3% moisture content in a drier with a perforated travelling metal belt. Temperature in the oven may be divided into 4 regions; for example, heated sections at 310°, 300° and 280°F, and an unheated section to partially cool the product. The decreasing temperature is said to promote the development of the desirable curling and blistering.

A recently patented process (Benson and Merboth 1967) describes the production of breakfast cereal flakes from a continuous sheet of dough, the improvement being the greater control achieved over the size and shape of the finished product. In the referenced patent, the thickness of the dough is reduced by running the conveyor belt faster than the extruded sheet, and tempering is accomplished by cooling the outer surface of the dough with currents of air.

Oat Flakes

Because of the comparatively high level of fat in oats, flakes made in the conventional way have a limited storage life. Lilly and Reinhart (1967) describe a process said to give products of satisfactory stability. Oat groats are pressure-cooked until the starch is gelatinized, and then kneaded while being held at a temperature of 150° to 212°F until a plastic dough is formed. The dough is formed into flakes about 0.011 in. thickness and these are flash dried to 2 to 10% moisture by contact with air having a temperature in the range of 400° to 800°F.

SHREDS

Shredded Wheat Biscuits

The most popular representative of this class is the shredded wheat biscuit manufactured by several US firms. This product differs from most other prepared breakfast cereals in that it is made from whole grain without the addition of any flavor and without the removal of the germ or bran. Cooking may be done at atmospheric pressure, in boiling water. After 1 hr or more of cooking, the moisture content of the wheat has reached 50 to 60% and the kernels are very soft. Some preliminary drying in louver ovens may be done at this time, but the whole wheat is not brought much below 45 to 50% moisture. The cooked and partially dried wheat is transferred to stainless steel bins and tempered for many hours before it goes to the shredding rolls.

The shredding rolls, as shown in Fig. 72, are from 6 to 8 in. in diameter and as wide as the finished biscuit is to be and thus are much smaller than flaking rolls. On one of the pair of rolls is a series of about 20 shallow corrugations running around the periphery. In cross section, these corrugations may be rectangu-

FIG. 72. SHREDDING ROLLS

lar, triangular, or a combination of these shapes. The other roll of the pair is smooth. Soft cooked wheat is drawn between these rollers as they rotate, and issues as continuous strands of dough.

Biscuits are built up by layering strands on a moving belt which passes under sets of rolls working in tandem. Ten to 18 rolls may be used for circular biscuits, while 22 rolls is a common number for rectangular biscuits. In the latter case, layered strands are separated into biscuits by passing them below blunt "knives" which fuse a thin line of the dough into a solid mass at regular intervals (Fig. 73).

Circular biscuits are formed quite differently. One end of the sheet formed from layered strands of whole wheat dough is caught up by a smooth roll which rotates just above the belt. This roll turns the layer of strands back upon itself, and the forward motion of the belt, combined with the reverse and upward motion of the cylinder surface, causes the layer to roll up into a circular biscuit. As the biscuit reaches the proper size, a knife chops down on the belt, severing the strands so that the formed biscuit is released. Automatic controls vary the speed of the belt as the diameter of the biscuit increases. The completed disk falls into a cup from which it is transferred to a belt leading to the ovens.

Since the biscuits are formed from dough of relative high moisture content, they are quite tender, and must be handled very

FIG. 73. CUTTING SHREDDED WHEAT BISCUITS

carefully to prevent distortion. In practice, this means that the transfer steps must proceed more slowly than is necessary with flakes or puffed products.

The wet biscuits are placed on a metal belt moving through a high temperature gas-fired oven. After 10 to 15 min, the outside of the product is dry and toasted while the interior is still wet. Then the biscuits are transferred to another hot air oven (or to a different section of the same oven) where they are dried at 250°F for 30 to 60 min through time depending upon the size and the air flow. Finished moisture content is about 11%. The combination of heat treatments causes the biscuit to assume the familiar oval cross section as a result of differential shrinkage of the layers.

A triple shredding mill is used for producing bite-sized break-fast cereals (see Fig. 74). Dough made from wheat, corn, or rice is fed to long, water-cooled shredding rolls. These rolls deposit a shredded dough sheet onto a constant speed conveyor to form a wide, three-layer ribbon. The rolls on the first and third shredding mills extrude dough sheets with a laced pattern, due to the presence of smooth and grooved rolls. The middle set of rolls revolves at higher speed than the other two sets. As a result, the dough sheet in the middle folds as it falls onto the relatively slow moving conveyor belt covered by the first sheet. Sugar is sprinkled over the middle dough sheet, and the top sheet is added. The combined structure passes between scoring rolls, and the baked cereal is finally broken along scored lines to form individual bite-sized pieces (Anon. 1965).

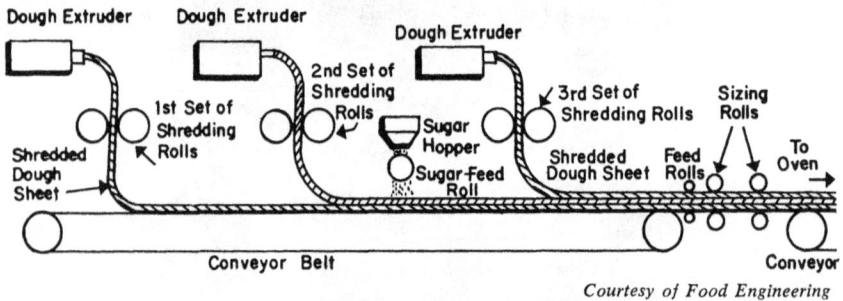

Courtesy of Food Engineering

FIG. 74. METHOD OF FORMING BITE-SIZED SHREDDED BISCUITS

Hale and Carpenter (1956) described the preparation of a biscuit having a lattice-like network of shreds. Their product may be puffed and therefore resembles not only shredded wheat biscuits, but also the product patented by Huber (1955) as described in the Puffing section of this chapter.

The flavor of shredded wheat differs markedly from that of whole flakes because the latter include added condiments and are subjected to much more heat in both the cooking and the toasting step. The more rigorous heat treatment applied to flakes results in considerably more caramelization in the finished product.

Rancid odors tend to accumulate if shredded wheat is stored in sealed containers. For this reason, the product is sold in "breather" boxes without outer or inner linings. When so packaged, the product is just as stable to storage deterioration as any other prepared cereal except that moisture absorption may occur in atmospheres of high relative humidity with a consequent loss of crispness.

CEREAL GRANULES

The only product of this type being marketed on a nationwide basis is Grape-Nuts made by Post Cereals Division of General Foods Corp. It is manufactured by a method quite different from that applied to any other breakfast cereal. In essence, this food is the toasted fragment of a loaf of bread—albeit an unusual type of bread. It bears a strong resemblance to some of the earliest types of precooked cereals.

The initial step in the manufacture of cereal granules is the preparation of a stiff dough from wheat, malted barley flour, salt, dry yeast, and water. A dough weight of 1600 lb is a common size. After mixing, the dough is dumped into troughs and stored at 80°F and 80% RH for 4.5 to 5 hr. During this stage, much hydrolysis of the starch to sugars occurs by virtue of the action of the malt enzymes and some leavening takes place as a result of the yeast fermentation.

At the end of the fermentation period, the dough is formed into loaves and transferred to the ovens without the intervention of a proofing period. The loaves are baked for 2 hr at 400°F and then depanned (Fig. 75).

The baked loaves are fragmented by shredding knives, or saws, and the pieces transferred to the secondary ovens after passing through a sizing step which removes fines. After 2 hr or more of baking at about 250°F, the pieces are broken up into small granules which are carefully sized before packaging. Fines from each stage of the operation are used in subsequent doughs.

PUFFED CEREALS

General Considerations

Puffing processes may be conveniently divided into two types: (1) atmospheric pressure procedures which rely upon the sudden application of heat to obtain the necessary rapid vaporization of water; and (2) pressure-drop processes which involve suddenly transferring superheated, moist particles into a space at lower pressure. In the latter case, the pressure-drop may be achieved by releasing the seal on a vessel containing a product which has been equilibrated with high temperature steam, or it may be se-

FIG. 75. REMOVING GRAPE NUTS LOAVES FROM PANS

cured by transferring the hot material from the atmosphere into an evacuated chamber. The former process is much more widely used.

The puffing phenomenon results from the sudden expansion of water vapor (steam) in the interstices of the granule. The particle is fixed in its expanded state by the dehydration resulting from the rapid diffusion of the water vapor out of it. Gun-puffing may result in an increase of apparent volume (bulk density decrease) of eightfold to sixteenfold for wheat and sixfold to eightfold for rice. Oven puffing causes a lesser increase, about threefold to fourfold for rice.

Puffed products must be maintained at about 3% moisture (or less) in order to have the desired crispness. Even at 5% moisture a definite toughness becomes evident (Carlin 1956). These levels are more critical and harder to maintain in foods which have been gun-puffed.

Oven-Puffed Rice

This product is prepared from whole kernels of domestic short-grain milled rice. Frequently the rice is parboiled and pearled. A batch of 1400 lb of rice is weighed into cookers such as are used in the preparation of corn or wheat flakes. About 53.5 gal. of sugar syrup with salt are added, and the mixture is cooked for 5 hr under 15 lb steam pressure. Sometimes nondiastatic malt syrup and enriching ingredients are added before cooking.

The lumps of cooked rice coming from the retorts are broken up and dried to approximately 25 to 30% moisture content in rotating louver driers. At this point, the moisture is allowed to distribute uniformly in the grain mass by storing the partially dried product in stainless steel bins for about 15 hr. Lumps form during the tempering process, and must be broken up before the rice is sent to the flaking rolls.

After the individual kernels are separated and again dried so that a moisture content of 18 to 20% is reached, they are passed under a radiant heater which brings the external layers of the rice to a temperature of about 180°F. The outside layers of the kernel are plasticized by the heat so they do not split when the grain is run through the flaking rolls.

The rolls used in preparation of oven-puffed rice are set relatively far apart so that the tremendous compression effect necessary in corn flakes manufacture is not achieved. In fact, the rolls contact only the central part of the rice kernel. The "bumped" grains are again tempered, this time for about 24 hr.

To secure the puffed effect, the cooled and tempered rice is passed through toasting ovens at 450° to 575°F. Transit time is about 30 to 45 sec.

A cereal called "Special K", manufactured by the Kellogg Co., is basically a rice kernel which is cooked, then coated while in a moistened condition with wheat gluten, wheat germ meal, dried skim milk, debittered brewers' yeast and other nutritional adjuncts. Finally the material is oven-puffed. A more complete description of the product has been given by Thompson and Raymer (1958).

Gun-Puffed Products

The manufacture of a composite cereal by the gun-puffing method will be described since it includes several concepts not previously treated in this chapter. Figure 76 gives a pictorial summary of the steps involved in making General Mills' Cheerios.

Corn cones, oat flour, and a flavor pre-mix consisting of sugar, coloring substances, flavoring compounds, etc., are combined in a screw conveyor with interrupted flights. The homogenous mixture is dumped through a rotary valve into a continuous steam-jacketed cooker. Water is added by a metering device so that the product going to the extruders is at 38 to 40% moisture content.

Auger-induced pressure extrudes strands of cooked dough around the periphery of a circular die. A knife travelling over the die surface cuts the strands into short pellets. The pellets

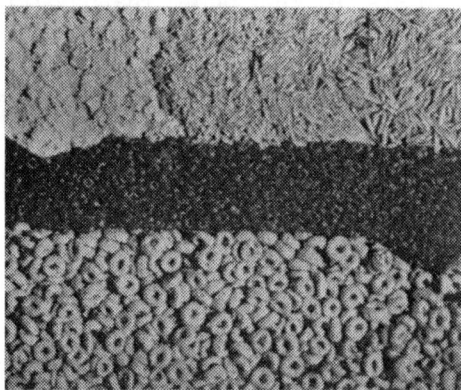

Courtesy of General Mills

FIG. 76. STEPS IN MAKING GENERAL MILLS'
CHEERIOS

may be solid or have a hollow center. The pellets go to a tumbling cooker which reduces the surface moisture and prevents the occurrence of agglomeration. The product is piled about 3 in. deep on the metal belt of a Procter and Schwarz oven.

Solid pellets at 15 to 16% moisture content may be bumped between rolls to make a disk-shaped particle with serrated edges.

The guns, so-called, are pressure vessels about 6 in. in diameter and 30 in. long. They are provided with a steam inlet, a bleed-off valve, and a means for heating the gun—usually by a gas flame. A charge of pellets at 11 to 12% moisture is dropped into the open end of the gun from a gravity chute leading from the storage bins on the floor above. The end of the gun is sealed by a trip-valve. All valves are closed, sealing the gun. If the gun is heated by gas flames, it is slowly rotated while the temperature, and consequently the pressure, is slowly built up. The temperature may reach 500° to 800°F, and the pressure at the end point may be 100 to 200 psi. This process may take 5 to 7 min.

When the pressure reaches the necessary level, the end of the gun is suddenly opened by a trigger mechanism. The contents explode (Fig. 77) into a cage or bin provided with a floor opening leading to a conveyor belt. The ejected material may be dried in a rotating heated cylinder, and then cooled. The material is then visually inspected for color and agglomeration and sent to the packing line.

Courtesy of General Mills

FIG. 77. DOUGH PIECES EXPLODING FROM AN EXPERIMENTAL PUFFING GUN

There can be almost endless variations on the preceding method. Cocoa can be added to the dough to make a chocolate flavored confection. Products may be made of any color for which there is a suitable edible dye. Shapes can be varied within a wide range. Raw materials may be almost any combination of cereals.

Some composite-dough, gun-puffed cereals are the trademarked Kix, Coco-Puffs, Jets, Trix, Cheerios, etc.

Wheat and rice kernels can be gun-puffed. Usually durum or a hard wheat is used as the raw material for the puffed wheat, and it is often pearled before puffing to reduce the amount of bran present on the finished product. This bran becomes loosened by the puffing process, and if an excessive amount is present, it may form an unsightly deposit in the bottom of the package.

Puffing by Extrusion

Many fancy shapes of puffed breakfast cereals, as well as many kinds of snacks, are being made by extruding superheated and pressurized doughs through an orifice into the atmosphere. The sudden expansion of water vapor as the excess pressure is released increases the volume several times. Apparent specific volumes can reach or exceed those attained by gun-puffing, and the process seems to have several advantages over gun-puffing.

The cereal pre-mix containing on the order of 60 to 75% expandable starch base is moisturized with water or steam. The resultant mash is compacted by a screw revolving inside a barrel which may be heated by steam. The thread of the screw has a progressively closer pitch as it approaches the discharge. The pressurizing and steam heating bring the dough to a temperature of around 300° to 350°F and a pressure of 350 to 500 psi at the die head. Under these conditions the dough is quite flexible and easily adapts to complex orifice configurations.

The die head may contain several orifices, and pieces of correct size are sliced off by revolving blades resting on the exterior die surface. Adjustment of the speed of rotation of the knife assembly controls the piece size. Figure 78 illustrates several features of a typical extrusion puffing device.

The dough pieces expand very rapidly as they leave the die orifice but the expansion may continue for a few seconds since the dough is hot and still flexible and water continues to boil off. Even so, the moisture content of 24 to 27% is too high for satisfactory stability, and the pieces are further dried on vibrating screens in hot-air ovens. Fines and agglomerates are removed at this time, and the products cooled and packaged.

According to Sanderude (1969) the following raw materials can be expanded satisfactorily in equipment described above:

Rice flour   excellent expander, white and bland tasting; colors and flavors can be added easily

Corn meal or flour   very good, expands but retains the corn flavor

Oat flour   high moisture needed to expand due to higher fat content and high temperature also needed

Wheat flour   high moisture necessary and high temperature needed

Potato flour   high moisture and temperature needed to properly puff

Tapioca flour   high temperature, moderate moisture, bland taste

Defatted soy flour   high temperature and moderate to high moisture

Full fat soybeans   3 to 5 min preconditioning with steam prior to extrusion; extruded at 250°F

Plain corn and wheat starches and acid modified starches   medium to high temperatures using either steam or water as moisturizer

Courtesy of Wenger Mfg. Co.

FIG. 78. COOKER-EXTRUDER FOR FORMING PUFFED-SHAPED CEREALS CONTINUOUSLY

As the fat content increases, there is a tendency to reduction in expansion, but the pieces become more uniform and their surface becomes smoother and brighter, while the cell size becomes smaller and more uniform. Monoglycerides seem to increase these effects. Sugars modify the flavor and texture, and may help to control shape and size of tough doughs.

Capacity of existing units range up to about 5,000 lb per hr. Figure 78 shows a small unit suitable for pilot plant or experimental work. Figure 79 is a flow diagram of a possible installation.

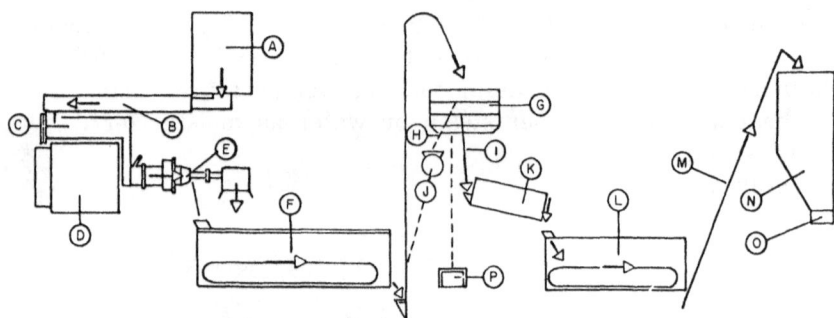

FIG. 79. FLOW DIAGRAM FOR EXPANDED CEREAL PRODUCTION

A—Circular bin discharger. B—Steam preconditioner. C—High speed liquid mixer. D—Short time/high temperature extrusion cooker. E—Variable speed knife. F—Drier. G—Two screen shaker (note). H—Overs spout to reel. I—Sized products spout to reel. J—Overs separator (crusher). K—Optional revolving reels (animal fats). L—Cooler. M—Inclined belt conveyor or pneumatic system. N—Packing bin. O—Packer. P—Cooler fan. Note: Spout fines from bottom pan or shaker to eye of cooler fan.

Alternate Puffing Methods

Processes depending upon contact of dough pieces with very hot embossed cylinders are in current use. Huber (1955) describes such a method in which roll temperatures of 350° to 800°F are used. Very likely in practice temperatures in the upper half of this range would be employed. The rolls may be heated by radiant (infrared) heat or by the circulation of high temperature fluid media inside the cylinder. In Huber's process, the dough has a moisture content of 8 to 18% going into the rolls and 6 to 7% after puffing. The dough may be preheated to temperatures below boiling: in this case lower roll temperatures are practicable.

Gates (1958) described a cereal puffing method which is somewhat analogous to gun-puffing except that the grain is steam-treated under less pressure than usual and is puffed by passing into a chamber having a considerable negative pressure with respect to the atmosphere. Gates' method also differs from gun-puffing in being continuous. It has been used to make a quick-cooking rice, but apparently there are no prepared breakfast cereals being made by this method.

Sugar-Coated Products

Spherical or disk-shaped products such as puffed wheat and rice can be coated by a method very similar to the pan-coating technique used in confectionery manufacture. The requisite apparatus somewhat resembles a cement mixer in having an open bowl rotating about an axis inclined to the horizontal (see Fig. 80). The very dry cereal particles are placed in the bowl, and, as it rotates, a molten (250°F) sugar syrup is slowly dripped upon the mass. Coconut oil may be added to decrease foaming in the sugar syrup and to promote separation of the coated particles. The tumbling action of the particles results in each of them remaining separate and being uniformly coated with a thin glaze of sugar

Courtesy of Post Cereals

FIG. 80. A COATING REEL—A DRUM IN WHICH SUGAR SYRUP IS COATED ONTO TOASTED FLAKES

TABLE 61

COMPOSITION AND NUTRITIVE VALUE OF BREAKFAST CEREALS, PER 100 GM[1]

Variety	Cal-ories	Protein Gm	Fat Gm	Total Carbo-hydrates Gm	Cal-cium Mg	Phos-phorus Mg	Iron Mg	Thiamin Mg	Riboflavin Mg	Niacin Mg	Fiber Gm	Sodium Gm
Corn flakes	372	7.50	0.300	86.2	5.3	42	1.80	0.420	0.070	2.10	0.600	1.00
Oven-crisped rice	377	5.50	0.300	88.7	24.8	117	1.80	0.390	0.035	7.10	0.400	1.00
All-bran	334	11.00	2.500	75.8	85.0	1240	10.30	0.390	0.318	17.70	8.200	1.20
40% bran flakes	359	10.20	2.000	79.8	57.0	602	4.60	0.350	0.180	8.50	3.500	1.30
Whole wheat flakes	374	8.50	1.300	83.7	37.2	382	2.84	0.885	0.106	3.90	1.700	0.80
Shredded wheat	370	10.00	1.500	80.8	48.7	410	3.50	0.288	0.100	4.40	2.300	...
Sugar coated puffed corn	378	4.00	0.200	92.3	10.6	28	1.77	0.424	0.159	2.12	0.300	0.30
Sugar coated puffed wheat	389	5.00	1.500	89.8	14.0	142	1.66	0.420	0.035	5.00	1.000	0.06

[1] Kellogg Co.

which hardens upon cooling. A stream of hot air is usually directed into the coating reel to speed drying.

Some details of the syrup preparation and its storage have been given by Massman and Rivers (1955). Other authorities have suggested a syrup formula of about 86% sucrose, 13% corn syrup, and 1% salt. Sometimes 0.01 to 0.05% sodium acetate may be added to prevent crystallization of the coating. From 25 to 60% of the weight of the finished product is due to the glaze.

BIBLIOGRAPHY

ANON. 1965. Let rolls work for you. Food. Eng. *37*, No. 2, 60–64.

BENSON, J. O., and MERBOTH, J. A. 1967. Process for making shaped cereals. US Pat. 3,332,781. July 25.

CARLIN, W. 1956. Uses I-R for fast tasty-crisping of moisture-sensitive food. Food Eng. *28*, No. 6, 72–73.

CARSON, G. 1957. Corn Flake Crusade. Rinehart and Co., New York.

CLAUSI, A. S., VOLLINK, W. L., and MICHAEL, E. W. 1967. Breakfast cereal process. US Pat. 3,318,705. May 9.

FAST, R. B. 1967. Process for preparing a coated ready-to-eat cereal product. US Pat. 3,318,706. May 9.

GATES, W. C. 1958. Puffing method and apparatus. US Pat. 2,838,401. June 10.

GUNDERSON, F. L., and BROWNLEE, H. J. 1938. Oats and oat products. Culture, botany, seed structure, milling, composition, and uses. Cereal Chem. *15*, 257–272.

HALE, D., and CARPENTER, E.J. 1956. Apparatus for manufacturing a cereal food product. US Pat. 2,743,685. May 1.

HUBER, L. J. 1955. Process of preparing a puffed cereal product and the resulting product. US Pat. 2,701,200. Feb. 1.

LILLY, E. F., and REINHART, R. R. 1967. Process for producing ready-to-eat oat cereals. US Pat. 3,345,183. Oct. 3.

MASSMAN, W. F., and RIVERS, R. W. 1955. How Post Cereals built syrup-coating efficiency—advance handling did it. Food Eng. *27*, No. 5, 70–72.

SANDERUDE, K. G. 1969. Private communication. Kansas City, Mo.

THOMPSON, J. J., and RAYMER, M. M. 1958. Production of ready-to-eat composite flaked cereal products. US Pat. 2,836,495. May 27.

CHAPTER 7

Charles M. Hoskins[1] | # Macaroni Production

INTRODUCTION

Macaroni is a generic term covering a wide variety of products sometimes termed alimentary pastes, which includes the common items of macaroni, spaghetti, and egg noodles, plus a whole range of other products of various shapes and sizes obtained by adding special ingredients or by using special forming techniques. Macaroni products are widely known and widely used. In the United States, they are one of the few products made from flour that have enjoyed an increasing per capita consumption. In the year 1939, per capita consumption was 5.1 lb; in 1947, 5.8 lb; in 1958, 5.96 lb and in 1967, 7 lb. Canada has a record of increasing per capita consumption that is greater percentage-wise than the United States over the same years. Reports from Italy indicate per capita consumption varying from 20 to 60 lb per person per year, depending on the part of the country. Consumption of the various macaroni shapes in the United States is approximately as follows: 40% long goods, 35% short cuts such as elbows, shells, etc., 20% noodles, and 5% specialties such as bow ties, rigatoni, tufoli, and mafalda.

In the United States, the Food and Drug Administration has published a Standard of Identity (Anon. 1966) for macaroni products which establishes ingredients and labeling requirements for the products. These Standards represent practices which are more or less common in other countries of the world as well as in the United States. According to the Standards, the basic raw material will be semolina, durum flour, farina, flour or any combination of two or more of these with water. Permitted optional ingredients are egg white solids from 0.5 to 2.0% of the weight of the finished food, disodium phosphate, onions, celery, garlic, bay leaf, salt or other seasonings. Gum gluten can be added in such quantities that the protein content of the finished food is not more than 13% by weight. Up to 2% glyceryl monostearate can be added. Egg white, gum gluten, and glyceryl monostearate are

---

[1] CHARLES M. HOSKINS is a consultant and marketing representative for De Francisci Machine Corp.

usually used to prevent disintegration and sticking together of canned products. Whole wheat macaroni, soy macaroni, and vegetable macaroni products are made in small quantities and consumed largely by the health food trade. Whole milk macaroni is permitted by the Standards, but rarely made because of possible rancidity of butterfat and poor texture of cooked product. Macaroni made with nonfat milk and with carrageenan added is permitted by the Standards and is an improvement over whole milk macaroni in cooked texture and keeping quality.

Eggs can be added to macaroni products in which case they become either egg macaroni or noodles. Noodles are defined as the product which is formed in ribbon shape and which contains not less than 5.5% by weight of the solids of egg or egg yolk as a percentage of the total solids of the noodle product.

Enriched macaroni and noodles are in widespread use. Enrichment is obtained by adding to each pound not less than 4 mg and not more than 5 mg of thiamine, not less than 1.7 mg and not more than 2.2 mg of riboflavin, not less than 27 mg and not more than 34 mg of niacin or niacinamide, and not less than 13 mg and not more than 16.5 mg of iron. Enrichment can be added through synthetic mixtures, yeast, or vital gluten. The required amounts of thiamine, riboflavin, and niacin are higher than the corresponding requirements for bread flour to compensate for loss in the discarded cooking water. Macaroni products are made in a great variety of shapes. Some manufacturers regularly manufacture more than 70 different shapes which obtain their value from providing variety and interest to meals.

There are so many shapes and sizes that no official standard dimensions are available for most. However, the Standard of Identity makes certain definitions that are of interest. Macaroni, according to the standard, is the macaroni product, the units of which are tube-shaped, hollow, and more than 0.11 in., but not more than 0.27 in. in diameter. Spaghetti is the macaroni product, the units of which are cord-shaped (not tubular) and more than 0.06 in., but not more than 0.11 in. in diameter. Vermicelli is the macaroni product, the units of which are cord-shaped (not tubular) and not more than 0.06 in. in diameter.

RAW MATERIALS

The ideal raw material for making macaroni products should lend itself to easy processing on macaroni presses and in driers to yield a smooth and mechanically strong product of uniform color. When the product is cooked in boiling water, it should maintain its shape without falling apart or splitting and should cook to a firm consistency free from a slimy, sticky surface film. The cooking water should be relatively free of starch and the product should be resistant to disintegration due to overcooking. It is generally believed that durum products, and particularly durum semolina, come closest to satisfying these requirements. This opinion is confirmed by the fact that most macaroni manufacturers will use durum when it is available and that the per capita consumption of macaroni decreased in the United States when durum crops were destroyed by stem rust and increased as soon as durum was again available in large quantities. Because macaroni products made from durum have a characteristic yellow color, this color has become associated with good quality and much effort in plant breeding and manufacturing procedure has been directed toward increasing the uniformity and intensity of the yellow color of macaroni.

Durum wheats were originally grown in Russia in a cold, dry climate. They were imported to the United States by Carleton, a US Dept. of Agr. scientist who has been called the father of durum breeding and improvement, and for many years the two varieties Kubanka and Mindum were widely used. The varieties Carleton and Stewart were developed from these varieties to resist the prevalent races of stem rust and were in use for many years. However, in 1954 the new race 15B of stem rust, which had begun its depredations in 1950, practically destroyed the durum crop. All-out effort on the development and distribution of rust-resistant varieties, enabled the growers to plant a full crop of durum resistant to the 15B variety of rust in 1957. The new varieties were chiefly Langdon and Ramsey although some Towner and Yuma were also planted. During the period when durum was not available, blends of durum and hard red wheat in percentages from 50 to 100% hard wheat were used to make macaroni. In 1960, the improved varieties, Lakota and Wells, replaced the varieties developed specifically to overcome the rust. In 1966, Leeds was planted and by 1968 new varieties were being developed to get back to the large kernels characteristic of the pre-rust varieties.

The durum wheat kernel is very hard and both the endosperm and the individual starch kernels are translucent. It is high in carotenoid pigments, particularly xanthophyll and taraxanthin.

This hardness and translucency is dependent on the durum being grown in a dry cool climate as in the Dakotas and in western Minnesota.

In commercially milled wheat flours, the percentage of starch granules damaged mechanically during milling to flour varies directly with the hardness of the grain. In soft wheat, 1 to 2% of the starch granules are damaged; in hard wheat flour, from 3 to 4%, while in durum flour, from 6 to 8% are damaged. The starch of durum wheat is more subject to amylase attack than the starch of common wheats. This may be due to the fact that starch granules are more damaged during milling than is the case with common wheats. The swelling capacity of durum starch is greater than that of hard red spring wheat. The sugar content of durum flours is somewhat higher than that of other wheat flours.

The gluten of durum wheat has different characteristics from those of bread wheat. In the dry state, this results in a very hard endosperm which is much harder than the common hard wheats. However, when a dough is made from durum semolina or flour, it is not so tough or elastic as dough made from hard wheat. The durum dough will extrude through a small hole at lower pressure than hard wheat dough. When durum is made into bread, the loaf volume is much less than the loaf volume with hard wheat flours. When durum gluten balls are dried in an air oven for testing the per cent gluten in flour, the volume will be much less than the volume of the gluten ball of hard wheat dried in this same oven. Hard wheat flour and farina will make an acceptable macaroni product, but the color is not so yellow as the color of the durum product; the product is not so resistant to overcooking, and it does not have so desirable a taste as durum. However, if it is cooked to precisely the correct consistency and served with a good sauce, it is a good food.

Durum semolina, durum granular, and durum flour are the three general classes of durum products used for making macaroni. Durum semolina is the purified middlings of durum wheat ground so that all of the product passes a No. 20 US sieve and not more than 3% passes through a No. 100 US sieve. The durum granular product is a semolina to which flour has been added so that about 7% passes through the No. 100 sieve. Flour is a product all of which passes through a No. 100 sieve.

Flour makes a dry macaroni which is mechanically very resistant to breakage, smooth, and of a clear yellow color. The dry semolina product is not quite so strong mechanically and not quite so uniform in color. It can be identified by the presence of a small number of bran specks which are visible to the naked eye. The semolina product takes longer to cook and is more resistant to overcooking than the flour product and causes less cloudiness in the cooking water. Durum granular and blends of semolina and flour have properties intermediate between flour and semolina. Granular and semolina are somewhat more desirable for use in a macaroni plant because they flow from bins into the continuous press more evenly and with less trouble from bridging in bin outlets. The water absorption of flour is greater than that of semolina so that flour products require more drying time than semolina products. There is more slippage in the extrusion screw of continuous presses when flour is used so that production is decreased by the use of flour.

Effect of Growing Conditions on the Raw Material

Growing conditions to which the crop is subjected can have an important influence on the macaroni-making qualities of the flour made from that wheat. While it is not necessarily true that the amount of protein in a wheat is a direct measure of the quality of the resultant macaroni, still the amount of protein and its quality are vitally important.

Proteins are the principal nitrogen-containing compounds of the wheat kernel, and consequently, of the flour. The primary factor causing the difference in protein content is the difference in environment affecting the nitrogen nutrition of the wheat plant. Any factors of soil, general climate or season which limit the amount of nitrogen available to the plant during the grain formation and maturation period reduce the protein content of the grain. Nitrogen utilized by the plant before heading and blossoming is reflected in total yield. It used to be thought that the wheat plant took up nitrogen early in its growth and that subsequent development depended on this supply. It is now known that the plant is actively functioning and that nitrogen taken up even after vegetative requirements are met is deposited in the grain itself.

The amount and timing of rainfall influences nitrogen available to the plant. If the rainfall is abundant in the early stages of vegetation and inadequate later on, there could be a deficiency in the protein content of the wheat kernel. Excessive rain at any time could result in leaching of nitrates from the soil. The location of nitrates in the soil is a factor in protein content. The time and quantity of rainfall can remove the accumulated nitrate supply to a point in the soil where it cannot be reached by plant roots.

Effect of Blight Damage

Blight and related forms of damage can have an influence on the quality of semolina and the macaroni made therefrom. Blight is particularly prevalent in wet, cloudy harvest seasons. Experiments have been performed in which carefully prepared experimental blends containing graduated proportions of light and heavily damaged kernels were milled and the resulting semolina processed into macaroni. It was found that 10% of lightly damaged kernels with discoloration evident only at the tip was without detrimental effect, while 25% did not greatly lower the color or increase semolina speckiness. Over 50% by weight would be extremely bad to use in the mill mix. The influence of heavily damaged kernels with visible injury in the crease and other portions of the kernel was more marked. Even 5% of such grains significantly increased the number of specks in the semolina and decreased macaroni color, while 10% was very detrimental.

Sprout Damage

Durum wheat is grown in rolling country where the entire field does not always become ripe at the same time. To correct this, it is cut in swaths which are laid on the ground until drying and ripening are complete. During this period when the durum is laying on the ground it is especially susceptible to damage by rain which causes sprouting. When the wheat germinates, alpha amylase is released which attacks the starch reducing the length of starch chains down to dextrins and sugars.

Some experimental work has been done on sprouted wheat to determine the effect on macaroni-making qualities. In 1 experiment, samples of sound, hard amber durum wheat were sprouted under approximately uniform conditions for varying lengths of time to obtain 3 distinct stages. These three "stages" were defined by length of sprout obtained. Each of the stages was then blended in various proportions by weight with the original sound wheat to obtain mixes for experimental milling.

Sprouting apparently had no effect on the ease of milling, but properties of the dough during macaroni processing were affected. Those made from blends containing a high percentage of badly sprouted wheat were crumbly and "short," but after the customary amount of kneading in batch equipment did have normal consistency. Semolina yield was reduced when more than 20% of sprouted wheat was included in the blend. Diastatic activity of the semolina was greatly influenced by the proportion of sprouted wheat in the blend and by degree of sprout, while absorption was generally lowered by sprout damage.

Macaroni color was markedly decreased by increased sprouting and there was a highly significant negative relationship between diastatic activity and color. Ten percent blends of the second and third sprouted stages had more effect on both these properties than 100% of stage one. Five percent of heavy damage reduced the color score 40%. Semolina from wheat at the first stage of sprouting noticeably affected color at 20% concentration. It appeared from the data secured in the study that the length of the sprout is more important than the amount of sprouted kernels present.

The first large crop of durum grown from the varieties resistant to 15B stem rust was harvested in 1957. Continuous wet weather toward the end of the crop year resulted in large quantities of sprouted wheat. Most of the wheat with a large mixture of sprout in it was made into flour because the yield of this wheat in granular and semolina was too low. This flour made the operation of the continuous mixers very difficult because very slight changes in moisture content would cause the dough to become sticky and form large lumps which would not feed into the extrusion screw properly. Noodles made from the sprouted wheat were so sticky that difficulty was experienced in drying due to the formation of lumps. Because the dough was very sensitive to temperature changes, the extrusion rates in continuous long good spreaders varied considerably across the length of the die, so that the extrusion pattern was uneven. Before the sprouted wheat came into use, a spreader could be made to yield a stick containing no short strands after 10% to 12% trimming. With the sprouted wheat flour, the amount of trim was increased to as much as 35%. These trimmings are returned to the mixer and reprocessed so that the production of the press is seriously reduced by the sprouted wheat. The difficulty may have been caused by the softening of dough due to diastatic activity, which is enhanced by heat, or it may have been due to the general softening of the dough which heating produces even in the absence of diastatic activity.

Eggs for Noodles

The only other major raw material besides flour or semolina used in macaroni products is eggs. United States Standards of Identity require that anything called noodles or egg spaghetti, egg macaroni, etc., must contain 5.5% of the solids of eggs as a percentage of the total solids in the finished product (Anon. 1967). These egg solids can be put into the product by the addition of frozen yolks, dried yolks, frozen whole eggs or yolks. However, it is the practice in the United States to use dark-colored yolks with an NEPA color score of 4.0 to 5.0 or a carotenoid pigment color of as high as 75 or 80 ppm. Frozen yolks are most commonly used for this purpose although a number of manufacturers are using spray-dried yolks.

Between 1945 and 1958, it became increasingly more difficult to obtain an adequate supply of dark-colored yolks. A tendency was noted in the United States for poultry men to grow larger flocks of chickens and to keep them inside buildings on prepared feeds. Part of the reason for this, besides the productivity factor, was that light-colored yolks were preferred for table use, by far the largest market for egg products. Dark color in eggs is obtained when chickens are fed outdoors on natural feed. It appears to be related to the consumption of large amounts of plant pigments. This meant that the dark-colored yolks were available only in the springtime when flocks were normally turned outside, and a diminishing number of farmers were following this practice.

Methods of spray-drying egg yolks have improved greatly in the period since World War II. Both taste and color retention have been good and this has resulted in a more widespread usage of spray-dried yolks for making noodles. One of the advantages of dried yolks is that they can be measured accurately by weight into a dry blender for blending with the correct amount of noodle flour. There is also the advantage of less bulky storage and the elimination of the need for refrigeration.

Fresh refrigerated, but not frozen, eggs are used in bulk in some large factories. The eggs are delivered in a tank truck to a sanitary stainless steel refrigerated tank and fed directly from the tank into the process.

Whole eggs are infrequently used by the manufacturer of noodle products. Despite the fact that egg white adds something to the strength of the product and its ability to withstand overcooking, the dilution of the natural yolk color by the white solids results in a poorer finished color of the noodle. At times, though, cost considerations encourage the use of whole eggs.

THE PRODUCTION PROCESS

Basically, the production process for macaroni products (without eggs) consists of adding water to flour or semolina in such quantity as to produce a mixture of 31% moisture, mixing these ingredients together for a short period of time, kneading the dough to obtain a plastic, homogeneous mass and then extruding the mass through dies under pressure in such a way that the product comes out in the shapes which are normally seen on store shelves. After extrusion, products are dried, packaged and sold.

Until about 1940, most macaroni products were made by a batch process. The semolina and water were weighed and combined in a mixer of about 300 lb. capacity. The mixer was operated for approximately ten minutes and then dumped into a kneader or gramola. The loose dough was compacted in the kneader by subjecting it to heavy corrugated rollers which bore down on the dough as it passed under the rollers in a rotating pan. The thick ribbon of dough was alternately turned up on edge by a plow and crushed down by the corrugated rollers. Slabs of plastic dough were cut from the kneader and placed in the chamber of a hydraulic press. Pressure was then brought to bear on the dough to force it through dies at the bottom end of the chamber. Pressures of 1500 to 3000 psi on the dough were used.

More recently, the functions of mixing, kneading, and extruding have been combined in the continuous screw press. The continuous press is now widely used throughout the world and only smaller isolated plants still use the batch process. The continuous press is normally equipped with volumetric feeders which provide a continuous flow of semolina and water to the press at rates of 200 to 5000 lb of flour per hour. The continuous mixers are equipped with horizontal shafts and blades that move the product slowly forward while mixing the dough. The dough is approximately 31% moisture and is rather dry, so that it remains in the form of small balls from ¼ to 1 in. in diameter and does not form a continuous smooth mass in the mixer as in a bread dough (see Fig. 81). At the end of the mixer, the dough drops into a specially designed auger which is in a tightly sealed cast housing. The auger moves the dough forward and at the same time compacts it, building up pressure and kneading the dough simultaneously.

Courtesy of DeFrancisci Machine Corp.

FIG. 81. VIEW INTO TWIN SHAFT DOUGH MIXER
SHOWING LOOSE CONSISTENCY OF DOUGH FORMED
AS SMALL BALLS

Vacuum tight cover is raised to show mixer action.
Trimmings from long spaghetti return through airlock at left. Flour is fed through airlock at right.

Operation of Auger

In an auger extruder flow occurs through a channel of approximately rectangular cross section. Two sides of the rectangle are
formed by the leading and trailing surface of the auger flight.
The bottom of the rectangle is the root of the auger while the top
is the inside surface of the barrel or cylinder in which the auger
revolves. The visualization of auger action is simplified if the cylinder wall is considered to be moving and the auger stationary.

The force causing the dough to move down the channel is the
component of the velocity of the cylinder wall parallel to the
channel (V_x). This causes a flow which varies linearly from 0 at
the root of the auger to V_x at the cylinder wall (see Fig. 82).
The rate of shear is the same from the top to the bottom of the
channel. However, the material near the shaft of the auger moves

much more slowly than the material near the cylinder wall and, therefore, remains in the auger for a long period of time and is greatly overworked.

If there is a restriction on the outlet such as a macaroni die, the pressure back along the channel will cause a counter flow against the "drag flow" caused by the forward motion of the cylinder wall. This is called pressure flow and is a maximum halfway up the channel and is 0 at the cylinder wall and the root of the channel. The action of the screw is greatly influenced by the ratio of pressure flow to drag flow, (a). For no back pressure (a) equals 0 and for a completely stopped off flow (a) equals 1. The distribution of flow velocity due to drag flow, pressure flow and combined flows are shown in Fig. 82 for (a) $= \frac{2}{3}$.

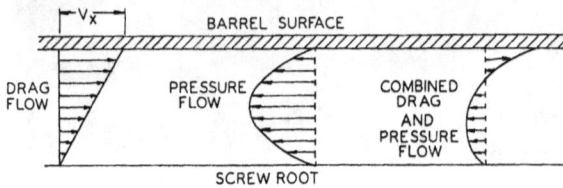

FIG. 82. VELOCITY PROFILE OF FLOW OF DOUGH DOWN CHANNEL FORMED BY BARREL, SCREW FLIGHTS AND SCREW ROOT

Drag flow is caused by movement of barrel relative to screw. Pressure flow is caused by back pressure of die and other obstructions. V_x is component of barrel velocity parallel to channel.

The component of the cylinder wall motion perpendicular to the flight causes dough to flow back toward the leading edge of the flight next to the cylinder wall and to flow away from the flight near the root of the screw. This causes the dough to spiral down the channel. This is the principal mechanism which causes mixing and kneading of the dough. The number of turns which the dough makes is proportional to the time the dough stays in the auger for a given auger design and rpm. For example, assume that the dough stays in the auger 1 min with no back pressure, (a) $= 0$, and makes 5 spiral circuits of the channel. If the back pressure is such that (a) $= 0.5$, the dough will stay in the auger 2 min and the dough will make 10 circuits of the cross section of the channel. Mixing will be doubled.

For an excellent and thorough mathematical treatment of the

action of an auger, see *Processing of Thermoplastic Materials* by Ernest C. Bernhardt.

Macaroni press augers are designed with deep flights which results in a large backward pressure flow. The ratio between pressure and drag flow in a macaroni press screw (a) is normally about 0.6. Thus, in a 1000 lb per hr press the total drag flow would be 2500 lb, the pressure flow would be 1500 lb and the net flow would be 1000 lb.

The result of this complex flow through the screw is that in the screw itself the dough in contact with the cylinder, the front and back surfaces of the flights and the root of the screw is worked more than the dough at the center of the channel. When the dough comes off the end of the screw, the cylinder of overworked dough around the shaft is reduced to a small rope of dough. The overworked dough against the cylinder wall is around the outside of the mass of dough and the material against the flights is a spiral screw shape in the mass of dough. In further flow the material against the flights is dispersed quite well throughout the mass of dough, but the material at the center of the cylinder of dough and at the outside keeps its integrity until it goes through the extrusion dies. This dough is softer than the less worked, cooler dough and it, therefore, extrudes at a higher rate causing an uneven extrusion pattern which must be trimmed off to length for long spaghetti and which results in uneven lengths of short cut product such as elbow macaroni. In a round die this generally results in higher extrusion rates around the outside of the die. The uneven extrusion rates from a long, narrow spaghetti die result in longer than normal strands at the two ends and center of the die and at the front and back edge of the die.

Superimposed on the pattern caused by the screw is a certain amount of heating due to flow through the distributing tubes leading to the long narrow die used in a long spaghetti extruder. The flow through these tubes is parabolic. That is, there is a zero flow at the surface of the metal and this increases parabolically to a maximum at the center of the tube. The highest rate of shear is at the surface of the tube and the lowest rate of shear is in the center of the tube. The longest amount of dwell time is at the surface of the tube and the shortest amount of dwell time is in the fast flowing dough at the center of the tube. Therefore, the material against the metal walls is worked more than the material at the center and this is readily apparent in the extrusion pattern on the long goods press.

Heat Generation in Extruder

A 1000 lb per hour press uses a 15 hp motor. Perhaps 90% of this 15 hp is converted into heat of friction in the dough. This heat is created in the auger, in the distributing tubes, and in the very high rates of shear obtained when the dough is extruded through holes in the die. The following calculation shows the temperature rise which would be obtained if no heat were taken from the dough by the cooling jacket and also shows the estimated amount of heat given up by the dough to the water in the cooling jacket, through the walls of the distributing tubes to the surrounding air, and through other exposed surfaces of the press and die.

Calculation of Frictional Heating of Dough in a Macaroni Press.—

Heat equivalent of horsepower $= 15 \text{ hp} \times 0.9 \times 2545 = 34,357$ Btu hr

$$Q = CW\Delta T \qquad \text{and} \qquad \Delta T = \frac{Q}{CW}$$

where

Q = heat absorbed by dough (Btu)

C = specific heat of dough at 31% moisture = 0.435 Btu per lb per °F

W = weight of dough per hour = 1000 lb at 12% moisture
$\qquad\qquad\qquad\qquad\qquad\qquad$ = 1255 lb at 31% moisture.

T = temperature rise of dough (°F)

$T = \dfrac{34,357}{0.435 \times 1255} = 63°F$

If dough enters the screw at 90°F, the leaving temperature would average 153°F. Actually it leaves at about 110°F so that about ⅔ of the heat is removed by the water in the cooling jacket and other heat losses. It should be kept in mind that most of this heat is generated near the metal surfaces so that local heating may result in temperature much higher than the average temperature.

Heating at the surface of spaghetti being extruded from a die sometimes causes the surface to be damaged to the extent that it peels off and forms a small collar around the spaghetti at the die opening. This collar breaks off periodically causing rings around strands of spaghetti which may be spaced as closely as 6 in. apart. This difficulty usually occurs in hot weather when all the temperatures of the press and dough are somewhat above normal.

Overworking of the dough has some negative aspects but it should be kept in mind that a certain amount of shear and working of the dough is necessary to achieve a homogenous dough which will give a uniform texture and color to the finished macaroni product. In fact, many presses have a kneader plate at the end of the screw which consists of a metal plate perforated with holes about ¼ in. in diameter with a finely perforated sheet metal plate on its upstream side. This breaks the flow into very small streams and recombines it to work out any inequalities in the dough and to filter out chunks of dry dough or extraneous matter so it will not plug the die.

Vacuum

In most macaroni presses vacuum is applied to the dough at some point in the process to remove air. This creates a very uniform and translucent dough which enhances the yellow color in the finished dry product.

Previous to the application of vacuum, the air incorporated in the dough in the mixer was dissolved in the dough by the pressure exerted by the screw. When the product was extruded, the air formed bubbles in the dough, especially near the surface where shear rates were high. These bubbles dispersed the light impinging on the product causing a white color which obscured the deep yellow of the durum semolina (Fifield *et al.* 1937).

Controlling Feeding and Mixing

The aim in the feeding and mixing functions of the press is to deliver to the auger a continuous flow of dough containing a constant percentage of moisture—about 31% on an as is basis. The control of the feed of the flour and water is not a simple matter. One of the factors that enters into this control is the design of the hopper above the press used to hold a reservoir of flour or semolina for the press. Semolina, being a coarse material, somewhat similar in flowing characteristics to granular sugar, flows rather easily and does not give too much of a problem. Flour, on the other hand, is very much inclined to bridge across an opening and the result is that there are interruptions to flow and occasional flushing. Using flour there is a varying head pressure on the flour feeder which can result in uneven feed.

The design of the hopper over the press is important. Sides of the coned portion of the feed hopper should be as steep as possible. One side should be straight up and down and the others should be sloped at not less than 60° to 70° from the horizontal. Hoppers are normally equipped with vibrators or impactors to keep the flour or semolina flowing smoothly. The discharge of the hopper normally is offset from the press feed inlet itself so that the direct pressure of the material does not bear on the feeder. This results in less compacting of the material and a more even rate of feed. The most common type of flour feeder is a simple volumetric device consisting of a short belt conveyor about 6 in. wide by 12 in. long passing under the feed inlet. An adjustable gate combined with the constant speed of the conveyor belt controls the volume of the flour or semolina being fed into the mixer. In conjunction with the volumetric feeder for flour, the water feed is taken care of by providing a constant head of water in a hopper equipped with a standpipe and a valve to set the rate of water feed.

Several manufacturers have employed feeders using a gravimetric principle. In these feeders a stream of material is fed onto a small continuous belt. The belt is mounted on a scale so that this scale can sense the amount of material on the belt at any particular time. The weight of the material on the belt operates to control the feed gates so that a uniform rate of flow is obtained.

Another method which is used in Europe to some extent uses small scales for both water and flour or semolina. The scales are set for a certain amount of material and come up to weight to be dumped on a timed cycle.

Control of the dough mix is kept within surprisingly close limits by both volumetric and gravimetric feeders. Actual press control is obtained by the operator by watching the ammeter attached to the auger drive and by regulating the size of dough balls in the mixer.

Size of dough balls is important. When flour or semolina and water are fed into the press, they gradually mix together as they are agitated and worked forward down the mixer. As they mix, the flour particles start to cling together forming lumps. Size of lumps is dependent on raw material used, type of mixer, amount of water, and rate of feed. The press must be controlled in such a way as to get the most possible water in the mix (for color), but not so much that large lumps result and interfere with the product feeding into the auger. Too high moisture will cause the wet extruded spaghetti to stretch on the stick and will cause noodles and short cut products to stick together and form lumps. Dough balls ½ to 1 in. in diameter seem to be about the right size for most presses.

Addition of Eggs

Egg yolks are combined with water in a mixing tank in the manufacture of noodles. Most often an attempt is made to combine just the correct amount of yolks with water so that the resultant mixture fed to the press will provide the right dough consistency and the correct amount of egg solids to give the correct solids content in the finished product. This results in a rather difficult feeding problem, especially where the flow is regulated by a valve or orifice. The normal practice is to use a constant head tank with a standpipe which discharges through a line having a valve to control the rate of feed. This valve is subject to a buildup of sediment from the egg solids and, therefore, some noodle manufacturers have substituted a constant displacement piston pump which assures a constant volumetric feed of the egg-water mix.

The egg ingredients in noodles cost nearly as much as the flour, even though they represent only 5.5% of the total solids in the product. Consequently, even a small deviation from the correct rate of feed of eggs can result in a substantial difference in the cost of the finished product. Being on the short side can bring danger of fines or confiscation of the product by regulatory agencies of the Federal government.

Dry blenders, such as ribbon blenders, have come into use in the manufacture of noodles and egg macaroni products because it is possible to weigh the dry ingredients into the blenders accurately and obtain the correct final solids content. Where dry blenders are used, there is little objection to the use of a volumetric feeder for flour-egg mix since the finished product will contain the correct solids content regardless of the amount of water used for moistening the dough.

Calculation of Rate of Liquid Addition

Flour or semolina normally contain about 86% solids, as used, and the moisture content of the product leaving the press varies 1 or 2% either side of 31% moisture on an as is basis. This means that for every 100 lb of flour used it is necessery to add approximately 24.8 lb of extra water to get the required dough consistency plus a small amount to replace evaporation while mixing. The addition of egg solids to flour to make noodles requires:

11.12 lb of egg yolks at 45% solids
19.25 lb of whole eggs at 26% solids
5.27 lb of dried yolks at 95% solids, or
5.27 lb of dried whole eggs at 85% solids

Table 62 indicates the pounds of eggs required to obtain 5.5% solids egg content in noodles with a varying egg solids content in the yolks and moisture content in the flour. The normal practice is to mix a 30-lb tin of frozen egg yolks (45% solids) with 60 to 65 lb of water to obtain the correct egg solids content in the finished product. The reason for the variation in the amount of water is that there is a variation in the final moisture of the product as it leaves the press due to the type of press used and the type of flour or semolina used.

TABLE 62

EGGS PER 100 LB FLOUR TO OBTAIN 5.5% EGG SOLIDS IN NOODLES

	Moisture Content of Flour		
Egg Solid Content	13	% 14	14.5
% 43	11.78	11.64	11.57
44	11.51	11.38	11.31
45	11.25	11.12	11.06
46	11.01	10.88	10.82

Where dried yolks are added to water in the egg dosing operation, an attempt is made to obtain approximately 15% egg yolk solids in the egg-water mixture being fed into the press. In order to get the same mixture as would be obtained with 30 lb of frozen yolks added to 60 lb of water, it would be necessary to add 14.2 lb of dried yolks to 75.8 lb of water.

Frozen or dried egg whites are sometimes used to improve the resistance to overcooking in the finished product. One of the principal uses in this connection is for canning where the product is subjected to long periods of blanching and retorting. Standards of Identity permit the addition of 0.5% to 2.0% of the solids of egg whites as a percentage of the weight of the finished food. Table 63 indicates the pounds of water to be added to a 30-lb can of liquid whites (12.5% solids) fed into a continuous press.

NOODLE PRODUCTION

Noodle production methods differ somewhat from other macaroni products. The problems involved in feeding the eggs required

TABLE 63
POUNDS WATER TO BE ADDED TO 30-LB TIN OF EGG WHITES[1]

Moisture in Goods Leaving the Press	Egg White Solids in Finished Product			
	%			
	0.5	1.0	1.5	2.0
% 30	152.0	63.2	33.6	18.7
32	181.0	77.5	43.0	25.9

[1] Assumes moisture content of 11 % in the finished product.

for noodles are described above. The principal difference in manufacturing, however, is that noodles are a flat product lending themselves to production from a sheet of dough.

A few manufacturers extrude noodles through a die in much the same manner in which short cuts are made. The majority, however, employ continuous processing equipment which forms a sheet of dough about 20 in. wide and varying from 0.070 to 0.125 in. in thickness. This sheet of dough is then fed into a cutter which consists of calibrating rolls which reduce it to the required final thickness, cutting rolls which cut the sheet lengthwise to the required width and cutters which cut the strips of noodle dough to the required length.

Basic manufacturing functions with noodles are very much the same as with macaroni in that the ingredients are combined (under the old method) in a mixer for 10- to 15-min mixing. The mixed products are then dumped into a kneader, or gramola, where the dough is compacted. At the end of this operation, the procedure differs somewhat in that the chunks of dough are fed into a machine called a "dough break." The slabs of dough are fed back and forth through rollers which are moved closer and closer together, gradually reducing the thickness of the dough sheet. The sheet is folded double on several passes to laminate the product. When a certain thickness has been reached, the dough is wound on a spindle for later feeding to the cutter.

The first step toward making this operation continuous was a continuous noodle sheeter made by the Clermont Machine Company of Brooklyn, New York. This unit performed the same basic functions as the batch process, but used a continuous mixer which fed dough into a pair of rolls which formed a continuous sheet. This relatively thick sheet was folded back and forth on itself, reduced in thickness through a pair of rolls, folded again and put through a third pair of rolls and then fed continuously through a noodle cutter (see Fig. 83). This resulted in a laminated sheet

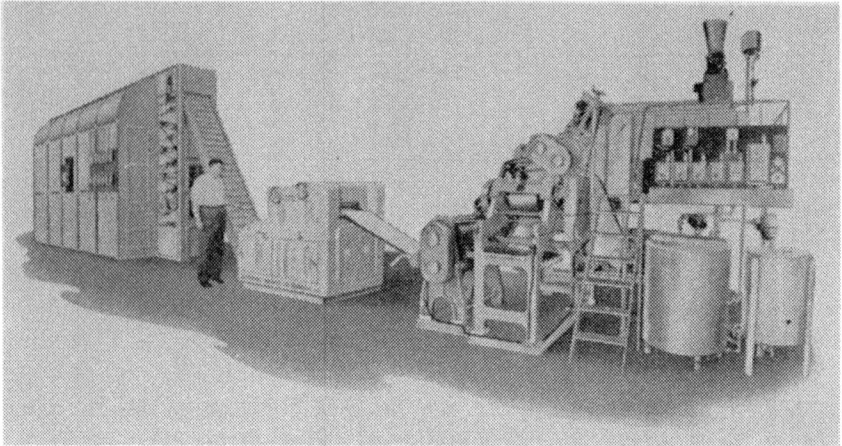

Courtesy of Clermont Machine Co.

FIG. 83. LAMINATED ROLL TYPE SHEET FORMER, NOODLE CUTTER AND PRE-LIMINARY DRIER

which was porous and easy to dry. The resulting product cooked through quickly because moisture could penetrate more easily than into an extruded product.

More recent developments use the standard macaroni press extruding a sheet through a circular or rectangular slotted die, forming a sheet which is fed automatically into a cutter (Fig. 84). The press exerts more pressure on the dough than is the case with the old dough break, or the continuous sheeter method, so that the product made on the press has a more translucent, deeper yellow color.

The batch process and the Clermont Sheeter both fold dough sheets in such a way that a large number of small air bubbles are entrapped in the finished product. This causes the product to have a whiter, more opaque appearance than when the product is made on a press, especially under vacuum. Color improvement of the extruded sheet can be obtained by using Teflon or other plastic liners in the dough slot so that the surface of the dough sheet is very smooth.

Bologna Style.—Many manufacturers, especially those catering to the Italian trade, make what is called "Bologna style." This classification would include such items as bow ties. Such shapes are formed out of a dough sheet made either continuously or by the batch process and fed into a stamping machine. These machines are somewhat like cookie cutters except that they have

Courtesy of Braibanti and Co.

FIG. 84. PRESS FEEDING SHEET OF NOODLE DOUGH DIRECT TO CUTTER

attachments to form special shapes and to do such functions as crimping the center of the bow tie.

Twisted Goods.—Spaghetti and vermicelli are sometimes sold in "biscuits," twisted clumps, which are dried on trays. These twisted products were formerly formed by cutting a handful of vermicelli or spaghetti in 8- or 10-in. lengths and putting this product in the form of a figure 8 on a tray. When the product dried, the "biscuit" held together and was packaged in this attractive form.

Automatic "twist" machines use an arrangement similar to a long goods spreader to extrude the product in clumps of the required number of strands to make the proper sized "biscuits." The clumps of strands are twisted into a figure eight after the product has been cut to length. These "biscuits" of twisted vermicelli are automatically placed in rows on drying trays with screen bottoms and then placed in dryers.

Nidis.—In Europe "Nidis" have become popular. These are birds' nests of noodles. The noodles are extruded in a cluster and hang down into a hollow metal cylinder. An air jet comes in tangentially near the top of the cylinder. When the noodle is cut off near the die, the air jet whirls it into the shape of the birds' nest which drops onto a drying tray. This has the advantage that very thin and very long noodles, up to 14 in., can be packaged without breaking.

DIES

The manufacture of macaroni and noodle products is essentially an extrusion function. The dough is prepared by the mixer, kneaded in the auger from the mixer to the die chamber, and then forced through the die under high pressure. The die performs

Courtesy of D. Maldari and Sons

FIG. 85. OUTLET SIDE OF SHORT CUT ELBOW MACARONI DIE SHOWING PINS IN HOLES TO FORM HOLLOW TUBE

the function of forming the dough into the characteristic, familiar shapes. Dies are normally made from bronze. However, stainless steel has been used and other materials and alloys are substituted from time to time, as well as a stainless steel frame with brass inserts which can be removed and replaced. Dies are about 1.5 to 2.5 in. thick. They are made in both circular and rectangular form, depending on whether they are to be used in presses for producing short goods (circular, Fig. 85) or used for an automatic spreader for long goods or noodle sheeting (rectangular, Fig. 86).

The simplest extrusion form, of course, is the familiar spaghetti strand. On the inlet side of the die the spaghetti hole is about $\frac{3}{16}$-in. diam. It tapers down to about the diameter of the spaghetti and the small diameter is maintained for only a short dis-

Courtesy of D. Maldari and Sons

FIG. 86. INLET SIDE OF RECTANGULAR DIE FOR LONG MACARONI SHOWING LARGE SIZE OF INLET HOLES TO ACCOMMODATE FINS TO SUPPORT PINS

tance called the gauging thickness. The tubular forms, such as macaroni, require a pin in the center as described in Fig. 87. Curved elbow macaroni is formed by drilling one side of the die hole deeper than the other so that the dough will flow faster on that side. In large elbow macaroni with thick pins a notch can be

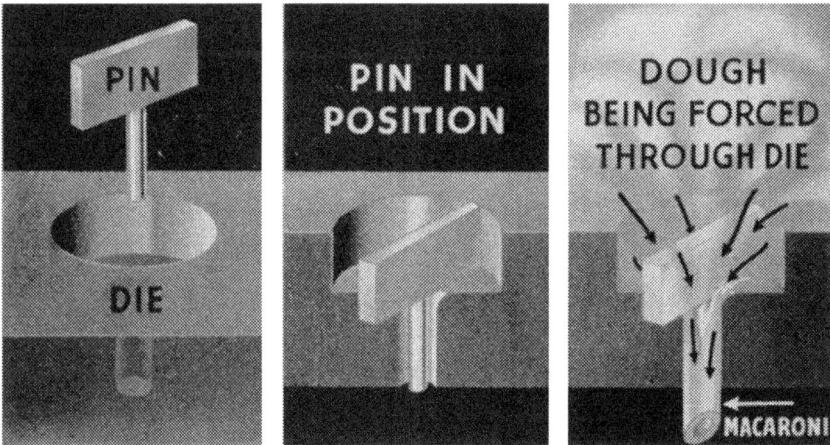

Courtesy of Glenn G. Hoskins Co.

FIG. 87. HOW DO YOU PUT THE HOLE IN MACARONI

(Left) Only two main parts shape the macaroni. One called the die, is made like a cup with a very thick bottom which has a hole through it of the same size as the outside of the macaroni which it is to make. The other is shaped like a pin with a square head. The long part of the pin is the same size as the hole inside of the finished macaroni. (Center) The pin fits inside of the cup. (Right) Pressure forces the dough around the pin and into the hole in the bottom of the cup. When the dough reaches the round part of the pin, the same pressure forces the dough together and out comes another length of delicious macaroni.

put into the pin to cause the dough to flow faster. Products with wavy edges can be made by allowing the dough to flow more rapidly at the edge than at the center of the strip.

The extrusion rate through a die hole in a bronze die is normally about 1 in. per second. Extrusion rates can be increased and the product made more smooth and more yellow by putting teflon inserts into the die so that the final extrusion surface is made of the very smooth and low friction teflon. Teflon dies have come into wide use in areas where products are packaged in transparent film. Bronze dies are used in areas where the product is put into cardboard boxes. It is generally considered that the bronze product cooks up in a more satisfactory manner. Water penetrates more slowly into the teflon product and there is a tendency for the surface to become soft before the interior has been thoroughly cooked.

Bronze has proved more satisfactory than stainless steel because the higher heat conductivity of the bronze takes the heat away from the die hole more rapidly than is possible with stainless steel dies. Stainless steel generally makes a rougher product.

The high pressure required to extrude dough through the dies subjects the dies to a substantial amount of wear. The first effect of the wear is to polish the dies smoothly so that after a brief initial period of running the product surface tends to become smoother. As production is continued, wear increases to the point where other production problems occur. Many manufacturers keep a set of standards on their product sizes so that the finished products can be checked frequently and dies repaired when product sizes get beyond certain limits of tolerance (Maldari 1956).

It is often difficult to tell just how die wear will affect production. Under normal circumstances, manufacturers will become conscious of die wear through the warning medium of packaging. Too heavy a product results in less volume per unit weight and there is a resulting slack fill in packages. Gradual wear of solid and tubular products can seldom be detected by visual inspection of the product, but must be determined by actual measurement. Fancy products tend to give some indication of wear by a change in physical appearance. For example, in sea shell production the flow of dough is at its maximum at the center of the shell, making this point more susceptible to wear than the ends. As wear increases, the dough flows faster at the center, thereby increasing curvature of the product. Another common warning of wear in shell dies comes in the form of checking either during or after drying. This checking can often be attributed indirectly to die wear and can be eliminated by reducing the thickness of the die outlet.

Wear in the wavy-type products, such as mafalda or wavy lasagne, becomes physically evident by more pronounced or closer curled waves. A cross section of this product should present a flat, noodle type appearance. The wave is the result of greater flow of dough on the ends of the slots in the die, making these ends the points of greatest wear.

Spiral products normally have a cross section like that of a noodle prior to wear. After wear has taken place, the outer circumference portion tends to become larger and rounded, thereby increasing the flow of dough in these points, resulting in a tighter curl or greater degree of twisting.

The problem of splits on short cut tubular products is not always directly traceable to the die, but can generally be blamed on grit that lodges in the die. The grit lodges between the pin and the outlet and prevents proper amalgamation of the dough before it leaves the extrusion opening. Sometimes pins are forced to one side by the grit, thereby increasing or decreasing the rate of curl on an elbow macaroni, or causing excessive curl on long macaroni products.

Die Cleaning and Storing

Many manufacturers have a large number of dies available due to the fact that some 100 shapes may be manufactured in a single plant on perhaps 2 or 3 presses. Some dies can make more than one product, but many are suitable for only a single shape. Consequently, the storage conditions under which dies are kept must receive careful consideration.

Cleaning of dies is quite difficult. A relatively dry dough is forced into the holes in the die under pressure of approximately 1000 psi. A small proportion of the dough remaining in the die at the time of removal from the press is exposed to any cleaning action that might be used. Dough cannot be left in the dies because eventually souring will take place and the resulting acids will attack the bronze surface to cause pitting which impairs quality of the finished product.

The normal procedure is to remove dies from the press and place them in a water bath in which the water is continually moving and kept fresh. Dies are often placed on wooden trays and isolated from the walls of the storage tank in order to prevent electrolysis and consequent pitting and corrosion from occurring. Overnight soaking causes the dough inside the dies to become softened and greatly eases the eventual cleaning operation.

Dies are usually cleaned by high pressure water jets in an automatic apparatus which either rotates the dies under stationary jets or moves jets across the surface of a stationary die. Dies can be cleaned in modern die washers in between 1 and 1.5 hr. Teflon dies are generally easier to clean than bronze dies.

It is vital that all of the dough be removed or pitting will occur. It is possible to clean dies with high pressure steam jets. However, in the normal thick die, approximately 2 in. thick, the high temperature steam tends to cook the dough inside the die and make it very difficult to remove. High pressure steam can be used to good advantage in cleaning dies which are relatively thin, made of stainless steel with brass inserts. After the dies are cleaned and dried, they are stored in clean neutral mineral oil or in air until the next use.

DRYING MACARONI PRODUCTS

Upon extrusion from the press, macaroni is a soft, plastic product containing approximately 30% moisture. This wet product must be dried to 12% moisture or less to obtain a hard product which will not support the growth of mold, yeast or other spoilage organisms. Too rapid removal of moisture will cause the product to "check." Too slow removal of moisture may permit stretching of long products on the sticks, or souring, or mold growth. To achieve proper drying, the rates of drying must be controlled by adjusting air circulation, temperature and humidity. The drying rate of a macaroni product is determined by the shape of the product, the temperature, humidity and velocity of the air, and the moisture content of the macaroni.

If a hygroscopic material such as macaroni is placed in a stream of air of given temperature and humidity, it will gain or lose moisture until it reaches a constant percentage of moisture which is called the equilibrium moisture. In the case of macaroni, the raw material is a complex, organic system containing starch, protein, and other materials which change their properties according to the variety of wheat used, growing conditions, milling procedures, percentage of protein, previous drying conditions, and many other factors. Therefore, the equilibrium data obtained by various investigators are not completely consistent. The data obtained by Earle (1948) are as representative as any. These data were obtained by drying the product to constant weight in a laboratory drier with close control of temperature and humidity and adequate air circulation at a temperature of 90°F. The data are summarized in Table 64.

TABLE 64

EQUILIBRIUM MOISTURE OF MACARONI AND EGG NOODLES AT 90°F

Relative Humidity %	Equilibrium Moisture	
	Dry Basis (Macaroni) %[1]	Dry Basis (Egg Noodles) %[1]
90	22	...
80	18.2	...
70	16	14
60	13.9	11.9
50	12.1	10
40	10.5	8.5
30	8.8	...
20	7.0	...
10	4.9	...

[1] Percent bone dry basis.

Drying Rates

On the basis of extensive drying tests run by the author on spaghetti with a diameter of 0.069 in., with an air velocity between 150 and 300 ft per min. and with temperatures varying from 90° to 170°F, drying rates can be calculated by means of the following formula

$$\log \frac{F_o}{F} = KAt \tag{1}$$

where

F = % free moisture = $M - M_e$

M = % moisture in macaroni (dry basis)

M_e = % equilibrium moisture (dry basis)

K = a constant experimentally determined to be 0.0406

A = sq ft of surface area of macaroni per lb of bone dry solids

t = time in hr

F_o = free moisture at zero time

log = logarithm to the base 10

$$A = \frac{0.61D}{D^2 - d^2} \tag{2}$$

where

A = sq ft of surface area per lb of dry solids

D = outside diameter (in.)

d = inside diameter (in.)

d = O for spaghetti and the equation reduces to

$$A = \frac{0.61}{D} \tag{2a}$$

Substituting in equation 1

$$\log \frac{F_o}{F} = \frac{0.0248Dt}{D^2 - d^2} \tag{3}$$

From equation 3, we can derive equation 4

$$t_a = 12.1 \frac{D^2 - d^2}{D} \tag{4}$$

where

t_a = the time in hr required to reduce the free moisture to half its initial value

D = outside diameter of macaroni (in.)

d = inside diameter of macaroni (in.)

Example: Plot a drying curve for spaghetti of .072 in. diam. with an initial moisture content of 31% (dry basis) dried in an atmosphere of 90°F and 80% RH.

Solution: From Table 64 we find that the equilibrium moisture is 18.2% (dry basis) at 80% RH.

$$t_a = 12.1 \times 0.072 = 0.87 \text{ hr}$$
$$F_o = 31 - 18.2 = 12.8$$

For every 0.87 hr, the free moisture is halved. A calculated drying curve is shown in Table 65.

TABLE 65
CALCULATED DRYING CURVE FOR SPAGHETTI

Time (hr)	F = Free Moisture[1]	M = Total Moisture[1]
0	12.8	31.0
0.87	6.4	24.6
1.74	3.2	21.4
2.61	1.6	19.8
3.48	.8	19.0
4.35	.4	18.6
5.22	.2	18.4

[1] Percent bone dry basis.

Checking

Macaroni drying is not a simple matter of removing all of the moisture as rapidly as it can be evaporated from the product. "Checking," or cracking of the macaroni piece will result unless the drying conditions are carefully controlled. Checking is caused by the differential expansion and contraction of the layers of macaroni dough under the influence of changes in moisture and temperature. In practice, cracking must be prevented by a relatively gradual removal of the moisture from the product. This is usually accomplished by drying the macaroni in three basic stages.

In the preliminary stage approximately 40% of the total moisture removal is accomplished in 30 to 40 min. This case hardens the product and a period of resting or "sweating" in high humidity air allows the moisture to distribute itself fairly evenly throughout the cross section in from 1 to 2 hr. From this point, slow drying removes the moisture at a rate which will not cause damage to the product.

Extensive studies of the physical properties of macaroni were carried out by Earle (1948) in order to establish a scientific basis for the understanding of the checking phenomenon. He found that the coefficient of thermal expansion of macaroni dough averaged 58×10^{-6}. The coefficient of expansion related to moisture content is 4×10^{-3}.

Earle compiled a considerable amount of data on the strength of macaroni. The breaking strength under tension varied all the way from 1500 psi for noodles which were overdried in the predrier to 7300 psi for a commercial macaroni product. He found that harsh preliminary drying decreased the strength of noodles from 3000 to 1500 psi. He also listed data which showed that macaroni with 14% protein had a breaking strength of 5117 psi while a 10.6% protein product had a breaking strength of 3978 psi.

Calculations based on breaking strength, modulus of elasticity, the coefficient of expansion due to moisture and the thermal coefficient of expansion showed that macaroni containing 12% water on a dry basis and at 90°F suddenly placed in a current of moving air of 70°F with a relative humidity such that no change could occur in the surface moisture content would develop a stress of 18.7 psi. This is a very small value in comparison to the 5000 to 7000 psi required to break the macaroni or check it.

Macaroni in equilibrium with air of 65% RH which was moved into 85% RH would develop a maximum stress of 4700 psi. This would come close to checking the macaroni and would probably actually cause check because of irregularities in the structure. This shows that the checking of macaroni is caused by differences in moisture content and not by differences in temperature. It should be kept in mind, however, that differences in air temperature have a very marked effect on the distribution of moisture so that temperature differences can cause checking indirectly. The physical properties of macaroni are listed below:

Coefficient of thermal expansion (avg) 58×10^{-6} per °F

Coefficient of moisture expansion 4×10^{-3} per per cent moisture (dry basis)

Modulus of rupture (breaking strength) at 13%, 5400 psi (This varies from 1500 to 7300 psi depending on the condition of the product and previous drying history)

Thermal stress set up by moving from 90° to 70°F air, 18.7 psi

Stress set up by moving from 65% to 80% RH 4700 psi

Specific gravity of spaghetti at 10% moisture content, 1.4

Modulus of elasticity at 10% moisture dry basis 1×10^6

Modulus of elasticity at 15% moisture content on a dry basis 0.6×10^6

Specific heat of dough (calculated to 0% moisture) 0.18 Btu per °F per lb

Specific heat of dough at 31% moisture (wet basis) 0.435 Btu per °F

Specific heat of flour, 10% water, 0.26 Btu per °F

Checking or deformation of macaroni products in the preliminary drier has increased since the vacuum process of extrusion has come into use. If the heat in a long spaghetti preliminary drier is arranged in such a way that the entering goods are dried very rapidly by a blast of hot dry air and then drying is continued at a slower rate to the end of the drying section, the spaghetti will case harden at the beginning. As moisture is slowly removed from the interior, it will contract, but the surface will not be able to follow it as it shrinks. This causes the surface to distort so that when it leaves the preliminary drier the spaghetti looks like rubber tubing flattened by vacuum. The same type of thing occurs with short cuts, but the product will have ridges from one end to the other, or will shrivel up like a prune.

Normally, overdrying in the preliminary drier will produce white spots or bubbles in the interior of the spaghetti. This is caused by the shrinking interior trying to pull away from the case-hardened surface and causing the dough to pull apart at the points where bubbles appear. Usually these preliminary drier "checks" do not cause trouble with respect to cooking quality.

If dried macaroni is moved into a very humid atmosphere, it will absorb moisture on the surface which will then expand. If this process is carried too far, the surface will pull away from the interior and cause serious checking. Before the vacuum process was adopted, this checking showed up in spaghetti as a crack under the surface which ran at about a 30° angle to the axis of the spaghetti at both ends of the checked place. The shape was very much like a boat. Since the advent of vacuum, this check often shows up in spaghetti as short lines perpendicular to the axis of the spaghetti inside the strand. This is the most serious kind of check.

The difference in vacuum and nonvacuum is caused by the fact that checks normally begin at a bubble where the stresses concentrate in a product not vacuum-treated while a vacuum-treated product is very homogeneous and there is no focal point for the check to start. Therefore, the checks in products not vacuum-treated take place perpendicular to the maximum stress which is along the length of the spaghetti.

If a product at 12 or 13% moisture is moved into hot, dry air, the surface will dry and contract. This will cause a very fine network of cracks to appear on the surface. Usually these cracks do not extend very deeply into the macaroni and do not usually cause trouble in cooking. This check is called "tension check" because the surface is in tension when it occurs.

If wet macaroni is dried rapidly, the surface will be dried, but no stresses will be set up because the product is plastic. Under these circumstances the moisture content at the surface will be small and in the interior it will be large. The solids content at the surface will be large and the solids content in the interior will be small. If drying is continued with this difference in concentration down through the plastic range into the brittle range, there will be no stress set up as long as the drying rate is fast enough to keep the moisture gradient in line with the solids content. When drying is stopped, the moisture will tend to distribute itself evenly. This will cause the surface to expand because it contains too great a concentration of solids and the interior will contract because it contains too low a concentration of solids. This will cause compression check. It is the most common check encountered in macaroni drying. Sometimes it takes several days for this type of check to appear and it may appear after the product has been packaged.

If short-cut macaroni has a small amount of stress trapped in it, which is insufficient to cause checking under normal circumstances, it will be more susceptible to checking due to moisture changes than a properly dried product. If such a product is put into a steel bin when it is warm and this bin is wheeled into an area where the air is cold, the layer of macaroni next to the surface of the bin will be cooled so that moisture will migrate from the hot interior of the bin to the cool product near the surface. The absorption of this moisture by the cool macaroni will cause check. This is quite common. Sometimes under these conditions the macaroni at the sides of the bin will check but the open top surface will not check because the moisture escapes to the air and is not absorbed by the cool macaroni.

In batch macaroni driers, there is often a part of the room which lags behind the rest of the room. Test runs were made on driers where one part of the room was at 15% moisture and there were parts of the room which were still at 21% moisture or more. When the greater part of the room has reached a moisture content of 15% or below, the wet bulb depression throughout the room will often increase rapidly because of schedules set on controls or because of the natural tendencies of uncontrolled driers. This will cause the high moisture part of the drier to lose moisture rapidly and perhaps dry from 21 to 15% in an hour. The product will have a very bad trapped stress-type of check. It is this mechanism that accounts for the fact that a certain spot in the drier will often yield macaroni which is checked and moldy at the same time.

If a strand of spaghetti is supported between two rods approximately 6 in. apart and a weight is hung on the center of the strand which is slightly lighter than the weight which will break the strand instantly, the strand of spaghetti will gradually bend more and more and after a period of ½ to 1 hr it will break. The same mechanism may account for the fact that it takes a long time for some macaroni to check. The continuously applied force due to the weight causes the macaroni to deform slowly in an attempt to relieve the stress. However, there are portions of the dough which will not flow and these portions gradually take on the entire load which was originally supported by all of the dough in the macaroni. Eventually, the rigid "skeleton" which will not flow has insufficient strength to support the stress which has been loaded onto it and the entire strand of spaghetti breaks.

Where stresses are trapped in the spaghetti, the unevenly distributed water is continually exerting pressure on the surface to expand and the interior to contract. The fraction of the dough which flows gradually accommodates itself to this pressure and the skeleton of the dough eventually cannot support the pressure of the water and checking occurs.

Long Spaghetti Production Systems

In the 1920's, spaghetti was extruded from a hydraulic press through a round die. The strands of spaghetti were cut off with a knife, manually draped over a wooden stick 54 in. long and trimmed to about 22 in. below stick. These sticks were placed on racks on casters in three tiers of about 15 sticks on each tier. This truck was pushed into a preliminary drier which removed about 35% of the moisture and then put into a finish drying room which usually had a large 6-ft fan or 2 fans in one end and held about 20 trucks. In most cases there was no heat in this drier and it took from 2 to 5 days to dry.

In the 1930's and 1940's, the continuous press was developed with an automatic spreader which extruded spaghetti from a rectangular die so that it hung straight down in two curtains. Two sticks were then pulled through the curtains of spaghetti on chains so that the spaghetti hung down on both sides of the stick and then the spaghetti was automatically cut off. The lower ends were trimmed by a reciprocating or rotary knife. The sticks then passed through the drying section of a preliminary drier on chains with the spaghetti hanging down and then passed through two levels of resting section where the moisture was allowed to distribute itself evenly in the strand.

The sticks of predried spaghetti were placed on racks which were put into the drying rooms very similar to the noncontrolled rooms. Glenn Hoskins applied wet and dry bulb temperature controls to these drying rooms from the 1930's on and reduced the drying time to about 36 hr. Later developments of better air circulation and higher temperature of operation have reduced drying time in batch rooms to 15 hr for spaghetti. Time schedule controllers automatically change the temperature settings throughout the drying cycle from a very high humidity at the beginning to a lower humidity at the end of the cycle when the goods are practically dry.

In the late 1940's, completely automatic continuous production lines for long spaghetti were developed and now most of the large factories are equipped with continuous lines. The integrated continuous long spaghetti production systems of each of the major manufacturers are described below.

Long Goods Continuous Production Lines

Each of the functions of a long goods continuous production line will be described in considerable detail in the description of the DeFrancisci machine. All of these functions are performed by the other continuous production lines, but only features which differentiate that particular machine will be described.

DeFrancisci Long Spaghetti Production Line.—Flour is fed into the mixer through a rotary airlock with teflon-lined pockets running at constant speed. Water is fed through a valve which gives constant flow while compensating for variations in upstream water pressure. The water is adjusted to give the correct proportion of water and semolina. A percentage timer turns the flour feed and water on and off for adjustable percentages of each minute in order to control the total rate of flow per hour.

The mixer is a twin shaft mixer with blades wiping down the sides of the mixer and up in the center. It is kept under a vacuum of about 20 in. of mercury in order to preserve the yellow pigment and to eliminate bubbles in the final product. This gives a more translucent yellow color.

The auger is stainless steel with bronze tipped flight crests to give a good bearing surface against the hardened steel cylinder. The cylinder has longitudinal grooves to keep the dough from turning as the auger turns. After leaving the auger, the dough passes through a sheet metal plate, perforated by small holes held by a stainless steel disk with approximately ¼ in. holes in order to knead the dough and strain out any pieces of hard dough or extraneous matter.

A network of tubes carries the dough to 12 equally spaced points along the long rectangular die head (see Fig. 88). The dough is extruded through a rectangular die containing more than 1000 holes. The spaghetti is extruded in two curtains of strands to hang in front of two aluminum drying sticks 80 in. long. A timer starts the stick and cutting mechanism in operation. The sticks are advanced to pick up the curtains of spaghetti and then a knife cuts the spaghetti against the surface of the die so that the spaghetti hangs on both sides of the two sticks. The sticks are then carried forward past a reciprocating cutter similar to the cutter of a mowing machine which trims the uneven lengths off. The trim is carried by a conveyor into a centrifugal chopper blower which cuts the spaghetti lengths into about ½ in. pieces and pneumatically conveys them back to an air lock feeding into the vacuum mixer.

The spaghetti hangs about 21.5 in. on each side of the stick. The sticks are carried by chains through a preliminary drier in which air is blown down through the spaghetti from the top and recirculated. Wet and dry bulb controls maintain a wet bulb de-

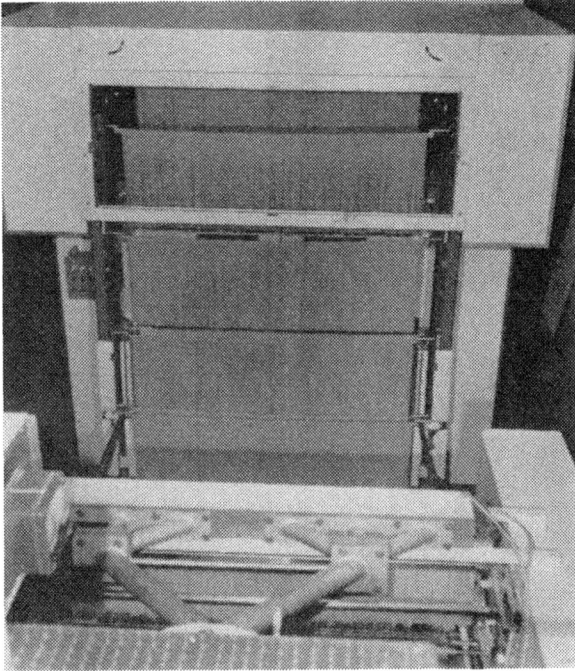

Courtesy of DeFrancisci Machine Corp.

FIG. 88. LONG SPAGHETTI SPREADER SHOWING DISTRIB-
UTING TUBES, RECTANGULAR DIE HOLDER, CURTAIN OF
SPAGHETTI EXTRUDING FROM DIE AND FULL STICKS OF
SPAGHETTI BEING CONVEYED INTO PRELIMINARY DRIER

pression of about 10°F in this drier and a dry bulb temperature approximately 110°F for a period of 30 to 45 min.

A chain with hooks picks up the individual sticks, carries them up to the ceiling and then lowers them down past the five tiers of chain carriers in the finish drier. When 5 sticks have been brought approximately to the level of the 5 chains, fingers reach out and grasp the sticks and deposit them on the 5 chains simultaneously.

The sticks are then carried forward at the same rate by the chains at 5 levels in the drier until they are dry after about 18 to 22 hr of drying at approximately 105°F. Wet and dry bulbs are both controlled by instruments which operate exhaust fans, intake dampers and automatic valves supplying steam to the coils which run the length of each of the control sections. There are either 2 or 3 separately controlled areas from the beginning of the finish drier to the end in order to vary conditions as the spaghetti becomes drier.

The air circulation is down through the top 3 tiers and up

through the bottom 2 tiers. There are fans on both sides of the drier which suck the air out between the third and fourth tiers of spaghetti. The air from these fans splits in the air chambers on the two sides of the drier so that part of the air goes up the air chambers and part of it goes down the air chambers. After drying, the sticks move into the accumulator still on the same level of chain carriers. The accumulator will usually hold 18 hr of production. The accumulator fills up during the 16 hr of operation at night and is emptied in 8 hr during the day. Sticks are taken from all five levels at once and placed on the inlet chains of the stripper and saw. This machine automatically lays the spaghetti on its side, removes the sticks and makes three cuts which remove the heads and the uneven lower ends of the spaghetti and cut the remaining strand into two pieces approximately 10 in. long. The spaghetti is then either taken directly to the packaging machines or placed in tote boxes for delivery to multiple packaging machines.

The sticks are automatically placed on the stick return which consists of two rotating cylinders the length of the drier with a helical groove cut in the surface. This carries the sticks back to the press. Since the sticks are normally not being used as fast as they are being stripped, it is necessary to accumulate about 24 hr of sticks on this stick return at the end of the packaging operation.

Buhler Long Goods Production Line.—Buhler makes two types of long goods production lines; the traditional Buhler line and the more recently developed Bassano line. Flour feed for the Buhler press is through a variable speed screw feeder. The mixer is a two-shaft mixer. A short cross-feed auger picks up the dough at the end of the mixer and forces it into a small vacuum chamber through which the main auger passes. A vacuum of 25 to 27 in. of mercury is pulled at this point. The cross auger acts as an airlock and is equipped with variable speed to vary the production of the press. The main auger operates at an rpm higher than necessary to extrude at the press capacity so that it is never overloaded by the cross auger.

Dough is distributed by tubes to the long die. The spaghetti is cut several inches below the die face by a pair of disk knives which move from one end of the curtain of spaghetti to the other to cut it off quickly and evenly without closing the opening of hollow products like macaroni.

The sticks of spaghetti are carried to the top pass of the preliminary drier. After they have traversed the length of the drier, they are lowered by chains to a second pass which moves back toward the press and then to a third pass which discharges out the end of the preliminary drier to the first finish drier. There are 3 passes in the first finish drier and 3 more in the second finish drier.

Air is alternately blown up and down through the spaghetti as the goods pass through the drier. The fresh air dampers and the heat are controlled by the wet bulb depression directly. The temperature rises to a point which will give the correct wet bulb depression. As the drier becomes empty, the operating temperature automatically lowers to keep the wet bulb depression at the correct level.

Buhler Bassano Drying Line without Sticks.—The Rolinox process developed by Bassano and manufactured by Buhler Brothers eliminates the use of drying sticks and the stripper and saw.

Products are discharged evenly from a long die onto a horizontal table. They are then lifted by suction to the screen of an overhead carriage and cut to length in this position by rotary knives. Excess trim is returned to the press. The cut goods are then placed automatically on preliminary drying trays. After the trays have passed through the two zone preliminary drier, the product is stripped off into perforated metal drums of the final drier and dried to the final moisture content (see Fig. 89).

Courtesy of Buhler Brothers

FIG. 89. PERFORATED METAL DRUMS CONTAINING FIVE BUNDLES EACH OF SPAGHETTI CUT TO LENGTH AND ROLLING THROUGH BUHLER ROLINOX FINISHING DRIER

These drums are perhaps 5 ft long and 8 in. in diameter with divisions separating the bundles on 10 in. spaghetti in each compartment. The drums are oscillated backward and forward and then rotated throughout their progress through the drier to insure straightness and uniform drying of the goods. All long shapes including thick macaroni, noodles, and lasagne can be produced on the Rolinox line.

The Rolinox system eliminates drying sticks, stripper, and saw and the trimmings from the dried product which must be reground and reused as raw materials in other systems.

Braibanti Long Goods Production Lines.—The Braibanti press (see Fig. 90) has four mixers in series. A vacuum is maintained on the last mixer which discharges into two augers running parallel to each other. Each of the 2 augers supplies ½ of the long narrow die through tubes which fan out downward at right angles to the auger. Trimming is done by a rotary knife like a lawn mower. After passing through the preliminary drier, the sticks of spaghetti are carried to the top pass of a five-tier drier. They are deposited on a pair of stationary bars running the length of the drier holding up each end of the sticks. A "comb" with

Courtesy of Braibanti and Co.

FIG. 90. COMPLETELY AUTOMATIC LINE FOR CONTINUOUS PRODUCTION AND DRYING OF LONG SPAGHETTI. AUTOMATIC SPREADER IS IN FOREGROUND

notches spaced properly for each stick moves back one stick length below the level of the stationary bars, moves upward to pick up the stick and then forward and down to deposit the stick one space forward. After the sticks have progressed the whole length of the drier, they are picked up by a chain with hooks and deposited on the second tier, conveyed by "comb" backward to the intake end of the drier and proceed thus through the five tiers of the drier. Air circulation is down through the spaghetti with alternate sections of air circulation and no air circulation. Reheating coils are placed at various points between tiers so that drying can continue as the air passes down through the five tiers. Without heating coils the air would become saturated and drying would take place only on the top tier. The net result of this is that the goods are alternately dried and rested to give the desired overall drying program. The sticks are carried to one tier at a time of the storage unit. When one level has been filled, the sticks are diverted to the next lower level. The accumulator is also unloaded one level at a time.

Pavan Long Goods Drying System.—The Pavan press which has a single mixer under vacuum is mounted with the mixer close to the floor (see Fig. 91). The flour feed is under the storage

Courtesy of Pavan Co.

FIG. 91. PAVAN LONG GOODS PRESS WITH INCLINED SCREW WHICH PERMITS INSTALLATION OF MIXER AT FLOOR LEVEL

bin which may be located at a remote point. The flour is sucked into the press by the vacuum in the mixer through a transparent plastic tube about one inch in diameter. The auger is slanted upward to gain the proper height for discharging directly into the rectangular chamber above the die without distributing tubes.

After passing through the preliminary drier, the sticks of spaghetti are conveyed into one drying tunnel at a time. These tunnels are placed one above the other and are completely separated from each other. When one tunnel is full, the doors at both ends are closed and the spaghetti is dried in a batch fashion. Very high velocity but low volume jets of controlled hot air are blown up through the spaghetti from underneath through a duct which runs from one end of the drying tunnel to the other. This duct is moved back and forth from side to side of the tunnel so that only a few inches of the spaghetti hanging on the stick are being dried at a given instant and the rest is being allowed to equalize its moisture. Moisture is removed from the air by refrigerating coils and run to the sewer and the air is then reheated by hot water coils.

The drying tunnels are used as storage units as well as driers. They are unloaded one at a time just as they are loaded one at a time. The net result is that the drier operates as a continuous drier even though the individual tunnels are used as batch driers.

Clermont Continuous Production Line.—The first continuous long goods lines in the United States were built by Clermont Machine Company. After leaving the spreader, the spaghetti passes through a 3 tier or 5 tier preliminary drier. If it is a 5 tier drier, it is divided into two sections as far as control and air circulation are concerned with a resting tier between these sections. Spaghetti passes through the top tier by means of of chains and then is lowered to the second tier and goes back to the intake end of the drier and so on through the five tiers.

The 2 finishing driers are each divided into 2 drying sections from top to bottom with suitable resting sections so that there are actually 4 finish drying sections with rest in between.

Air circulation varies from one unit to another according to the needs of spaghetti.

Essonica Press.—Essonica has produced a press with 11 short vertical augers directly above the rectangular die. These augers are short and their purpose is to improve the spreading pattern and reduce the working of the dough.

Short Cut Production Lines

Originally short cuts were dried on trays in rooms similar to long goods drying rooms or in drawers with screen bottoms in enclosed cabinets with forced air circulation. These processes have been out of use for many years except for very thick-walled or special products such as egg barley and Kluski noodles.

DeFrancisci Short Cut Production Line.—The mixer, vacuum, and auger are the same in the short cut production press (see Fig. 92) as in the long goods press. The dough is extruded through an elbow into a cast steel bell which spreads out to the

Courtesy of Glenn G. Hoskins Co.

FIG. 92. CONTINUOUS DRIER FOR SHORT CUT PRODUCTS

diameter of the round 2.5 in. thick by 15.5 in. diameter die. A vertical shaft holds up the center of the die so it will not bend under pressure and also supports the rotary knife which cuts on the surface of the die. This knife may have from 1 to 4 blades depending on how short the product should be cut. The short product drops onto a shaking conveyor with a perforated bottom through which air is blown to dry the surface of the product so that the pieces will not stick together and so that large shapes will have the strength to maintain their form until they reach the preliminary drier. Product is conveyed by trough belt conveyers to a distributor which spreads the produce out across the top drying screen. In the preliminary drier the product is spread across the 5-ft width of the top screen to a depth of less than $\frac{1}{2}$ in. It passes rapidly to the end of the preliminary drier and drops to the second screen where it is piled a little bit deeper and so on through 5 screens with gradually increasing depth of piling until the bottom is about 1.5 in. deep. Forty percent of the moisture is removed in about 35 min in the preliminary drier.

The product is piled deeply on a resting screen which is either at the bottom of the preliminary conveyor or the top of the first finisher to allow the moisture to distribute itself evenly. The product is then conveyed to the first fininshing drier where it is distributed across the width of the top screen and passes through approximately five screens and is then conveyed to the second finishing drier where drying is completed.

Air circulation is down from the top through the goods. Temperature and humidity controls are similar to the controls of long goods dryers.

The conveying screens are of nickel-plated steel wire mesh with roller chain at the edge and cross braces every 6 in. Material is prevented from falling under the chains and air is prevented from overdrying the edges by aluminum baffles with conveyor belt edging which touches the screen about 2 in. inside the edge of the screen.

Uninsulated driers housed in plastic coated plywood require 12 to 14 hr to dry short cut macaroni. By insulating the driers and raising the temperature of operation to 130°F the drying time has been cut to 5.5 hr for elbow macaroni (see Fig. 93).

Buhler Short Cut Drying System.—The Buhler drier uses "S" shaped cross section extruded aluminum slats for coveying the product through the driers. These slats normally are not perforated, but the air passes through the space between the slats and is forced through the material. These driers operate at high temperature so that drying time is in the neighborhood of 6 hr. There is a short preliminary and one finishing drier which is vertically divided into three temperature-humidity control zones. Air is forced in between the slat conveyors and passed through the goods at each level. There are reheating coils between conveyors.

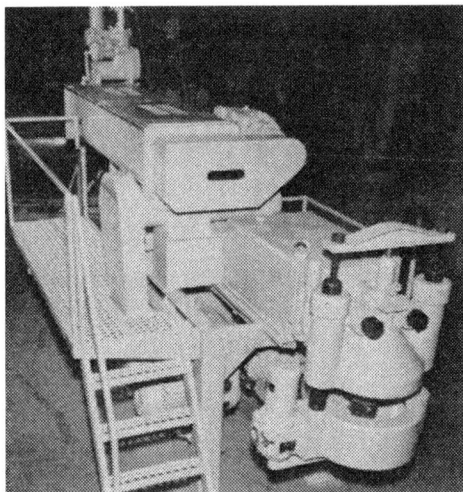

Courtesy of Buhler Brothers

FIG. 93. SHORT CUT EXTRUSION PRESS CONSISTING OF FLOUR FEED AT BACK, TWIN SHAFT MIXER, AND CROSS FEED SCREW AND AIRLOCK UNDER MIXER. VACUUM IS DRAWN AT DISCHARGE OF CROSS SCREW AS MATERIAL IS PICKED UP BY MAIN EXTRUSION SCREW FEEDING SHORT CUT DIE HEAD

Braibanti Short Cut Drying System.—The Braibanti short cut drier consists of a preliminary and one finishing drier. Conveying screens are of nylon mesh without chains at the side. Tension is kept on these conveying screens by gravity takeups on the drive pulleys. There are cross braces with rollers at each end about every 6 in. of screen. A flap of the screen rides up on side baffles so that the macaroni will not fall over the edge of the conveying screens. Air circulation flows in on top of one screen and passes down through that screen and out through the screen below.

Pavan Short Cut Production System.—The circulating fans and heating coils in the Pavan system are underneath each screen between the top flight and the return flight of the conveying screens. On a long finishing drier there may be from 2 to 3 separate fan chambers along the length of the screen so that the drier can be divided into a large number of controlled sections. Conveying screens are of nylon mesh with cross braces.

Garbuio Short Cut Drier.—The Garbuio system dispenses with conveying screens. The macaroni is placed in the first of a series of troughs across the width of the drier. These troughs are approximately 10 in. by 5 ft. They are formed of screen mesh through which air is blown. The first trough is hinged and dumps into the second trough which then dumps into the third. The product is left at rest except when it is dumped into the succeeding trough.

Noodle Production Lines

Noodle driers are essentially the same as short cut driers except that the material is more bulky and more subject to breakage and formation of lumps due to the noodles sticking together. This means that there must be more space between screens and in general, the top screens should not be piled as deeply as the lower screens as material passes through the drier.

The method of forming the noodles is different from the short cut system. This has already been described in an earlier section.

Ravioli Production

Ravioli is a pillow formed of two sheets of dough with a pocket of meat, cheese or vegetable filling. Traditionally, dough was made on the mixer, kneader, and dough break combination and then fed into hoppers at the top of the ravioli machine which rolled the dough into two sheets. These two sheets were formed into pillows by a pair of rolls and the filling was put into the pockets formed by the roll just before the roll closed the pocket.

In Europe, the product is made in sheets of a number of pillows only partially cut through so that they can be broken off by the user and these products are dried (see Fig. 94 and 95).

In the United States, dried raviolis are not sold, but the product is sold in cans or in frozen form.

Direct Canned Spaghetti

DeFrancisci has developed a machine which extrudes bundles of spaghetti from a long rectangular die, trims them to length

FIG. 94. SHEET OF DRIED RAVIOLI

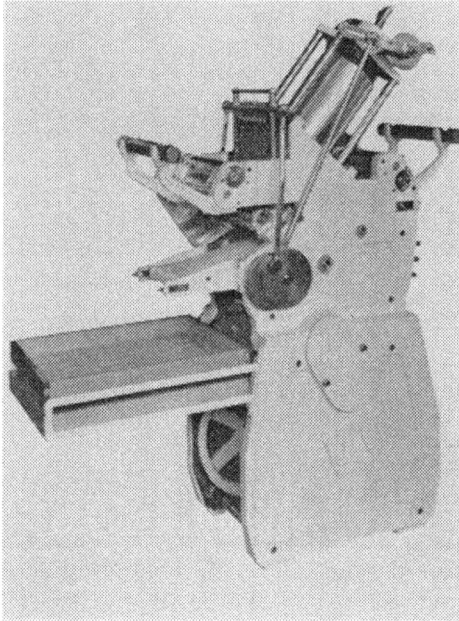

Courtesy of Giacomo Toresani S.p.a.

FIG. 95. RAVIOLI MACHINE

Rolled dough sheets are placed on spindles at right and left. Filling is placed in the cylinder at top. Filling is measured by reciprocating pistons in front of large filling cylinder and the pillows are formed by two rolls with pockets directly underneath the filling pistons.

and drops them directly into cans (see Fig. 96). Meat balls and sauce are then added and the cans closed and sterilized in a continuous rotary sterilizer made by FMC. This sterilizer rolls the cans so that the strands of spaghetti are agitated and do not stick together.

Courtesy of DeFrancisci Machine Corp.

FIG. 96. PRESS FOR EXTRUDING AND MEASURING LONG SPAGHETTI DIRECTLY INTO CANS

Special Drying Processes

Jacob Zwick in Hamburg, Germany developed a method of drying short cuts in which the product was put through a steam jacketed drum drier and dried in 1 hr. Scotland (1966) has developed a process for drying elbow macaroni in ½ hr using radiant heat and vibratory conveyors. This product cooks in about 1 min and has been tested by the US Army as a quick cooking field ration. Pavan has used a "Roto-therm," a drier in which long spaghetti is heated to a high temperature between plates and this is thought to reduce the drying time. W. K. Kellogg Company has developed a process in which the macaroni is first cooked in boiling water and then dried in ½ hr without checking (Poole 1955). These processes are not in commercial use with the exception of the Pavan Rototherm.

Reground Finished Product

There are various unusable products after the process is complete. These include the trim from the long spaghetti, checked macaroni products, and deformed products which may occur at the beginning or end of the production run (see Fig. 97).

This product is usually ground in a hammer mill to approximately the granulation of semolina and mixed with the raw material within the range of 1 to 10%.

In an automatic system, long spaghetti, short spaghetti, and the trim from the saw are dropped into a paddle-type screw conveyor which breaks the product to about 2 in. long. This feeds into a centrifugal blower with a material handling blade which breaks the pieces to about $\frac{1}{8}$ in. long and conveys them back to the flour room. This crushed material is stored in a surge hopper and fed through a hammer mill back into the stream of raw material at a metered rate to give the desired percentage of regrind in the mix.

COMPOSITION OF MACARONI

Table 66, adapted from US Dept. of Agr. Handbook No. 8 (Anon. 1967) compares properties of macaroni and noodle products with other common foods.

Courtesy of Clermont Machine Co.

FIG. 97. AUTOMATIC SPAGHETTI CUTTER FOR REMOVING SPAGHETTI FROM STICKS AND CUTTING TO LENGTH

TABLE 66

COMPOSITION OF MACARONI, NOODLES AND RELATED FOODS

(Based on 100 gm of edible portion of each food)

Food	Food Energy Cal	Pro-tein (Gm)	Carbohydrates Total (Gm)	Fiber (Gm)	Fat (Gm)	Water (Gm)	Ash (Gm)	Cal-cium (Mg)	Phos-phorus (Mg)	Iron (Mg)	Vitamin A (IU)	Thi-amine (Mg)	Ribo-flavin (Mg)	Niacin (Mg)	Ascorbic Acid (Mg)
Macaroni, (unenriched)															
Dry	377	12.8	76.5	0.4	1.4	8.6	0.7	22	165	1.5	0	0.09	0.06	2.0	0
Cooked	149	5.1	30.2	0.2	0.6	60.6	3.5	9	65	0.6	0	0.02	0.02	0.5	0
Macaroni (enriched)															
Dry	377	12.8	76.5	0.4	1.4	8.6	0.7	22	165	2.9	0	0.88	0.37	6.0	0
Cooked	149	5.1	30.2	0.2	0.6	60.6	3.5	9	65	1.1	0	0.17	0.10	1.4	0
Noodles (containing egg)															
Unenriched, dry	381	12.6	73.2	0.4	3.4	9.6	1.2	22	199	2.1	200	0.20	0.11	2.3	0
Unenriched, cooked	67	2.2	12.8	0.1	0.6	83.8	0.6	4	35	0.4	30	0.03	0.02	0.4	0
Unenriched, cooked (with 3:1 water absorption)	124	4.1	23.8	0.2	1.1	69.9	1.1	7	65	0.7	56	0.06	0.04	0.7	0
Enriched, dry	381	12.6	73.2	0.4	3.4	9.6	1.2	22	199	2.9	200	0.88	0.37	6.0	0
Enriched, cooked	67	2.2	12.8	0.1	0.6	83.8	0.6	4	35	0.5	30	0.14	0.06	1.0	0
White rice															
Raw	362	7.6	79.4	0.2	0.3	12.3	0.4	24	136	0.8	0	0.07	0.03	1.6	0
Cooked	119	2.5	26.2	0.1	0.1	70.5	0.7	8	45	0.3	0	0.01	0.01	0.4	0
Precooked, dry	382	8.8	83.3	0.4	0.2	7.6	0.1	4	66	0.8	0	0.02	0.02	0.1	0
Potatoes															
Raw	83	2.0	19.1	0.4	0.1	77.8	1.0	11	56	0.7	20	0.11	0.04	1.2	17
Baked	98	2.4	22.5	0.5	0.1	73.8	1.2	13	66	0.8	20	0.11	0.05	1.4	17
Boiled, peeled before cooking	83	2.0	19.1	0.4	0.1	77.8	1.0	11	56	0.7	20	0.09	0.03	1.0	14
French fried	393	5.4	52.0	1.1	19.1	19.6	3.9	30	152	1.9	50	0.18	0.11	3.3	28
Corn															
Raw	92	3.7	20.5	0.8	1.2	73.9	0.7	9	120	0.5	390	0.15	0.12	1.7	12
Cooked	85	2.7	20.2	..	0.7	75.5	0.9	5	52	0.6	390	0.11	0.10	1.4	98
Grits, unenriched, cooked (degermed)	51	1.2	11.0	0.1	0.1	87.1	0.6	1	10	0.1	40	0.02	0.01	0.2	0
Grits, enriched, cooked	51	1.2	11.0	0.1	0.1	87.1	0.6	1	10	0.3	40	0.04	0.03	0.4	0
Minimum daily requirements (Adults) (FDA)	3500	65.0	..	..	..	..	..	750	750	10.0	4000	1.0	2.0	16.0	30

In the opinion of the authors, the figures given for cooked noodles are not fairly representative. Apparently, the tests were made on canned noodles or something of the kind, because the samples must have gained much more water than normal.

Information, released by the National Macaroni Manufacturers Association in 1958, indicates a calorie level for cooked macaroni of 115 to 118 Cal per 100 gm, somewhat lower than the figures in Table 66. It should be pointed out that macaroni products continue to absorb water for quite a long time during cooking. Therefore, differences in length of cooking would have an important bearing on the calorie level of a cooked product, as well as on other qualities, such as percentage of protein.

MACARONI PRODUCTS QUALITY CONTROL

A few simple tests are needed to control the quality of macaroni products and raw materials. These are described in detail in *Cereal Laboratory Methods, Seventh Edition American Association of Cereal Chemists 1962, Revised,* and *Methods of Analysis of the Association of Official Agricultural Chemists, Current Edition.*

Sifting Tests

The tailings from the sifter should be examined to see if the flour or semolina is infested with insects.

Granulation Test

A rotary stack of sieves should be used to determine the particle size distribution. All of the semolina should go through a US No. 20 sieve and not more than 3% should go through a No. 100 sieve. A typical granulation quality standard for a mill might show the following percentages on the various screens:

$$
\begin{array}{l}
\text{On 20— 0\%} \\
\text{On 28— 0.5\% max} \\
\text{On 35—24.0\% max} \\
\text{On 42—22} \pm 5\% \\
\text{On 60—33} \pm 5\% \\
\text{On 100—18} \pm 5\% \\
\text{Through 100— 3.0\% max}
\end{array}
$$

Ash Test

The standard of identity specifies that durum flour should have an ash content calculated to a moisture-free basis of not more than 1.5%. Semolina standards specify an ash content of not more than 0.92%, although it is generally agreed that a good semolina should not run more than about 0.65% ash.

Brabender Farinograph Test

Irvine and Anderson (1951) have found that macaroni making qualities of semolinas and flours can be predicted by a Farinograph test conducted with a 50-gm mixer and with an absorption equivalent to 31% moisture in the dough. A desirable raw material will rise to a peak fairly rapidly and will break down slowly. A material which has a wide and uneven trace will have a tendency to form a rough product. A raw material which drops off from the peak rapidly will break down in the auger and extrusion tubes of the press and would be expected to give an uneven extrusion pattern.

Speck Test

The semolina is spread over a flat surface with a 1-in. grid and the number of black specks and bran specks per square inch is counted.

Grit Test

The semolina is floated off on carbon tetrachloride and the grit which is sand and other stony material settles to the bottom. It is collected on filter paper and weighed.

Moisture Test

The most important test in a macaroni plant is the moisture test which is generally performed by drying ground macaroni at 266°F for 1 hr. There are also electrical methods which measure the resistance of the product and high speed drying methods using infrared and other means for quick evaporation.

This test is used for analyzing the operation of the driers. Under ordinary conditions the final moisture of macaroni products should be approximately 12%. Normally, they will dry down to 10% or less in the package unless it is moisture-vaporproof.

Evaluation of Dry Macaroni Products

Figure 98 shows a sheet for evaluating the quality of a dry macaroni product. Demerits are assigned to each fault of the finished products. All of the demerits are added and subtracted from 100 to give the quality score. The total score makes it possible to rate samples relative to each other and the breakdown of demerits for each product shows exactly what is wrong so that the cause can be determined.

Code No. _____   Panel Member _____
Description _____

	Range of Demerits	Demerits
Serious flaws		
Checked	(0–20)	( )
Split	(0–10)	( )
Deformed	(0–10)	( )
Color		
Gray or brown (10 = very dark)	(0–10)	( )
Yellow (5 = no yellow)	(0–5)	( )
Appearance		
Large bubbles (poor predrying)	(0–5)	( )
Small bubbles (poor vacuum)	(0–5)	( )
White specks	(0–5)	( )
Dark specks	(0–5)	( )
Rings	(0–5)	( )
Streaks	(0–10)	( )
Roughness	(0–10)	(_____)
Total Demerits	(0–100)	( )
	Score (100-demerits)	( )

FIG. 98. MACARONI PRODUCTS DRY PRODUCT EVALUATION

Cooking Test Profile

Fifty grams of sample are boiled in 1 liter of distilled water (see Fig. 99). In the case of spaghetti the strands are cut to a uniform length of 4 in. A portion of the sample is removed 1 min

Code No. _____   Date of Test _____
Description _____
Job No. _____ P.O. No. _____ Sample No. _____ Panel Member _____

Dimensions: Diameter or Width _____ Wall Thickness _____
Cooking Time (T) _____ How Determined: Glass Plate _____ Bite _____
Appearance of Cooking Water after _____ Minutes of Cooking:
Cloudiness: None ( ) Slight ( ) Moderate ( ) Extreme ( ) Odor of
                                                                        Cooking
Yellow Color: None ( ) Slight ( ) Moderate ( ) Extreme ( ) Water _____

No.	Time	Range of Demerits	Cooking-time Periods ____ Minutes
1	Gray or brown color (5 = very dark)	(0–5)	
2	Yellow color (5 = no yellow)	(0–5)	
3	Surface irregularity	(0–5)	
4	Splitting or breaking	(0–10)	
5	Stickiness	(0–10)	
6	Slime	(0–10)	
7	Odor	(0–10)	
8	Taste	(0–10)	
9	Too soft	(0–10)	
10	Too firm	(0–5)	
11	Sticks to teeth	(0–10)	
12	Doughiness or lack of elasticity	(0–10)	
	Total Demerits	(0–100)	
	Score (100-demerits)		

General Observations:

FIG. 99. MACARONI PRODUCTS COOKING TEST PROFILE REPORT

after it has cooked for the correct time. In the case of spaghetti the correct cooking time is determined by smashing a strand between two microscope slides or other glass plates until the uncooked line of dough at the center of the strand of spaghetti has disappeared. Demerits are assigned for each of the quality defects. These are totaled and subtracted from 100. This is repeated at cooked time plus 6 min and cooking time plus 11 min. This gives a profile according to the type of flaw in a particular sample and also gives a measure of resistance to over cooking.

Borazzio Cooking Test

The Italian professor Borazzio (as quoted by Hummel 1950) developed a cooking test which is widely used in Europe and for which equipment is made by Buhler Brothers, Uzwil, Switzerland. Samples of 250 gm of dried products are cooked in 1 liter of 1% salt solution in each of 2 cooking vessels, heated in a constant temperature oil bath at 218°F. This causes the product to cook at a temperature of 208°F, so that disintegration of the product caused by the turbulence of boiling is kept to a minimum. The cooking time in one vessel is 18 min and the cooking time in the other vessel is 28 min. After the macaroni products have been cooked, the samples are drained for 5 min and then weighed. The difference in weight between the cooked sample and the dry sample gives the water absorption during cooking. The volumes of the dry samples and the cooked sample are measured by water displacement and the increase in volume recorded. The water in which the samples have been cooked is placed in a graduated glass tube and the suspension of starch and other materials is allowed to settle for ½ hr. The height of the milky portion which settles out at the bottom is measured. In some cases, a portion of the cooking water is evaporated to dryness to measure the amount of solids dissolved in the water. It is stated that less than 6% dissolved in the cooking water is very good and that 8% is about average.

Radley Cooking Test

Radley (1954) describes an interesting cooking test which might have wide application in research work. Approximately 3 gm of the product to be tested are placed in a wire basket with a lid. This is weighed in cold water at 68°F and is then plunged into rapidly boiling water. At a given time the product is removed from the boiling water and reweighed in cold water. It is then slung to remove the moisture and the outside of the basket is wiped dry. The basket is weighed in air. Then the product is returned to the boiling water. This is repeated until the product has completely disintegrated and the basket is then weighed in water, empty. The curve plotted from the weight values determined in air indicate the rate of absorption of water. The curve obtained from plotting the values obtained by weighing in water shows the rate at which the product is disintegrating since the water in the macaroni does not contribute to the observed weight when the product is submerged.

Egg Solids

This test is made to see whether the actual solids content of the yolk, whole egg, dried egg or egg white meets the guarantee furnished by the supplier. This test is made with an air oven and an analytical balance and is performed by measuring the weight loss of a sample placed in an air oven for 16 hr at 212° to 217°F.

Frozen dark-colored egg yolks will normally contain about 45% solids. Frozen whole eggs contain about 26% solids. Dried yolks and dried whole eggs and dried whites all contain approximately 95% solids. Frozen egg whites normally contain about 12.5% solids.

Color

Color of macaroni made from a given raw material can be predicted by:

(1) Determination of pigment content and lipoxidase activity. (2) The manufacture of a disc of macaroni dough which is evaluated by: (a) A reflecting photometer (Matz and Larsen 1954). (b) Visual comparison with whirling paper disks, consisting of white, black, yellow, and red. (c) Visual comparison with other samples. (3) The manufacture of a noodle about 50 mm wide and comparing its color with the whirling disk. (4) The manufacture of macaroni on small laboratory equipment.

Method (1) was worked out by Irvine (1955) of the Grain Research Laboratory of the Board of Grain Commissioners in Canada. It is based upon the fact that lipoxidase activity destroys the pigment in semolina during mixing. This is the fastest method, requiring only 30 min to obtain results. The pigment is determined by extraction with n-butyl alcohol solution and measurement of light transmittance at selected wave lengths.

Lipoxidase activity is measured by Warburg techniques. The

predicted macaroni color score is obtained by substituting the top values of light transmittancy and lipoxidase activity into an equation which can be found in the original reference. Extensive research shows good correlation between predicted and natural color scores.

Method (2) uses macaroni disks formed by mixing, kneading, sheeting, pressing and drying a small quantity of semolina to form a disk approximately 50 mm in diameter 3 mm thick. This is an accurate method, but drying of the disks takes as much as two days.

Method (3), using a rolled noodle in place of the disk, is faster because the porous noodle dries faster than a disk without checking, but the opacity caused by air bubbles obscures the yellow color somewhat.

Method (1) is recommended because it is simple and fast. However, it does not predict the amount of grayness in the macaroni. This can be determined by studying the color of the gluten washed out of the semolina.

BIBLIOGRAPHY

ANON. 1965. Official and Tentative Methods of Analysis, 10th Edition. Assoc. Offic. Agr. Chemists, Washington, D.C.

ANON. 1967A. Composition of Foods. US Dept. Agr. Handbook 8.

ANON. 1967B. Definitions and Standards for Foods. Title 21, Part 16. Macaroni and Noodle Products. US Dept. Health, Education & Welfare, Washington, D.C.

BERNHARDT, E. C. 1965. Processing of Thermoplastic Materials. Reinhold Publishing Corp., New York.

CUNNINGHAM, R. L., and ANDERSON, J. A. 1941. Micro tests of alimentary pastes I. Apparatus and method. Cereal Chem. 20, 171–185.

EARLE, P. L. 1948. Studies in the Drying of Macaroni, Factors Affecting Checking. Ph.D. Thesis, Univ. of Minnesota, St. Paul, Minn.

EARLE, P. L., and ROGERS, M. C. 1941. Drying macaroni. Ind. Eng. Chem. 33, 642–647.

FIFIELD, C. C., SMITH, G. S., and HAYES, J. F. 1937. Quality of durum wheats and a method for testing small samples. Cereal Chem. 14, 661–673.

HUMMEL, C. 1950. Macaroni Products Manufacture, Processing and Packing. Food Trade Press, London.

IRVINE, G. N. 1955. Some effects of semolina lipoxidase activity on macaroni quality. J. Am. Oil Chemists' Soc. 32, 558–561.

IRVINE, G. N., and ANDERSON, J. A. 1951. Air bubbles in macaroni doughs. Cereal Chem. 28, 240–243.

IRVINE, G. N., and ANDERSON, J. A. 1954. An improved wheat prediction test for macaroni quality. Cereal Chem. 32, 88.

MALDARI, C. D. 1956. Tracing production problems to die wear. Hoskins Macaroni Production Manual. Glenn G. Hoskins Co. Libertyville, Ill.

MATZ, S. A., and LARSEN, R. A. 1954. Evaluating semolina color with photoelectric reflectometers. Cereal Chem. *31*, 73–86.

POOLE, R. S. 1955. Production of precooked alimentary paste products. US Pat. 2,704,723. March 22.

PORTESI, G. 1957. The Alimentary Paste Industry. Editrice Molini D'Italia, Rome. (Italian)

RADLEY, J. A. Apr. 1954. Chemistry and Physics of Macaroni Products. British Magazine, "Food Manufacturing."

SCOTLAND, B. S. 1966. Processing for preparing a ready-to-eat alimentary paste. U.S. Pat. 3,251,694. May 17.

TRESSLER, D. K., VAN ARSDEL, W. B., and COPLEY, M. J. 1968. The Freezing Preservation of Foods, 4th Edition, Vol. 4, Freezing of Precooked and Prepared Foods. Avi Publishing Co., Westport, Conn.

CHAPTER 8

Samuel A. Matz

Starch and Oil Production from Cereals

INTRODUCTION

Starch was separated from other grains and from root vegetables such as potatoes long before corn was used as a raw material. Doubtless the development was based on the observation that a white, insoluble granular material settled to the bottom of the pan when quantities of cut tubers were washed. John Biddis set up the first starch factory in the United States at Hillsborough, N.H. in 1802. He used potatoes. In 1842, Thomas Kingsford founded the cornstarch refining industry. He was the first person to extract starch from maize on a commercial basis. This advance occurred at the factory of William Colgate and Co., in Jersey City, N.J. Kingsford's process used the same basic principles still in effect, but many technical improvements have since been made.

A few years later, three corn starch plants were erected, notable among them being the Kingsford plant at Oswego, New York and the Duryea plant at Glen Cove, Long Island.

A booklet by Jeffries (1942) contains much information about the operations of the Duryea plant. It was entirely self-contained and made all its own equipment, boxes, and wooden tanks. It had a press for printing its labels and manufactured its own illuminating gas. There was not an electric motor or electric light in the plant in 1891.

The corn was placed in wooden, flat bottomed tanks, covered with warm water and allowed to stand. After the corn was sufficiently softened, it was ground in stone mills and sieved and washed on silk screened shakers powered by reciprocating engines. The slurry that was washed through the sieves was put in wooden tubs, treated with caustic soda and allowed to settle. After settling, the water was sent to the sewer, taking all of the gluten and soluble materials with it. This settling process was repeated three times for each batch.

Yield calculations were not made in those days, but it seems probable that the starch recovery could not have been much above 50% of the total starch available. No chemical tests were run in the plant, but an effort was made to control the alkalinity of the settling tanks. Key employees were trained in testing the slurry in the settlers and they either approved the batch or caused more caustic soda to be added. When the starch was hydrolyzed to convert it to "glucose," tasters who had no other responsibility, judged the completeness of the conversion.

From these crude beginnings have come the modern efficient corn refining plants. They do not empty their valuable products down the drain, they have excellent control laboratories, and they use the most modern equipment available to produce quality materials economically and maintain good material balances (Goodwin 1959).

Although some other grains are wet-milled, corn far outweighs the combined total of other cereals given this treatment. For this reason, the following discussion will be primarily concerned with corn.

THE CORN KERNEL

The composition of the corn kernel on a dry basis is shown in Table 67. These figures are subject to slight variations depending on weather conditions during the growing season, type of seed and soil, amount and type of fertilizer, and other factors, but represent good averages for yellow dent corn. The water content of the corn which comes into the wet-milling plants varies from 10 to 25%.

TABLE 67
COMPOSITION OF THE CORN KERNEL

	%
Starch	73
Sugars	3
Pentosans	4
Protein	10
Oil	4.5
Fiber	3.5
Minerals	2

THE WET-MILLING PROCESS[1]

Raw Material

The corn wet-millers use only shelled corn and primarily No. 2 yellow dent corn. These kernels are shipped into the plants at a typical rate of about 260 carloads per day. The corn is transferred from boxcars to cleaners and then to temporary storage bins to await further handling.

---

[1] Based in part on GOODWIN (1959).

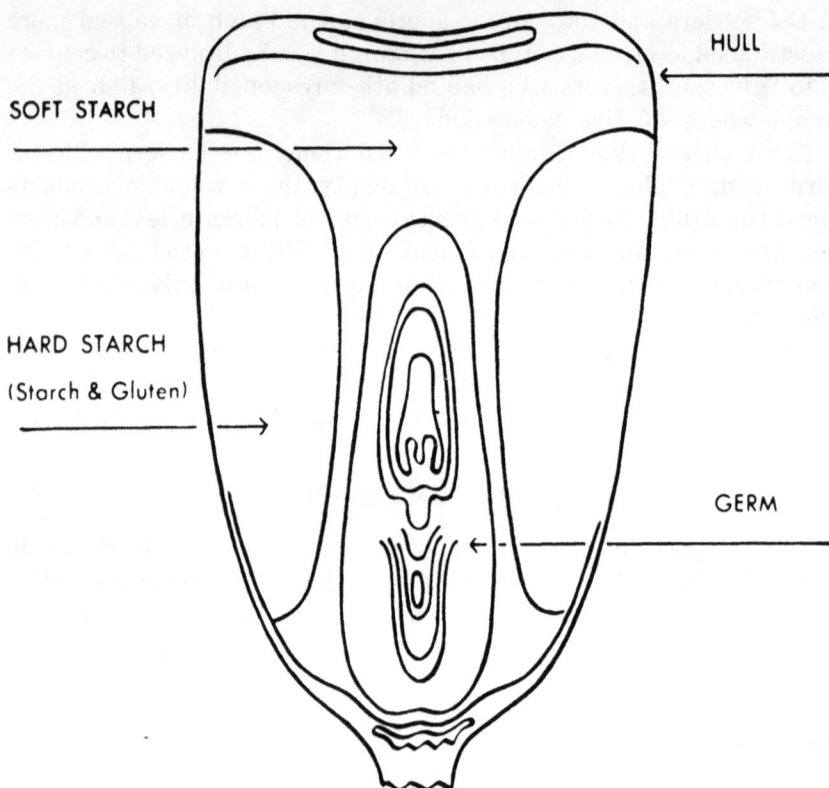

FIG. 100. DIAGRAM OF THE CORN KERNEL

The cleaning operations involve passing the corn past powerful magnets which remove scrap iron, steel, or nails which may have been introduced in the corn by previous handling. After passing the magnets, the corn is weighed on scale hoppers and sampled for quality.

After sampling and weighing, the corn is further cleaned by passing over perforated screens. The upper screen has holes just enough to let corn and smaller particles through and the lower screen holds back the corn but lets smaller pieces of chipped corn, cobs, sticks, and stones through. Next the corn is blown with high-pressure air. This removes low density materials and dust which are delivered to cyclone dust collectors. Because the corn itself is considerably agitated by the air, a further separation of higher density materials occurs and these materials are removed from the bottom of the flowing stream of corn.

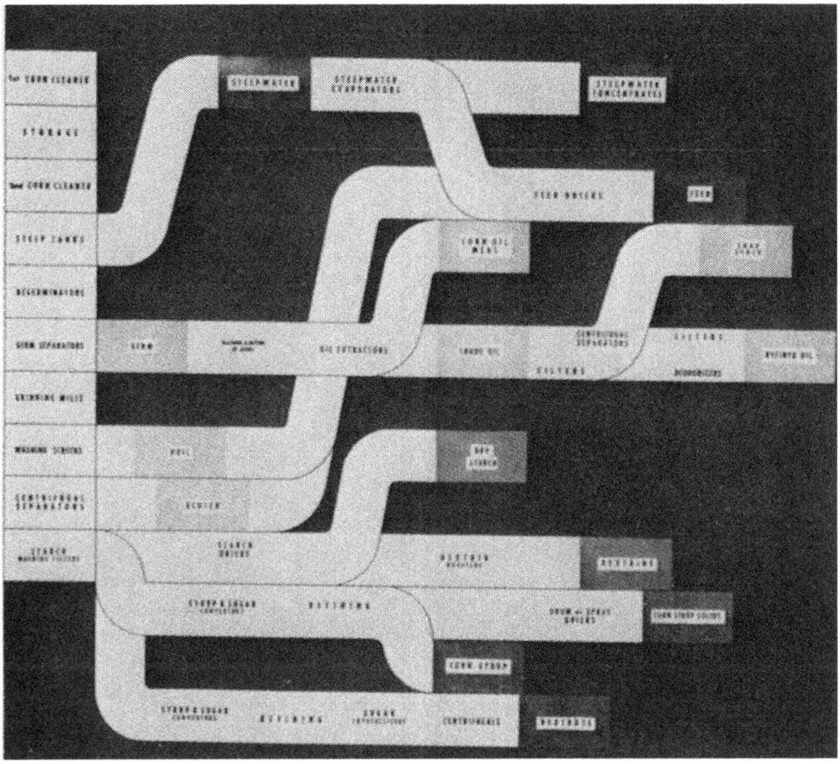

Courtesy of Corn Industries Research Foundation, Inc.

FIG. 101. FLOW DIAGRAM OF THE WET-MILLING PROCESS

Steeping

The cleaned kernels are transferred to "steep" tanks holding 2,000 to 3,000 bu and soaked for 36 to 48 hr in water at about 120° to 130°F. The water contains a slight amount of sulfur dioxide which prevents germination and keeps down fermentation and other undesirable microbiological changes but permits growth of lactic acid producing bacteria. This acid buffers the pH and aids in softening the kernel. The corn refiners manufacture their own sulfur dioxide by burning sulfur in rotary burners.

In the early days of wet-milling of corn, batch methods were used in steeping. In the other steps of processing, water was discarded. The result was a large demand for water, and large amounts of product were lost in the waste water discharged from the plant.

FIG. 102. CORN BEING DROPPED OUT OF ONE OF THE STEEP-
ING TANKS

It has been softened by soaking or "steeping," and carried
by a vibrating type conveyor to the grinding department.

From the early days of the industry, corn wet-millers had been
concerned with the waste water problem. They recognized that the
wastes produced pollution and odor problems and that part of their
raw material was being lost. In addition, there was a desire to
cut down on the water requirements, which sometimes tended to
outrun the available supply.

In the period 1925 to 1937, several inventors obtained patents
dealing with the "bottling up" or "corking up" of the corn
refining process. Among the inventors issued such patents were:
Jeffries (1930, 1937), McCoy (1931), Moffett (1928), Sherman
(1925), and Widmer (1926). These changes increased the yield

from 90 to almost 100%, greatly improved the efficiency of the entire process, reduced the waste disposal problems, and saved water, thus lowering the costs.

As a result of these developments, wet-milling is now conducted as a countercurrent system. The freshest water passes over the corn which has been steeped the longest, assuring removal of the maximum amount of soluble material. About 10 gal. of fresh water per bushel of corn are used to wash the separated starch at the end of the process. This water then flows countercurrently back through the separation system and finally goes to the steep tanks at about 1% solids. About 40% of the initial intake, or about 4 gal. per bu, are absorbed by the corn in the tanks, while the remainder is withdrawn as "light" steepwater containing about 5% dry substance—solubles leached from the corn. This steepwater is subsequently concentrated by evaporation to a slurry of about 54% dry matter. There are about 3 lb of solubles per bu of corn.

In the early days of the industry (1856 to 1890), steepwater was discarded, but the nutritive value was ultimately recognized and the wet-millers then started to concentrate it and add gluten and bran to make corn gluten feed. Steepwater is excellent for growing acid-producing bacteria and these bacteria are beneficial when fed to ruminants.

Degermination

The corn which has been softened by the steeping process is now ready for the first milling operation. This step, degermination, separates the oil-rich germ from the starch, gluten, hulls, and fiber.

The softened corn is passed over shakers in a second cleaning operation which removes any foreign material which has escaped the first series of cleaning operations or has been picked up in subsequent handling.

The corn is then ground in attrition mills. These attrition mills consist of two circular plates with projecting teeth on their facing surfaces. Some models have the 2 plates rotating in opposite directions while others have 1 stationary plate and 1 rotating plate. The spacing and other conditions are adjusted so that the rubbery germ is torn from the rest of the kernel, leaving the germ intact.

The slurry from the mills, which consists of endosperm, germ, and fiber, is diluted with a carefully controlled amount of process water. In the older style of factory, this mixture was transferred to a series of long troughs which served to separate the lighter

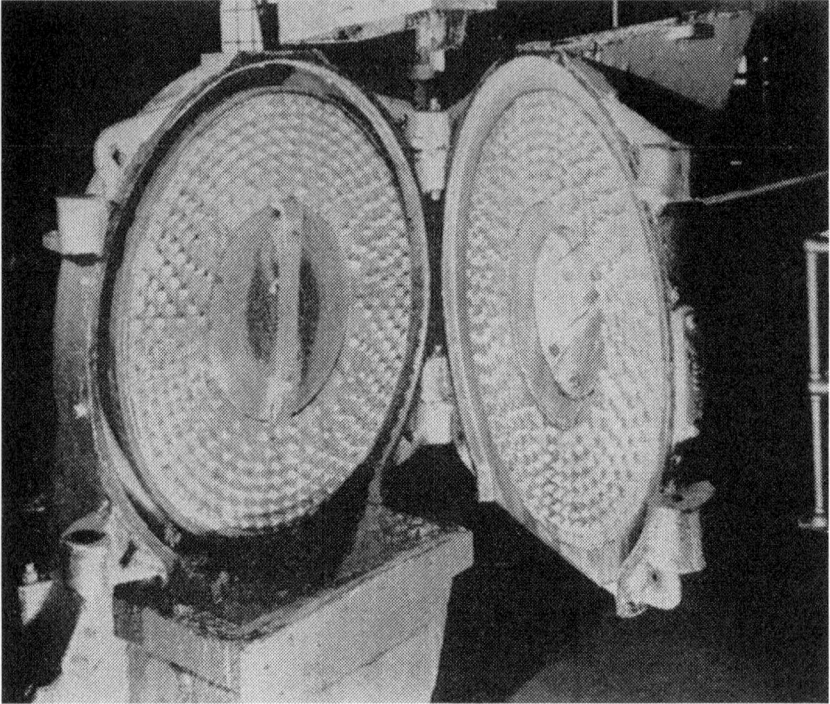

Courtesy of Corn Industries Research Foundation, Inc.

FIG. 103. VIEW OF AN OPENED DEGERMINATING MILL

oil-bearing germs from the rest of the material. There was an agitator in the lower part of the troughs which kept the products mixed and helped to free any germs which might have been entrapped. In the more modern plants, the mixture is fed to a battery of hydroclones. The germ, being lighter, spins off the top, and the heavier endosperm and fiber flow out the bottom. The germ is washed free of starch, dewatered, and dried.

In the old method, germs were removed from the top of the tanks along with some of the slurry. The germ-slurry mixture was then transferred to reels or shakers where the slurry was washed off and returned to the starch-gluten stream. At the same time that the germ was leaving the top of the tank, the mixture of starch, gluten, hulls, and fiber was leaving at the bottom, and it was sent to reels and shakers where most of the water and soluble materials were removed. Quality control tests were run to insure that the germ was completely separated from the starch and gluten, and, if not, the degermination process was repeated on the solids remaining after the starch-gluten slurry was removed and passed over the reels and shakers.

The germs are dried and sent to the oil extraction and refining plant. Since the extraction and refining process will be completely described later in this chapter, it will not be discussed here.

Separation of Hulls and Fiber from Starch and Gluten

The wet mash of fiber, hulls, gluten, and starch which remains on the reels and shakers is fed to mills which grind the materials to a very small particle size. The hulls and fibers are not reduced in size as much as the starch and gluten in the milling process.

The industry used hard, roughened flat mill stones for many years. These buhr mills were circular and rotated at high speeds over stationary stones. This process reduced the charge to a fine slurry. In recent years, the buhr mills have been replaced by stainless steel vertical mills which do the same job, require less maintenance, and last longer (Fig. 104).

In the older plants, the hull and fiber particles were removed

Courtesy of Corn Industries Research Foundation, Inc.

FIG. 104. BAUER ATTRITION MILL FOR FINE GRINDING OF CORN

One of these mills will do the work of several of the older type buhr mills and do it much more efficiently.

from the mill effluent by a series of reels (rotating hexagonal screens of copper) and shakers (flat nylon sieves). Modern plants separate fiber and hull from starch and gluten by screening over a series of D. S. M. (Dutch Slate Mines) screens. These have replaced all shakers and reels.

Gluten-Starch Separation

The corn refining industry has always used gravitational methods for separating starch and gluten. Initially, large wooden tubs were used to hold the slurry until the starch settled to the bottom. These were replaced by long, narrow slightly inclined wooden troughs called starch tables which permitted a constant flow of the starch-gluten slurry.

The starch tables were about 2 ft wide and 100 ft long. They had a pitch of about 5 in. which caused the slurry to flow slowly down them. The tables were almost always wooden although other materials such as concrete and slate have been used. For each 1,000 bu of corn that were ground, about 3,000 sq ft of table area were required. Since many plants mill upwards of 50,000 bu per day, the floor space requirements for tabling starch were enormous. In fact, the table house was usually the largest building in the refinery.

The starch gluten slurry flowed slowly down the tables and sedimentation occurred with the heavier starch depositing on the bottom and the gluten being removed at the top of the lower end of the tables. The first tabling operation produced starch containing small percentages of gluten and gluten containing larger percentages of starch. Further purification of the starch was accomplished by repeating the tabling operation or by washing the starch on rotary vacuum filters.

The tabling operation played a major role in improving the yield of products and lowering their cost. It did cause the loss of some starch in the effluent, and centrifuges were first introduced to recover additional starch and to raise the protein content of the gluten-containing fraction. In the 1940's centrifuges began to replace tables for the primary starch-gluten separation. Large, highly efficient continuous centrifuges now take the starch-gluten slurry, handle large volumes of materials, and yield products of high purity (Fig. 105). The tremendous floor space required by the tables is no longer needed. The entire operation is now inclosed and a major source of air contamination in the plant has been eliminated.

Courtesy of Corn Industries Research Foundation, Inc.

FIG. 105. INTERNAL VIEW OF PRIMARY SEPARATION CEN-
TRIFUGALS

In these machines gluten is removed from starch by centrifugal action.

The starch stream from the centrifuges contains about 1 to 2%
protein. It is further purified to less than 0.3% protein by passing
it through many small hydroclones. Fresh water is added only at
this point in the entire wet-milling process (subsequent to the
steeping operation). Its purpose is to wash out the last traces of
the solubles in the starch. Normally, it takes 7 to 10 gal. of water
per bushel of corn.

The wet starch is dewatered on rotary vacuum filters, moving
belt filters, or basket centrifuges. The final drying takes place in
tunnel (kiln) driers, continuous belt hot air driers, or in spray
driers.

The gluten after centrifugation is a thin slurry that is either separated by sedimentation in large tanks or, more generally, is "dewatered" and "destarched" in another centrifuge, filtered, and dried. It is then ready to be used as corn gluten meal or corn gluten feed, or processed to recover the protein, zein, which has many nonfeed uses.

Sorghum milo can be processed in much the same way as corn to yield a starch that is about the equal of cornstarch in most of the important characteristics. Sorghum starch is assuming increasing importance in the market place, especially for nonfood uses.

PREPARATION OF WHEAT STARCH

Although wet-milling of the whole wheat kernel has been used to produce wheat starch, it is believed that most current processing methods are based on the initial formation of a dough from low-grade wheat flour. This dough is washed in large volumes of water while being kneaded to release the starch. The dough may be formed in conventional batch mixers or in some type of continuous equipment.

In the Martin process, a pair of reciprocating rolls knead the dough under a spray of water. In the so-called batter method, the dough is chopped up in a cutting pump with additional water, and the gluten curds which form are separated from the much finer starch particles on shaker screens.

The crude starch suspension is passed through fine sieves to separate small fibers and small pieces of gluten. Further refining follows the same principles as in corn wet-milling, with the starch being separated on tables or in centrifuges. The damp starch can be dried in the same way as corn starch or in almost any other conventional drier, that is, in tunnel, tray, roll, spray, or flash driers (Anderson 1967).

Gluten is a valuable by-product, and the characteristics of the finished materials are very dependent upon the drying conditions. High temperatures lead to denatured gluten, which is useful as a protein supplement in foods and feeds. If the physical properties of the native gluten are desired (i.e. cohesiveness and elasticity), relatively gentle drying conditions must be used, such as spray-drying a 10% dispersion or vacuum-drying of small pieces.

Since the raw material for wheat starch production is generally a flour produced by conventional dry milling techniques, the starch system does not yield hulls or germ, although these fractions as well as some endosperm are recovered from the flour milling step. There is also no wheat equivalent of corn steepwater. The effluent from the washing process with about 10% solids is discarded.

An alternate method of producing wheat starch is based on dissolving away the gluten in an alkaline wash, while the starch granules remain relatively unaffected and are separated off.

PRODUCTION OF RICE STARCH

Rice starch is currently being made in Europe but apparently there are no commercial producers in the United States. The high cost of the raw material makes the starch uneconomical except for a few very specialized uses such as a component of face powder.

Simple washing methods will not release rice starch granules from their protein matrix. Chemical treatment is necessary to soften and disperse these proteins. Caustic soda solutions are generally used to effect the dissolution. For example, broken rice may be steeped for 24 hr in 5 times its weight of a 0.3% caustic soda solution as the preliminary step in rice starch production. The solution may be at ambient temperature or heated to about 120°F.

The caustic treated granules are washed and then dried in bags before being ground into flour. The flour is mixed with about ten times its weight of caustic soda solution. After a 24 hr treatment, the starch is allowed to settle and the supernatant solution, which contains most of the protein, is removed. Washing with water, settling, and decanting are employed to remove most of the remaining soluble materials from the starch granules.

The washed starch is dewatered by filtering or centrifuging and further dried in ovens or rotary driers. The starch cake is ground to the desired size and sieved.

Protein in the combined effluents is precipitated by the addition of hydrochloric acid. The supernatant fluid is discarded, and the precipitated material is partially dewatered in a filter press and finally dried by heat. It can be used as a protein supplement for cattle feed (Hogan 1967).

STARCH CONVERSION

Acid Hydrolysis

About half of the cornstarch that is produced is further processed to make corn sugars and syrups. This process, called "conversion" by the industry, involves catalytic hydrolysis of the starch. In the conversion process as well as in dextrinization,

starch modification, and starch derivation, the nature of manufacturing operation undergoes a pronounced change. The milling
process is entirely mechanical in nature and resembles other
milling operations. It differs from most of the others mainly in
the large amounts of water which are used. Conversion and the
other processes mentioned above are largely chemical in nature.

Starch is made up of amylose, a linear fraction, which in turn is
comprised of many dextrose molecules linked chemically in the 1–4
position and amylopectin, a branched molecule, in which branching
occurs at the 1–6 position. There is about 1 branch per 25 to 30
dextrose units in amylopectin. Since the ratio of amylopectin to
amylose in cornstarch is roughly 3:1, there is about one 1–6
linkage, for every 30 to 40 of the 1–4 linkages. Wolfrom and
Thompson (1956) report finding very small amounts of di- and
trisaccharides containing the 1–3 linkage.

There is an excellent discussion of the theoretical and experimental work which has been done in an attempt to clarify the
mechanism of the acid hydrolysis of starch in the books by
Whistler and Paschall (1967). A great deal of speculation and
hypothesizing has been done as to whether the two principal types
of linkages have different rates of hydrolysis. The observed results
lead to the conclusion that, from the practical standpoint, the
rates are not sufficiently different to affect significantly the nature
of the products formed. In enzymatic hydrolysis, which will be
discussed later, a different set of conditions prevails.

When starch is hydrolyzed with water and an appropriate
catalyst, the elements of water are added to the starch molecules.
This means that a given amount of starch will produce more than
that amount of hydrolytic products. The extreme case is the conversion of starch to the ultimate product, dextrose. The relationship is shown in the equation

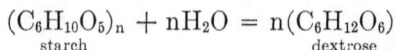

$$\underset{\text{starch}}{(C_6H_{10}O_5)_n} + nH_2O = \underset{\text{dextrose}}{n(C_6H_{12}O_6)}$$

In this case, 162 grams of starch react with 18 grams of water
to produce 180 grams of dextrose. Lesser degrees of hydrolysis,
which produce maltose and other polysaccharides result in smaller
weight gains.

The rate of hydrolysis of starch is increased by raising the
temperature and by increasing the acid concentration. Because
dextrose may react under these conditions to produce a disaccharide
according to the equation

$$2\ C_6H_{12}O_6 \rightleftarrows C_{12}H_{22}O_{11} + H_2O$$

and because this is an equilibrium reaction, the overall rate of hydrolysis if measured by dextrose produced in a given time, appears to be slower at higher starch concentrations.

The reaction indicated by the equation above as well as the Maillard browning reaction and oxidation which results in the production of furfural-type products are all undesirable. These reactions all lead to decreased yields of desirable products and to the production of materials which have a deleterious effect on the flavor and color of the products. For these reasons, the hydrolysis conditions are usually chosen and controlled to minimize the undesirable side reactions.

The commercial hydrolysis of starch is carried out in such a way that two main classes of materials are obtained; the first class includes the more completely hydrolyzed products which are crystallizable. These are the corn sugars, mainly dextrose and low molecular weight polysaccharides, and highly purified, crystalline dextrose. The other type of material is the noncrystallizable group of products called corn syrups and maltodextrines containing dextrose, maltose, and other polysaccharides. Whistler and Hickson (1955) have published data on the chromatographic separation and identification of the carbohydrate components in corn syrup. These data indicate that the closer the hydrolysis is carried to completion, the more nearly the products approach dextrose in composition. It is interesting to observe that the dextrose and maltose components increase at the expense of the higher polysaccharides while the amounts of the carbohydrates containing from 3 to 9 dextrose units remain relatively constant. These data are shown in Table 68.

The hydrolysis proceeds slowly at temperatures below 212°F. For this reason, starch hydrolysis is carried out in pressure vessels so that higher temperatures may be maintained. These vessels hold a charge of between 3,000 and 4,000 gal. of starch suspension. Typical operating conditions for syrup and sugar manufacture are given in Table 69.

The progress of the hydrolysis may be followed by measuring the dextrose equivalent (DE) of the solution. This dextrose equivalent is a measure of the reducing sugars present expressed as dextrose. Since the technique for directly measuring reducing sugars is time-consuming, a derived relationship between dextrose equivalent and specific rotation is used. The specific rotation starts

TABLE 68

DETERMINATION OF COMPONENTS IN CORN SYRUP

Syrup	Dry Solids[1] %	Syrup DE[2]	Mono[3]	Di	Tri	Tetra	Saccharides Penta	Hexa	Hepta	Octa	Nona
1	94.10	18.0	5.1	4.8	1.1	4.8	1.2	5.1	5.0	4.1	3.6
2	83.92	26.3	8.3	7.8	1.0	6.2	1.1	7.0	6.3	5.9	5.2
3	79.00	32.6	11.4	9.2	1.3	10.5	1.3	8.6	7.8	5.8	5.1
4	81.80	43.3	19.4	14.4	1.1	10.6	1.2	9.7	8.6	6.2	5.1
5	81.7	49.7	26.1	15.1	2.9	11.2	2.3	9.4	8.4	··	··
6	82.50	55.6	30.9	15.2	3.5	10.0	3.5	9.4	7.2	7.0	5.0
7	82.60	59.9	34.6	15.6	3.7	10.6	3.2	8.7	6.4	3.7	··
8	70.29	63.0	38.9	22.0	3.8	9.4	3.4	8.2	5.5	··	2.4

[1] Dry solids determined by Filtercel method (Cleland and Fetzer 1941).
[2] DE-determined by modified Lane-Eynon method.
[3] Glucose values A corroborated by glucose dehydrogenase and Sichert-Bleyer determination.

TABLE 69

STARCH HYDROLYSIS CONDITIONS

	Syrup	Sugar
Starch, lb	14,480	5,270
Water, gal	2,340	2,930
Hydrochloric acid, 20 Be	50	90
Steam pressure, psi	30	45
Conversion time, min	30	40
Temperature, °F	284	302

at $+200°$ and gradually decreases during the course of the hydrolysis. Pure dextrose has a specific rotation of $+52.5°$ and the solution approaches this value as a limit. When the specific rotation reaches the desired value, the acid is neutralized with soda ash and the hydrolysis stops. Acid-enzyme syrups will vary considerably in specific rotation from acid hydrolyzed syrups of the same dextrose equivalent because of differences in their carbohydrate composition.

The neutralized syrup or sugar solution is filtered, passed over a charcoal adsorbent or an ion-exchange resin to remove color bodies and other impurities, and concentrated to the desired solids value by evaporators.

The starch hydrolysis process is now being performed in some plants by means of continuous pressure converters. The quality of the syrup is superior because the conditions are more uniform. Patents have been issued dealing with this subject to Sipyaguin and Shoemakher (1943) and Horesi (1944); Dloughy and Kott (1948) have reported the details of a continuous process for the manufacture of dextrose. However, most of the materials prepared today are being produced by a batch-continuous process.

Enzyme Hydrolysis

It was mentioned earlier that at high starch and high acid concentrations, hydrolysis results in the occurrence of undesirable secondary reactions. The products from the reactions have an adverse effect on the use of such materials in many food applications. These effects are particularly important in the more highly converted materials.

Dale and Langlois (1940) have patented a process which overcomes most of the problems. In this process, an acid hydrolysis is carried out until the desired dextrose equivalent has been reached. The acid is then neutralized, the syrup is filtered and concentrated. An enzyme is added and the syrup is further hydrolyzed. This type of process produces syrups which are sweeter and less viscous than the acid-converted products. In addition, the secondary reactions are reduced to a minimum and the off-taste products are eliminated.

Dextrose

Dextrose, the end product of starch hydrolysis, is given a special place in this discussion because of its large usage. Of all the corn sweeteners produced, including corn syrups, crude corn sugar, and dextrose, the latter amounts to approximately one-third of the total, or about one billion pounds annually.

The manufacturing information given above indicates that, when it is desired to produce sugars by the hydrolysis of starch, lower concentrations of starch in the slurry should be used. It is also customary to use slightly higher acid concentrations and higher temperatures for longer times. The secondary reactions are not so important because the dextrose is finally purified by crystallization and the undesirable flavor and color constituents end up in the mother liquor. Nearly all commercial dextrose is now produced by the enzymatic hydrolysis of starch, although acid hydrolysis may be used as an initial step.

Newkirk (1936) in a series of publications and patents has described the manufacture and problems involved in the purification of dextrose. The procedure involves introducing a solution of the starch hydrolyzate, which has been concentrated by evaporation, into large vessels which are equipped with agitators. The solution is "seeded" with dextrose crystals and held at a temperature just above ambient for several days. After the dextrose has crystallized, the mother liquor is removed by centrifugation. The practical details for obtaining the final product, crystalline dextrose, reworking the mother liquor and controlling the crystallization are all contained in the writings and patents of Newkirk. Probably the most important factor in Newkirk's work was the discovery that, if the crystallization was allowed to take place while the system was being agitated and if the concentration of the sugar solution was high, dextrose would crystallize and a nearly chemically pure product would be obtained.

As this work was continued, conditions were worked out which favored the production of beta-dextrose. This particular crystalline form of dextrose has the property of dissolving easily and rapidly in water. Its use in beverages and other products requiring ease of solution followed immediately.

PRODUCTION OF OILS FROM CEREAL GRAINS[1]

Oils derived from cereal grains constitute important products in commerce, but their total world production is considerably below levels of output for many other vegetable, marine, and animal oils and fats, These oils, corn, rice, and wheat, are produced over a world wide area. Corn oil is produced in its largest quantities in the United States (Jamieson 1943) although in the Union of South Africa, USSR, Argentina, and Canada small amounts of it are manufactured. Of edible oils from vegetable sources produced in the United States, those from the soy bean and cottonseed each greatly exceeded the total output of corn, rice, and wheat oils.

Were it not for the fact that during the preparation of by-products from corn, the germ is almost completely separated from the rest of the products, corn oil would not have become an important oil from a commercial standpoint (Jamieson 1943). Wheat germ oil has no significance as a food and is only of very minor economic importance (Levin 1958). Although of greater significance in other parts of the world, principally in the orient, rice oil has not yet reached a level of great commercial importance in the United States. It can be readily seen from the foregoing facts that corn oil is the only oil produced from the true cereal grains which is of major economic importance in the United States.

SOME CHARACTERISTICS OF THE CEREAL OILS

Classification of Cereal Oils

Vegetable oils and fats are difficult to classify according to a plan which takes into account the many factors which would simply and completely describe them. In Table 70 are set forth 2 attempts at classifications of the 3 oils obtained from cereal grains produced in quantity in the United States.

Although the cereal grain oils are sometimes classified as semi-drying type oils on the basis of their iodine values, such a system of classification fails to take into account a number of rather important distinctions. These cereal oils contain principally oleic, linoleic and palmitic fatty acid esters, in that order (except for wheat oil, where the linoleic acid content slightly exceeds that of oleic acid), while linoleic acid or other fatty acids of greater unsaturation than linoleic are entirely absent from corn and rice oil.

---

[1] This section was written by RICHARD I. MEYER and JOHN IVERSON.

TABLE 70

CLASSIFICATION OF THREE CEREAL OILS[1]

TABLE 70

CLASSIFICATION OF THREE CEREAL OILS[1]

Oil[2]	Chemical Group or Type[3]	Botanical Name[4]
Corn	oleic-linoleic acid	*Zea mays*
Rice	oleic-linoleic acid	*Oryza sativa*
Wheat	linoleic acid	*Triticum sativum*

[1] All of these oils fall into the Hilditch Classification Group IVa, mono- and di-ethylenic and saturated fatty acids (containing principally oleic, linoleic and palmitic acid components). See Dean (1938).

[2] All of these oils are derived primarily from plants cultivated or processed in the main for products other than the oil.

[3] Bailey (1951).

[4] Williams (1950).

Table 71 indicates the disposition of the principal fatty acid components in corn, rice, and wheat oil. It has been thought that, since these oils do not contain linolenic acids or other highly unsaturated fractions, they are quite free from the tendency toward flavor reversion which is found in soybean and similar oils.

Corn Oil

Crude corn (or maize) oil has a dark reddish amber color. Even after refining, it is considerably darker than many other vegetable oils. By applying strong bleaching action, however, it can be lightened to a golden yellow color. This oil, unless subjected to considerable deodorization, retains the strong taste and odor characteristic of the original corn kernel. The unrefined oil contains relatively large amounts of phosphatides and other nonoil sub-

TABLE 71

PRINCIPAL FATTY ACID COMPONENTS IN CORN, RICE AND WHEAT OILS

	Saturated Acids		Unsaturated Acids			
Oil	Palmitic %	Stearic %	Oleic %	Linoleic %	Linolenic %	References
Corn	7.8	..	46.3	41.8	..	Dean (1938)
	12.0[1]	..	45.0	42.0	..	Hilditch (1949)
	11.0	2.9	48.8	34.0	..	Longenecker (1939)[2]
	8.1	2.5	30.1	56.3	..	Bauer and Brown (1945)
Rice	18.0	2.8	48.2	29.4	..	Dean (1938)
	17.1[1]	..	46.3	33.1	..	Murti and Dollear (1948)
	11.7	1.7	39.2	35.1	..	Jamieson (1926)[3]
	16.5	1.7	43.7	26.5	..	Jamieson (1943)[4]
Wheat	..	16.1	28.2	52.2	3.5	Dean (1938)
	13.8	..	30.0	41.1	10.8	Dean (1938)
	16.4	5.6	11.5	57.3	29.2	Gunstone and Hilditch (1946)
	..	15.5	25.5	52.6	6.3	Radlove (1945)

[1] Undesignated saturated fatty acids.

[2] Data are in mol. %.

[3] Rice plant variety: American.

[4] Rice plant variety: Hambus.

TABLE 72

OIL CONTENT OF THE WHOLE SEED AND THE VARIOUS PORTIONS OF CORN, RICE AND WHEAT

	Portion of the Grain		
Oil	Whole %	Bran %	Germ %
Corn	3–5.5[1]	. . .	30–35[1]
	3–6.5[2]	. . .	40–50[3]
Rice	14[3]	8–16[2]	. . .
Wheat	2[2]	5– 6[2]	12–18[2]

[1] Williams (1950).
[2] Jamieson (1943).
[3] Bailey (1951).

stances (often in excess of 2%) and its free fatty acid content (usually above 1.5%) is higher than that found in most vegetable oils. The refined oil contains small amounts of waxes which cause the oil to cloud at refrigerator temperatures unless it is selectively removed by a process called winterization. This process will be described in a later portion of this chapter. The keeping quality of refined corn oil is fairly good. The crude oil, however, is rapidly hydrolyzed, unless processing proceeds without delay. Such degradation is hastened when the oil remains in the presence of corn meal impurities. The germ portion contains from 30 to 50% of the oil of the corn kernel. Table 72 shows the oil content of various portions of the corn, rice and wheat plant. Table 73 indicates the many physical and chemical properties peculiar to corn, rice, and wheat oils.

Rice Oil

The color of rice oil, in either the refined or the crude state varies to a considerable extent. Oils which have attained a comparative

TABLE 73

CHEMICAL AND PHYSICAL PROPERTIES (WITHIN USUAL LIMITS) FOR CORN, RICE BRAN, AND WHEAT GERM OILS

Property	Corn	Rice Bran	Wheat Germ
Solidifying point, °F	5–14	2	. . .
Refractive index (Zeiss at 104°F)	58.5–60.5	61–68	72–80
Specific gravity 59.9/59.9°F	0.922–0.926	0.918–0.928	0.928–0.938
Titer	57–61[1]	77.4[2]	. . .
Iodine value	105–125	92–109	115–125
Saponification value	189–193	179–193	185–192
Unsaponifiable matter, %	0.8–2.0	2.7[3]	2–5
Free fatty acids, % as oleic	1.0	5–50	. . .

[1] Hilditch (1949).
[2] Murti and Dollear (1948).
[3] Williams (1950).

degree of acidity are difficult to refine, and, in particular, to bleach to a color considered acceptable for an edible oil. It is held that the oil extracted from rice is invariably high in its free fatty acid content (Bailey 1951). This high acidity is attributed to the activity of an unusually active lipase. It has been claimed, however, that if caustic refining of the rice oil is accomplished soon after extraction, much of this lipolytic activity is retarded (Jamieson 1943). Once processed the oil is reputed to possess unusual stability. Bailey (1951) suggests that this property is probably due to the presence of some potent antioxidants naturally present in the rice oil. Tables 71 and 73 show some physical and chemical properties of rice oil. Fuege and Reddi (1949) have shown that a tasteless and odorless oil can be produced through conventional refining, bleaching and deodorization of the oil.

Wheat Oil

As shown in Table 70 wheat oil, unlike corn and rice oil, does contain appreciable amounts of linolenic acid. The presence of this substance suggests potential problems with the keeping quality of the oil. Williams (1950) has found that wheat oil becomes rancid easily. The problems of palatability have not been a serious obstacle to its use because wheat oil is marketed, in the United States at least, principally as a specialty product for nutritional supplementation. Table 71 shows the principal fatty acid components of wheat oil. Table 73 indicates some chemical and physical properties peculiar to wheat oil.

COMMERCIAL PRODUCTION OF CEREAL OILS

The separation of oils and fats from vegetable materials comprises a distinct and specialized branch of fat and oil technology. All extraction processes have certain objects in common which may be mentioned briefly here: (1) to obtain the oil undamaged and as free as possible of undesirable impurities; (2) to obtain the oil in as high a yield as is consistent with the economic realities of the process; and (3) to produce an oil cake or residue of the greatest possible value as a by-product (Bailey 1951).

Production details of rice and wheat oil manufacturing will not be given within this brief survey, because essentially the same steps are used in extracting and refining of rice bran and wheat germ oils as are used for corn oil.

The raw material for corn oil extraction is the germ after it has been separated from the corn kernel. Following this step, the germs are washed and dried. They are then heated prior to their delivery into oil extraction presses. The oil may be expressed through hydraulic presses, but the usual practice in the United States is to employ semicontinuous or continuous screw-type presses. Such presses are called Anderson expellers. Fig. 106 illustrates these presses operating during production.

Courtesy of Corn Industries Research Foundation, Inc.

FIG. 106. ANDERSON EXPELLERS USED IN CORN OIL EXTRACTION

In utilizing the Anderson expeller, the germs are first passed through flaking rolls until they are ground into a coarse meal. This material is then passed into steam-heated temperers and then into the expeller unit. Here, the germs are forced under high pressure through a slotted barrel made up of steel sections in the form of a rotating screen called the "oil-reel." During this stage of processing, most of the oil is pressed out through the slots, while the fibrous portions (called the "foots") are discharged at the end of the barrel. This rejected material is usually returned through the expeller unit for repressing, but ultimately this residue material still retains about 5 to 8% of its original oil content.

Upon leaving the expeller units the crude oil is usually pumped through a lengthy filter press. Figure 107 illustrates an installation of corn oil clarifying presses. This filtration constitutes only

FIG. 107. FILTER PRESSES FOR CLARIFYING CORN OIL

the first of several such steps throughout the entire refining process (Jamieson 1943; Anon. 1957).

Some refiners employ a combination of mechanical presses and solvent extraction for removing oil from the grain germs. Solvent extraction is a highly efficient means of recovering the oil for it is capable of reducing the oil content of the residue to about 0.5% as compared with 5% in mechanically-expressed residues. Although solvent extraction is particularly advantageous in the processing of raw materials of extremely low oil yield, this process has been extended to the processing of corn germs (Bailey 1951). It is a common practice to "pre-press" seeds of high oil content in low-pressure screw presses to about ten per cent residual oil content prior to solvent extraction.

The usual solvents employed in this process are the so-called extraction naphthas such as light petroleum fractions consisting chiefly of n-hexane.

The solvent-oil mixture from the extractor is filtered to remove any solid matter, and then the solvent is separated from the oil in an evaporator or stripping column (Anon. 1957). Here the solvent boils off, and the vapor is condensed and collected for reuse. The germ flake from the extractor contains a certain amount of solvent and this is usually recovered by a steaming and heating process. The extracted flake, now stripped of almost its entire oil content, may be used in livestock feeds. Sometimes, in place of filtering the oil from the solvent extractor, it is simply run into large tanks where the solids settle to the bottom and are subsequently drawn off. These solids, known collectively as "foots," are utilized in soap manufacture.

Further refining of the extracted oil must be undertaken to remove fatty acids, phosphatides, and other gummy or mucilaginous materials. The first step in fatty acid removal is to expose the oil to a strong caustic aqueous alkali solution (Bailey 1951A). This process involves three operations: (1) emulsification of the oil with a considerable excess of alkali material; (2) heating to break the emulsion; and (3) separating the refined oil from the precipitated soapy substances and miscellaneous associated impurities. The precipitated material is again termed "foots" or "soapstock" and has commercial by-product value, for it contains about 30 to 50% free and combined fatty acids. Separation of the refined oil from the impurities is accomplished by centrifugation. Figure 108 illustrates an installation of centrifuges and flowmeters used in this stage of the refining process. By means of the alkali treatment for fatty acid removal, crude oils which seldom would contain

Courtesy of Corn Industries Research Foundation, Inc.

FIG. 108. CENTRIFUGES AND FLOWMETERS USED IN CORN OIL REFINING

less than 0.5% free fatty acids are reduced to a free fatty acid level of about 0.01 to 0.03%. Significant improvement in production efficiency has been obtained in recent years through the introduction of a continuous method of caustic soda refining to replace the more conventional batch method. This newer technique, first introduced into the United States in 1933, prevents to a large extent the so-called neutral losses which are common in the batch technique. These losses are reduced about 25 to 30% by reducing the time of contact between the corn oil and the caustic material and through efficient separation of the oil and the soapstock in very high-speed centrifuges.

A bleaching treatment of corn oil is usually required, since the alkali-refining process does not produce a sufficiently light-colored oil for marketing purposes. Bleaching is usually accomplished by treating the refined oil with an adsorbent material in powder form. Both natural and acid-activated bleaching earths (e.g., Fuller's earth or clay) are used as adsorbents. The acid-treated powders are more expensive, but have a greater adsorptive power for fat pigments, particularly for chlorophyll or related pigment compounds. The bleaching is usually effected at a temperature of 220° to 240°F under atmospheric pressure. Slightly lower temperatures may be utilized when the process is done under vacuum. When a good activated adsorbent earth is employed, an amount of only about 1% is required to bleach a good quality of oil. However, considerably larger amounts of earth material may be required if a natural earth of lower activity is used, or if the oil is of a poor grade or highly colored. Equilibrium between unadsorbed pigments in the oil and adsorbed pigments in the earth is usually established within a 5-min period if the oil and adsorbent are vigorously mixed.

In the common practice of bleaching, on a batch basis, the oil and adsorbent are mixed together in a kettle of about 30,000 lb capacity. To avoid oxidation of the oil and improve bleaching efficiency this procedure is usually performed under a vacuum, although the use of open kettles for the operation is not uncommon. The charge is pumped from the kettle through a filter press. The press cake of spent adsorbent is blown with air, or a mixture of steam and air, to reduce its entrained oil content to a level of 30 to 40%. Following this effort to recover oil, in some commercial operations solvent extraction is carried out on the press cake to recover more oil. Ultimately, the press cake of adsorbent earth material is discarded.

Bleaching as a distinct step in the oil-refining process is usually followed by a chilling treatment of the oil called "winterization." Winterization or destearination are the terms generally applied to a process of removing from the oil certain constituents which, though soluble at medium or higher storage temperatures, crystallize out and make the oil turbid in appearance under conditions of cold (Andersen 1953). Particularly when edible oils are used as salad oils, a small fraction of solid glycerides must be removed, as their presence makes the oil appear cloudy. In corn oil, however, it is not so much the presence of high-melting glycerides as a very small portion of waxes which produces this undesirable turbidity.

The process of winterizing consists in cooling the oil at temperatures some degrees below that at which the oil is required to remain clear. The oil is cooled with a cold water or brine coolant, or by cooling the air in a storage room where the oil is maintained. The cooling must be slow enough to enable the higher melting components to form definite well-built crystals. Small crystals resulting from too rapid cooling make filtration difficult. Frequently, a suitable filter aid is added to the oil before cooling, so that the particles can act as crystallization centers. To ensure maintenance of the low temperatures during filtration, the filters are often placed in the cooling room or else the filter presses are provided with internal channels for cooling. No wholly satisfactory continuous process for winterizing (Schwitzer 1951) has been perfected.

The final step in the refining process is usually the deodorization of the oil. Practically all of the vegetable oils used as edible fats products in the United States, as well as in most western nations, are subject to a deodorizing treatment.

Deodorization is a process of steam distillation, in which the relatively nonvolatile oil is held at a high temperature, under reduced pressures, while it is stripped of the relatively volatile constituents responsible for flavors and odors in the crude or partially refined oil. Concomitant with the deodorization process, the free fatty acid content of the oil is usually reduced to a level of from 0.01 to 0.03% and bleaching or loss of color of the oil also will occur (Bailey 1951).

Modern deodorization equipment operates in the temperature range of about 425° to 475°F and at very high vacuum levels. Many processors deodorize salad and cooking oils at somewhat lower temperatures than they do hydrogenated oils. The rigorous exclusion of oxygen during this process may be regarded as essential to the effectiveness of the operation. Figure 109 illustrates the

Courtesy of Corn Industries Research Foundation, Inc.

FIG. 109. TOP OF OIL DEODORIZING KETTLE SHOWING STEAM JETS FOR PRODUCING VACUUM

steam jets which produce the vacuum at the top of a typical deodorizer vat.

The stability of vegetable oils of good quality is usually improved considerably by the deodorization process (Bailey 1951). Deodorization destroys any peroxides in the oil and removes aldehydes or other volatile products which may have developed through atmospheric oxidation, but it should be noted that strongly rancid oils cannot be completely reclaimed by deodorization.

Batch-type processes are yielding to newer continuous processes for deodorization. In the former, the steps of heating, deodorizing, cooling, and filtering may take from 5 to 8 hr depending on the actual conditions of operation, and the size and quality of the equipment (Schwitzer 1951). Development of semicontinuous or continuous units with automatic controls have resulted in complete deodorization being effected in 2.5 hr. If deodorization has been properly conducted, the removal of odoriferous ingredients from the oil is substantially complete. Well deodorized oils of different identity, when fresh, cannot be easily distinguished from one another by odor and taste. They are quite bland and merely give one a sensation of oiliness in the mouth.

Corn oil, following deodorization, is usually cooled and finally packaged. Only a small portion of the total edible corn oil produced in the United States is subjected to the process of hydrogenation, and thus, as a processing step in the general refining of vegetable oils, it will be mentioned here only briefly. Primarily, hydrogenation is a means of converting liquid oils to semi-solid plastic fats suitable for shortening or margarine manufacture. By means of this process, enhancement of the oil stability and lightening of color is achieved. In the hydrogenation process, hydrogen is added at the double-bonds in the fatty acid chains. Hydrogenation as a process is more than a means of producing substitute products, for hydrogenated fats may be tailor-made to produce products superior in some important aspects to any of the natural plastic fats.

PRODUCTS AND THEIR USES

The fractions resulting from the normal corn wet-milling operation are: (1) steepwater; (2) starch; (3) gluten; (4) germ; and (5) hulls.

There can be various combinations and forms of these. For example, steepwater is evaporated to yield concentrates, or added to hull and gluten and dried to give a common feed ingredient. The germ is pressed and extracted to yield crude oil and corn oil meal. Much of the starch is sold in substantially unchanged form, by far the largest consumers being the paper and textile industries. Some goes into building materials, and food uses are also very substantial. Cornstarch is normally one of the cheapest food ingredients available, and it is used as a filler whenever possible. Whenever a gel structure is needed, as in gum drops, cornstarch (natural or modified) is used if possible. It is also indispensable in some special uses such as molds for candy pieces (e.g. chocolate cream centers).

Starch is modified in many ways to produce derivatives which have special advantages for certain food uses. The corn syrups described in an earlier section are widely used as sweeteners and texturizers. Since they are considerably cheaper than sucrose, they are used as replacements for that ingredient whenever possible. Corn syrups are sold at retail for use as sweeteners in home cooking (especially as a candy ingredient) as a component of baby formulas, and for uses on pancakes, waffles, etc. Most mixed syrups for consumer use, such as imitation maple syrups, contain substantial percentages of corn syrup. The baking industry uses corn syrup to make icings and frostings, fruit fillings, pies, jelly rolls, bread, etc. It is used to a minor extent in cookies.

TABLE 74

ANALYSIS OF TYPICAL CORN OIL CAKE[1]

	%
Moisture	9.47
Oil	12.23
Albuminoids	20.36
Digestible carbohydrates	46.13
Fiber	9.38
Ash	2.43

[1] Williams (1966).

The uses of cornstarch, corn syrups, and corn sugars in the food industry are so numerous that only a few of the largest ones will be mentioned.

Satisfactory jams, jellies, and preserves can be made when some of the sugar solids are replaced with corn sugar. The percentage substitution is limited by Federal Standards. The same situation applies to canned fruits.

Manufacturers of carbonated beverages utilize vast quantities of corn syrup, although complete replacement of sugar is not feasible in these products.

Dessert sauces such as ice cream toppings nearly always contain substantial percentages of corn syrup. These ingredients are useful as viscosity modifiers and "cling" agents in toppings. Corn syrup is a valuable ingredient in ice creams, sherbets, and the like. It modifies the initial texture and also retards the development of crystals of ice, lactose, and sucrose.

Most producers of caramel color use corn syrup as a raw material. Treatment of the syrup with alkali or acid under rigidly controlled temperature conditions can yield colors having widely different properties.

Dried corn syrups and dextrose are used mainly by bread bakers and cereal manufacturers. They are also used in canned, bottled, frozen, and preserved foods, carbonated soft drinks, candy, ice cream, "cream" fillings for cookies, mayonnaise, salad dressings, catsup, cheese spread, mustard, pickles, and prepared meats.

Almost all of the corn oil available is made into a salad or cooking oil (Bailey 1951). For these two purposes the oil is identical, except that when destined exclusively for salad use, the oil is winterized. The large use of corn oil for these purposes is due in part to the fact that oil requires little or no winterization treatment, and partly because the natural color of the corn is rather darker than is desirable for use in hydrogenated oils. Corn oil reputedly displays a high resistance to rancidity, which makes it especially valuable as an ingredient in products like mayonnaise or semisolid salad dressings (Anon. 1957). Despite properties of corn oil which seem to show its advantage in these latter products, recent surveys conducted under joint auspices of industry and government indicate that corn oil comprises less than 5% of the total vegetable oil usage in the United States for mayonnaise, salad dressings, and related products. It should be pointed out that this situation is governed almost entirely by the spread in price between corn oil (on the high end) and soybean and cottonseed oils (on the low end). In the highly competitive food market a price differential of 10% or more in a major ingredient like edible oils for these high fat products can and does favor the use of less expensive sources of oil.

In the United States, corn oil is used directly as a liquid fat, or is sometimes blended with harder fats. Bakers use the oil for greasing tins or for brushing baked goods. Actually the commercial bakery use for corn oil is quite small in terms of the total annual production of corn oil. In recent years, there has been an increasing utilization of corn oil in the liquid state by homemakers in contrast to use of hydrogenated fat sources.

A high smoke point (see Table 73) makes corn oil a preferred medium for deep-fat frying. The relatively high temperatures that may be used here, coupled with the other favorable characteristics make this oil particularly good for frying potato chips and doughnuts. The corn oil may be used repeatedly for frying, since it does not tend to absorb flavors from one food and transmit them to another.

The extensive use of the by-product of corn oil production referred to as soap stock, in the manufacture of soap and associated products, represents the most important nonfood application of corn oil. Smaller quantities of either crude or refined corn oil are used in ammunition, chemicals and insecticides, paint and varnish substitutes, rust preventive compounds, and textiles. Sulfonated corn oil has important applications in tanning and processing of leather. Refined corn oil has found wide usage as a carrier for vitamins and in other medicinal applications.

As previously mentioned, steepwater is usually dried in combination with one or more of the other by-products of starch manufacture, and the solids used as a feed component. On a dry-solids basis, the steepwater coming from a modern plant contains about 48%proteins, 26% lactic acid, 18% ash, and much smaller amounts of other substances. The amino acids of steepwater are not balanced, making it inadequate as the sole protein source in animal feeding. It contains nine vitamins; those present in largest amounts are inositol, choline, niacin and pantothenic acid. It was originally used for yeast culture, and then in the World War II period became an important part of the penicillin production method. It has also been used as a nutrient medium in the production of chloramphenicol, chlortetracycline, erythromycin, polymixins, tetracyclines, and streptomycins. It is also used in the fermentative production of vitamins, and inositol—a member of the B-complex—may be produced from steepwater.

The gluten fraction may be mixed with steepwater, corn oil meal, hulls, or any combination of these fractions to give corn gluten feed for cattle. Usually these high protein supplements are mixed with carbohydrate sources or fibrous materials. A sweetened corn gluten feed is made by adding hydrol or corn sugar molasses, a by-product derived from the corn sugar refining process. Zein can be extracted from the gluten as a more or less pure protein and used for certain specialty applications, such as a coating and anti-oxidant carrier for nut pieces.

QUALITY CONTROL TESTS

Anderson (1963) described a bench-scale method for determining the yield of products to be expected from wet-milling a given lot of grain. Most of the steps followed in a modern plant were duplicated in the laboratory procedure. Although the results of the small-scale processing did not correspond exactly to the results obtained in commercial milling of grain from the same batch, the differences were constant and relatively minor. Table 75 shows the comparison.

The standard tests listed in Table 76 have been approved by an analytical committee of the wet-milling industry and are run on the materials indicated. In addition to the tests listed in the table, many other tests which apply only to corn syrup have industry-

TABLE 75
COMPARISON OF LABORATORY AND COMMERCIAL STEEPING OF CORN
(Moisture-free basis)

Item	Laboratory-Steeped Corn %	Commercially Steeped Corn %
Composition of Corn		
Starch	72.1	73.0
Protein	10.3	9.8
Fat	4.7	. . .
Processing Data		
Steepwater: yield of solids	3.5	. . .
Protein content	35.9	. . .
Fibers: yield[a]	11.4	12.8
Protein content	16.2	12.7
Starch: yield	62.8	64.3
Recovery[b]	87.2	88.0
Protein content	0.29	0.32
Gluten: yield	14.3	14.5
Protein content	38.7	35.8
Squeegee starch: yield	1.0	1.4
Protein content	11.2	20.8
Process water: yield of solids	4.4	3.9
Protein content	33.8	47.0

[a] Combined coarse and fine fibers, and germ.
[b] Recovery based on total starch present in grain.

wide standard methods. Some of these are: Baumé, calcium, chloride, color, copper, dextrose equivalent, heavy metals, iron, fermentables, refractive index, specific rotation, and sulfate. There have been several published methods for automating the determination of dextrose equivalent of corn starch hydrolysates. A recent proposal was that of Ough and Lloyd (1965), who used the Technicon Auto Analyzer to make consecutive reducing sugar tests on samples before and after they were hydrolyzed with hydro-

TABLE 76
STANDARD ANALYTICAL PROCEDURES

	Corn	Cornstarch	Corn Syrups	Feedstuffs
Ash	x	x	x	x
Moisture	x	x	x	x
Protein	x	x	x	x
Insect infestation	x			
Starch	x			x
Acidity		x	x	x
Crude fat		x		x
Total fat		x		
pH		x	x	x
Solubles		x		x
Sulfur dioxide		x	x	

chloric acid. Reducing sugar content was determined by reagents made from potassium ferricyanide, potassium cyanide, and sodium hydroxide. Overall agreement between automated and Lane-Eynon results was good, there being no divergences greater than 4 DE. In the following discussion, very brief descriptions will be given of the most important test methods applied to cornstarch and oil. More detailed descriptions of cornstarch and oil tests can be found in Smith (1967) and, e.g., Kirschenbauer (1960), respectively.

Moisture is commonly determined on grains and starches by drying to constant weight in a vacuum oven held at 212°–248°F. Azeotropic distillation is the basis of the referee procedure of the corn wet-milling industry but it is probably too time-consuming for routine analyses. The Karl Fischer titration method is sometimes applied to starch products suspected of containing substantial amounts of volatiles other than water. Nuclear magnetic resonance (NMR) techniques for moisture in cornstarch were developed by the Corn Industries Research Foundation, and resulted in the manufacture of a commercial instrument which can determine moisture in cornstarch, flour, and similar materials in one minute with better precision than is obtained with standard methods in much longer times.

Inorganic materials are customarily determined as the residue (ash) after ignition at a specified temperature. The initial incineration may be over an open flame, but the final heating is usually at 525°F for about 2 hr or until the residue is free of carbon. The amount of ash in most commercial starch batches is quite small. To determine calcium, the sample is ashed and the residue dissolved in dilute hydrochloric acid before titrating with a standard solution of EDTA. The indicator is hydroxynaphthol blue.

Sulfur dioxide (as well as sulfites) may be carried over in free or combined form from the steeping operation or other steps in which it is added as a preservative or conditioner. The Monier-Williams procedure can be used to estimate the amount of this substance after it is stripped by a carbon dioxide stream from an acidified suspension of the sample. Alternatively, an extract can be colorimetrically titrated with iodine.

Acids present in the sample are determined by titration of a slurry or extract. The pH is usually determined on a slurry. Most unmodified starches are adjusted to a pH of about 5.0 during the final stages of processing.

The well-known Kjeldahl procedure is used to determine the nitrogen in grains and starches. The factor of 6.25 is applied to convert nitrogen to protein.

Measurements of the physical properties of starches are highly important to the buyer, being often much more indicative of their quality in a given use than are chemical tests. Determining the viscosity of a suspension cooked in a given way is a valuable test of utility for many applications. Several different instruments have been developed over the years to give empirical results which are useful for certain purposes. Among the most successful of these have been the Brabender Amylograph and the Corn Industries Viscometer. These instruments record the change in apparent viscosity of a starch suspension as it is taken through a heating and cooling program. The Brookfield Synchro-Lectric Viscometer and many other "viscosity"-measuring devices have been applied to control or research testing of starch gels.

Business transactions in crude and refined oils in the United States are conducted with the aid of uniform standards, specifications, and trading rules established by various trade organizations, government and associated scientific or technical societies. The trading rules generally set forth specifications for different grades of each oil and in some cases include systems for the establishment of premiums or penalties based on the contract price for oil which is better or poorer in quality than the standard for a common basis grade.

Although not directly concerned with oil trading, the American Oil Chemists' Society has for many years been active in the development of analytical methods which are generally the basis for trading rules utilized by the industry. Official and tentative test methods (AOCS) commonly used in examining and evaluating the various properties of corn oil or other vegetable oils include:

Free Fatty Acids and Acid Value

These are quantitative measurements of the amount of uncombined free fatty acids present in the oil. The determination of the residual free fatty acids and acid value serves as a measure of the effectiveness of the total refining procedure. The acid value is defined as the number of milligrams of potassium hydroxide necessary to neutralize 1 gm of the oil.

Color

Color is important not only in the appearance of the oil itself, but also as it affects the color of the product to which the oil may be added. The bleaching step in the refining process is varied according to the color in the oil which is actually desired. Although there are many methods for measuring color, it has been customary to report color in edible oils in terms of Lovibond units. This method expresses the color in a sample by comparison with red and yellow Lovibond glasses of known intensity.

Cold Test

This test measures the ability of an oil to remain sparkling clear at refrigeration temperatures. It is a check on the effectiveness of the removal of solid glycerides and wax during the winterization process. The test is made by actually holding a sample of the oil for 5.5 hr at 32°F and then observing for any cloudiness or turbidity.

Peroxide Value

This as a measure of the degree of oxidation that has taken place in the oil. Oxidation is related to rancidity and the test is indicative of the age of the oil in terms of probable handling conditions that have ensued (including temperature and exposure to light and air).

Smoke Point

This value represents the temperature at which the oil first gives off a continuous stream of smoke. The value is fairly constant for each particular type of oil and is indicative of the performance of an oil for deep fat frying applications.

Table 77 shows typical analytical data covering the basic properties of refined corn oil (Anon. 1957).

TABLE 77

REFINED CORN OIL ANALYTICAL DATA[1]

Property	Value
Acidity (free fatty acid as oleic)	0.020 to 0.050
Acid value	0.04 to 0.10
Color (Lovibond)	20 to 25 yellow
	2.5 to 5 red
Cold test	Clear
Saponification value	189 to 191
Iodine value	125 to 128
Hehner value	93 to 96
Titer	64° to 68°F
Melting point	4° to 12°F
Smoke point	430° to 500°F
Solidifying point	−4° to 14°F
Flash point	575° to 640°F
Fire point	590° to 700°F
Specific gravity	0.918 to 0.925
Pounds per gallon	7.672 at 70°F

[1] Anon. (1957).

BIBLIOGRAPHY

ALEXANDER, D. E., SILVELA, L., COLLINS, F. I., and RODGERS, R. C. 1967. Analysis of oil content of maize by wide-line NMR. J. Am. Oil Chemists' Soc. *44*, 555–558.

ANDERSEN, A. J. C. 1953. Refining of Fats and Oils for Edible Purposes. Academic Press, New York.

ANDERSON, R. A. 1963. Wet-milling properties of grains: bench-scale study. Cereal Sci. Today *8*, No. 6, 220–223.

ANDERSON, R. A. 1967. Manufacture of wheat starch. *In* Starch: Chemistry and Technology. R. L. Whistler, and E. F. Paschall (Editors). Academic Press, New York.

ANON. 1957. Corn Oil. Corn Industries Research Foundation, Washington, D.C.

ANON. 1958. Official and Tentative Methods of Analysis. The American Oil Chemists' Society, Chicago, Ill.

ANON. 1964. Critical Data Tables. Corn Industries Research Foundation, Washington, D.C.

BAILEY, A. E. 1951. Industrial Oil and Fat Products, 2nd Edition. Interscience Publishers, New York.

BALDWIN, A. R., and SNIEGOWSKI, M. S. 1954. Fatty acid composition of lipids from corn and grain sorghum kernels. J. Am. Oil Chemists' Soc. *31*, 24–27.

BAUER, F. J., JR., and BROWN, J. 1945. The fatty acids of corn oil. J. Am. Chem. Soc. *67*, 1899–1900.

BEADLE, J. B., JUST, D. E., MORGAN, R. E., and REINERS, R. A. 1965. Composition of corn oil. J. Am. Oil Chemists' Soc. *42*, 90–95.

BENNETT, C. D. 1966. Wet maize milling. Process Biochem. *1*, 318–320.

CLELAND, J. E., and FETZER, W. R. 1941. Determination of moisture in sugar products. Anal. Chem. *13*, 858–860.

DALE, J. K., and LANGLOIS, D. E. 1940. Syrup and method of making the same. U.S. Pat. 2,201,609. May 21.

DANIELSON, C. V., and KNAPE, S. D. 1958. Salad Dressing, Mayonnaise and Related Products, 1957. US Dept. Comm., Washington, D.C.

DEAN, H. K. 1938. Utilization of Fats. Chemical Publishing Co., New York.

DLOUGHY, J. E., and KOTT, A. 1948. Continuous starch hydrolysis. Chem. Eng. Progr. *44*, 399–402.

FUEGE, R. O., and REDDI, P. B. V. 1949. Rice bran oil. III. Utilization as an edible oil. J. Am. Oil Chemists' Soc. *26*, 349–353.

GOODWIN, J. T., JR. 1959. Corn wet-milling. *In* The Chemistry and Technology of Cereals as Food and Feed. Avi Publishing Co., Westport, Conn.

GUNSTONE, F. D., and HILDITCH, T. P. 1946. The use of low temperature crystallization in the determination of component acids of liquid fats. II. Fats which contain linolenic as well as linoleic and oleic acids. J. Soc. Chem. Ind. (London) *65*, 8–13.

HILDITCH, T. P. 1949. Oils (Fatty) and Fats. Thorpe's Dictionary of Applied Chemistry, 4th Edition, Vol. 9. Longmans, Green and Co., London.

HOGAN, J. T. 1967. The manufacture of rice starch. *In* Starch: Chemistry and Technology. R. L. Whistler, and E. F. Paschall (Editors). Academic Press, New York.

HORESI, A. C. 1944. Continuous conversion of starch. U.S. Pat. 2,359,763. Oct. 10.

JAMIESON, G. S. 1926. Chemical composition of rice oil. Oil and Fat Inds. *3*, 256–261.

JAMIESON, G. S. 1943. Vegetable Fats and Oils, 2nd Edition. Reinhold Publishing Corp., New York.

JEFFRIES, F. L. 1930. Manufacture of starch. U.S. Pat. 1,750,756. Mar. 18.

JEFFRIES, F. L. 1936. Manufacture of starch from corn. Improved mill house process. U.S. Pat. 2,050,330. Aug. 11.

JEFFRIES, F. L. 1937. Manufacture of starch. U.S. Pat. 2,088,706. Aug. 3.

JEFFRIES, F. L. 1942. Grinding Corn as I Have Seen It. Corn Products Refining Co., Chicago, Ill.

JENSEN, L. S., ALLRED, J. B., FRY, R. E., and McGINNIS, J. 1958. Evidence for an unidentified factor necessary for maximum egg weight in chickens. J. Nutr. *65*, 219–232.

KERR, R. W. 1950. Chemistry and Industry of Starch, 2nd Edition. Academic Press, New York.

KIRSCHENBAUER, H. G. 1950. Fats and Oils. Reinhold Publishing Co., New York.

LEVIN, E. 1958. Personal Communication. Viobin Corp., Monticello, Ill.

LOFLAND, H. B., QUACKENBUSH, F. W., and BRUNSON, A. M. 1954. Distribution of fatty acids in corn oil. J. Am. Oil Chemists' Soc. *31*, 412–414.

LONGENECKER, H. E. 1939. Deposition and utilization fatty acids. II. The non-preferential utilization and slow replacement of depot fat consisting mainly of oleic and linoleic acids; and a fatty acid analysis of corn oil. J. Biol. Chem. *129*, 13–22.

McCOY, R. O. 1931. Manufacture of starch. U.S. Pat. 1,832,229. Nov. 17.

MOFFETT, G. M. 1928. Manufacture of starch. U.S. Pat. 1,655,395. Jan. 3.

MURTI, K. S., and DOLLEAR, F. G. 1949. Rice bran oil. II. Composition of oil obtained by solvent extraction. J. Am. Oil Chemists' Soc. *25*, 211–213.

NEWKIRK, W. B. 1936. Development and production of anhydrous dextrose. Ind. Eng. Chem. *28*, 760–766.

OUGH, L. D., and LLOYD, N. E. 1965. Automated determination of the dextrose equivalent of corn starch hydrolysates. Cereal Chem. *42*, 1–15.

RADLOVE, S. B. 1945. A note on the composition of wheat germ oil. Oil and Soap *22*, 183–184.

SCHWITZER, M. K. 1951. Continuous Processing of Fats. Leonard Hill, London.

SHERMAN, R. F. 1925. Manufacture of starch. U.S. Pat. 1,554,301. Sept. 22.

SIPYAGUIN, A. S., and SHOEMAKHER, S. O. 1943. Apparatus for the continuous saccharification of starch. U.S. Pat. 2,337,688. Dec. 28.

SMITH, R. J. 1967. Characterization and analysis of starches. *In* Starch: Chemistry and Technology. R. L. Whistler, and E. F. Paschall (Editors). Academic Press, New York.

WHISTLER, R. L., and HICKSON, J. L. 1955. Determination of some components in corn syrups by quantitative paper chromatography. Anal. Chem. *27*, 1514–1517.

WHISTLER, R. L., and PASCHALL, E. F. (Editors). 1967. Starch: Chemistry and Technology, 2 Vols. Academic Press, New York.

WIDMER, J. M. 1926. Manufacture of starch. U.S. Pat. 1,585,452. May 18.

WILLIAMS, K. A. 1950. Oils, Fats and Fatty Foods, 3rd Edition. Blakiston Co., Philadelphia.

WILLIAMS, K. A. 1966. Oils, Fats and Fatty Foods, 4th Edition. American Elsevier Publishing Co., New York.

WOLFROM, M. L., and THOMPSON, A. 1956. Occurrence of the $(1 \rightarrow 3)$-linkage in starches. J. Am. Chem. Soc. *78*, 4116–4117.

CHAPTER 9

Ernest B. Kester

Samuel A. Matz

Rice Processing

INTRODUCTION

Nearly all rice consumed as food undergoes some type of milling operation during its preparation. Rice milling differs considerably from the milling of other cereals, because the preferred form of rice for culinary purposes is the whole grain rather than a flour or meal. Pulverized forms of rice are used in the kitchen infrequently, and then only as relatively minor ingredients in sauces and the like. Fairly large amounts of broken kernels and small pieces are sold for manufacturing purposes, as for brewing and the manufacture of puffed breakfast cereals or snacks. In the aggregate, however, the demand for whole kernel rice far exceeds that for smaller piece sizes, and the market value of the former is correspondingly greater.

The earliest type of rice milling relied on a mortar-and-pestle type of operation, although the "mortar" may have been merely a hole in the ground. The pounded grain was poured from one vessel to another, allowing gravity and the wind to separate fragments of hulls, chaff, bran, and germ from the kernels. In the natural progression of improvements, the mortars and pestles became larger, and arrangements were made for moving the pestles by animal power, windmills, or water power. Such simple methods of milling persisted for many centuries and are still used in many parts of the world, but a mechanical semiautomatic system using different principles was developed in Europe, starting in the eighteenth century.

The belief is widespread that rice is of major importance as a crop only in countries where its consumption is high and where it is the basic energy food for the inhabitants. It is, however, one of the principal grain crops of the United States, with an annual production that will soon reach 10 billion pounds of rough rice if the present trend continues (see Table 78). Our consumption is only about 7 lb per person per year. The United States is one of the largest exporters of rice, accounting for about 20% of the rice moving in world trade.

Vitreous long-grain rices are preferred in the domestic market for cooking and processing into canned formulations such as soups. The softer short-grain or Pearl types are sold to some extent for home use but are raised mainly to supply trade requirements in off-shore markets. They are the only type used for making puffed rice. Both kinds may be manufactured into other dry breakfast cereals and both are used in parboiling operations about which more will be said later. The short-grain varieties are not so easily cooked without becoming cohesive as are the long grain rices, but this disadvantage is overcome for the most part when precooked rice is canned or made into dry quick-cooking products. In some instances, the soft characteristics of short-grain rice are an advantage. For example, the brewing industry, which uses broken grades of rice as a source of carbohydrates, prefers short-grain types because they disperse more readily in the kettles.

TABLE 78

PRODUCTION OF ROUGH RICE IN THE UNITED STATES[1]

	Millions of Cwt
1957–59 average	45.7
1960	53.7
1961	54.6
1962	54.2
1963	66.1
1964	70.3
1965	73.2
1966	76.3
1967 preliminary	85.1

[1] National Food Situation, May 1968.

QUALITY OF RICE

Rice trading in this country is based on Standards promulgated by the US Dept. of Agr. (Anon. 1968B). Under these Standards, each lot of rough, brown, or milled rice can be provided with a designation which helps to fix its commercial value. In general, the designation will consist of the letters "US" followed by the number of the grade or the words "Sample Grade," the name of the class, and the name of the special grade (if any). For example:

"US No. 3 Medium Grain Brown Rice Parboiled."

There are 7 grades, 6 numerical plus "Sample," each separately defined for rough, brown, and milled rice. The grades are determined by limitations on the presence of damaged kernels, chalky kernels, red rice, and certain other undesirable forms or conditions. The grade requirements for milled rice are shown in Table

TABLE 79
GRADE REQUIREMENTS FOR MILLED RICE

Maximum Limits Of—

Grade[1]	Seeds, Heat-Damaged, and Paddy Kernels (Singly or Combined)		Red Rice and Damaged Kernels (Singly or Combined)	Chalky Kernels		Broken Kernels				Rice of Other Classes[3]
	Heat-damaged Kernels and Objectionable Seeds	Total		In Long Grain Rice	In Medium Grain or Short Grain Rice	Total	Removed by No. 5 Sizing Plate[2]	Removed by No. 6 Sizing Plate[2]	Through 6/64 Sieve[2]	
	Number in 500 Gm	Number in 500 Gm	%	%	%	%	%	%	%	%
US No. 1	2	1	0.5	1.0	2.0	4.0	0.04	0.1	0.1	1.0
US No. 2	4	2	1.5	2.0	4.0	7.0	.06	.2	.2	2.0
US No. 3	7	5	2.5	4.0	6.0	15.0	.1	.8	.5	3.0
US No. 4	20	15	4.0	6.0	8.0	25.0	.4	2.0	.7	5.0
US No. 5	30	25	[4]6.0	10.0	10.0	35.0	.7	3.0	1.0	10.0
US No. 6	75	75	[5]15.0	15.0	15.0	50.0	1.0	4.0	2.0	10.0

US Sample grade shall be milled rice of any of these classes which does not meet the requirements for any of the grades from US No. 1 to US No. 6, inclusive; or which contains more than 14.0% of moisture; or which is musty, or sour, or heating; or which has any commercially objectionable foreign odor; or which contains more than 0.1% of foreign material; or which contains live or dead weevils or other insects, insect webbing, or insect refuse; or which is otherwise of distinctly low quality.

[1] Color and milling requirements: US No. 1 shall be white or creamy and shall be well-milled. US No. 2 may be slightly gray and shall be well-milled. US No. 3 may be light gray and shall be at least reasonably well-milled. US No. 4 may be gray or slightly rosy and shall be at least lightly milled. US No. 5 and US No. 6 may be dark gray or rosy and shall be at least loosely milled. These color requirements are not applicable to Parboiled Milled Rice.

[2] Sizing plates shall be used for Long Grain Milled Rice and may be used for Medium Grain Milled Rice and sieves shall be used for Short Grain Milled Rice and may be used for Medium Grain Milled Rice; but any device which gives equivalent results may be used.

[3] These limits do not apply to the class Mixed Milled Rice.

[4] Milled rice in grade US No. 5 of the special grade Undermilled rice may contain not more than 10% of red rice and damaged kernels either singly or combined, but in any case not more than 6.0% of damaged kernels.

[5] Milled rice in grade US No. 6 may contain not more than 6.0% of damaged kernels.

79. This table is not applicable to the classes "Second Head Milled Rice," "Screenings Milled Rice," and "Brewers Milled Rice."

The classes for rough and brown rice are long grain, medium grain, short grain, and mixed. Milled rice includes these four classes in addition to head, screenings, and brewers.

The special grade designation, forming the final term in the usual description, are:

Rough Rice	Brown Rice	Milled Rice
Parboiled	Parboiled	Undermilled
Weevily		Parboiled
		Coated
		Granulated Brewers

MILLING

Figure 110 represents a grain of rice as it exists on the panicle of a rice plant ready for harvest. Threshing with a combine separates the grains from each other and yields what is known as rough rice or paddy. Each grain is encased in an easily removed protective hull. Inside the hull is a kernel of brown rice, so-called because of the dark bran layers or pericarp covering the endosperm. In a depression at the end of the grain is the germ. Underneath the bran coating is a layer of protein-rich cells called the aleurone layer.

Rice is milled to remove bran, germ, and the aleurone layer. The coarse outer bran coatings, the germ, and small bits of endosperm constitute the rice bran of commerce. The white inner bran and the aleurone layers are removed in the final stages of milling and together comprise the fraction known as rice polish. White rice is the final product. It is usually most valuable when perfectly white, admixed with a minimum quantity of broken kernels, and free of weed seeds, damaged kernels, and other objectionable materials. It is high in starch and so is an excellent energy source, but it is low in those life- and health-giving substances known as vitamins.

Rough rice yields approximately 20% hulls, 8% bran, 2% polish, and 70% milled rice. Although raw white rice is the commonest form of rice sold and used by consumers throughout the world, various practices have come into being for improving rice as a convenience item, for increasing its nutritional value, and for giving it improved cooking and processing characteristics. The processes for parboiling, enriching, canning, and converting

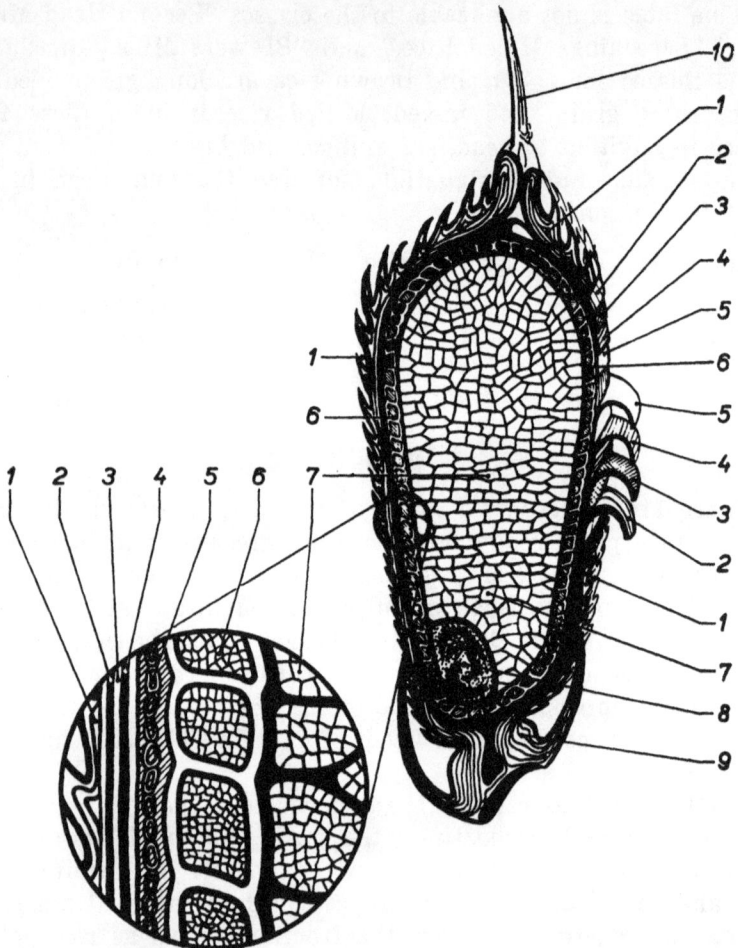

Courtesy of FAO

FIG. 110. STRUCTURE OF THE RICE KERNEL

(1) Hull (glume and palea), (2) epicarp, (3) mesocarp, (4) cross layer, (5) testa, (6) aleurone layer, (7) starchy endosperm, (8) embryo, (9) non-flowering glumes, and (10) apex or beard.

rice into quick-cooking forms will be discussed in detail in subsequent sections of this chapter.

STORAGE, TRANSFER, AND DRYING

The conditions under which rice is stored and the treatment it receives in transfer can have a pronounced effect on the yield of head rice and other responses to milling.

Rice is harvested by combines, large self-propelled machines that can cut, thresh, and clean a swath 10 to 18 ft wide at a speed of 2 to 6 mph. The best moisture for combining is too high for safe storage, so the grain is usually transferred quickly to drying installations which reduce the moisture content to 14% or below. Large column-type, continuous flow driers using heated air as well as batch-type units have been widely used.

The column-type methods of drying rice involve passing air heated to 140°F or less through falling or stationary columns of rice for short periods of time. Usually two or more passes through the drier are required to bring the moisture content down to 12.0 to 13.5%, which is usually considered a safe range for storage.

Most column-type units are in relatively large commercial plants and frequently dry over 120,000 cwt of rough rice annually. Considerable storage space is required to handle such quantities of rice, even though a major part of the grain may be shipped from the drier as soon as it has been dried. In order to make more efficient use of drying facilities, most commercial drying plants now have ventilated holding bins that allow high-moisture rice to be held for longer periods before drying.

For batch drying, both round bins and quonset-type bins have been used with equal success. Large volumes of air must be forced through the rice when it is to be dried in this way. The minimum airflow rate recommended by Sorenson and Davis (1955) was 9.0 cfm per bbl (162 lb) forced through an 8-ft depth of rice. The rice is dried to 13% moisture or less. It is of interest to note that this method of drying can be successfully used in the Gulf Coast areas of Texas and Louisiana where periods of relatively high humidity are common. Under these climatic conditions, the air is heated a few degrees to increase the drying rate.

Virtually all US rough rice is stored and transferred in bulk, although the products are frequently handled in bags. Wood, steel, or concrete bins of many different designs and sizes are used for storing the rough rice. If the bins are too large, breakage of kernels may become excessive and there will be a tendency for broken kernels to segregate toward the edges.

Transfer to the milling operation is by belt or screw conveyors, with bucket elevators for lifting. There has been some interest in pneumatic transfer, but, unless these systems are carefully designed, they may lead to added kernel breakage.

Some evidence exists that lower moisture contents in the rough rice lead to a greater yield of head rice (on a dry weight basis).

FIG. 111. DRYING PLANT WITH CONCRETE STORAGE BINS

Speed of drying, time elapsed between drying and milling, and other factors apparently interact with final moisture content in controlling the results, and the practical value of the process is somewhat uncertain at the present time.

The cooking and processing behavior of rice improves with storage; that is, the cooked texture of old rice is considered preferable by most people to that of freshly harvested rice. These aging or storage changes can be measured by objective tests as well as by panel evaluations. The obvious disadvantages in having to hold rice for weeks or months before it reaches its optimum consumer quality have motivated numerous investigations into possible methods for artificially accelerating the process. Sreenivasan and Giri (1939), in a study of rice quality changes after harvest, observed that increased temperature and reduced air supply during storage quickly improves the cooking quality of rice whereas paddy stored in the cold did not improve much after several months. Well-stored rice grains were found to swell on cooking to about four times their original volume, while freshly harvested rice scarcely swelled to double its size when cooked. These authors further concluded from their observations that fresh rice contains an active amylase which causes it to become pasty on cooking but is inactivated during storage. Desikachar (1956) reached similar conclusions—found that fresh rice not only showed a greater tendency to become pasty but also lost more solids into the cooking water than did stored rice. Moreover, its capacity to imbibe water was greater and the capacity of its amylose to bind iodine was lower. Normand *et al.* (1964) found that a change comparable to aging can be accomplished by heating milled or rough rice for about 1 hr in a tight container.

MODERN CONVENTIONAL MILLING

A modern rice mill is an elaborate assembly of devices for handling the rough grain as it comes from the warehouse end, with a minimum of manual labor, turning it into finished white rice, sacked and ready to market. All operations are mechanically performed, including transfer of the rice from one machine to another for the next operation. Excellent descriptions of rice milling and processing equipment have been published by the Food and Agriculture Organization of the United Nations (Aten and Faunce 1953; Borasio and Gariboldi 1957).

Equipment for milling rice is fairly well standardized in the United States. Rough rice is delivered to the mill from the drier or warehouse, usually admixed with various kinds of debris, such as straw, loose hulls, bran, weed seeds, pebbles, and granules of dirt. It also contains some broken rice. This mixture must first be cleaned with shaker screens to remove small and large heavy impurities and then aspirated to get rid of hulls and other chaff. The cleaned rice is then dehulled in a shelling device. There are several types of these devices, but the most commonly used is the stone sheller (Fig. 112), which consists of two horizontal disks separated by about the length of a rice grain. The apposed surfaces are coated with an abrasive emery-cement mixture. When in operation, the upper stone remains stationary while the lower revolves rapidly. A stream of rice is fed through the opening in the center of the upper stone. Centrifugal force throws the rice outward between the stones and, as the grain passes to the periphery of the disc, then it tumbles end-over-end allowing the ends of the hulls to be cracked and abraded by the surface of the stones. The rice grain then falls away from the loosened hulls, and the comparatively light hulls can be removed from the shelled grain when the mixture is aspirated through a cyclone separator. The brown rice produced in the shelling process includes some small unshelled rice grains which must be separated. This operation is performed in an ingenious device known as a paddy sepa-

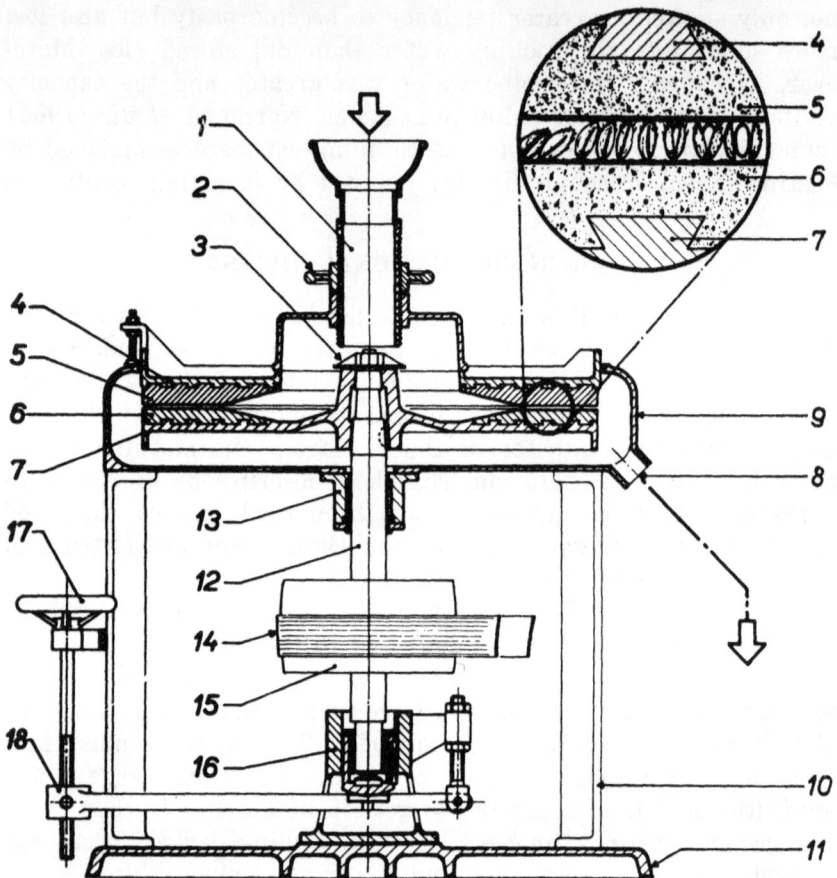

FIG. 112. STONE SHELLER

(1) Feed hopper, (2) feed regulation handwheel, (3) disk nut, (4) stationary iron disk, (5) stationary abrasive surface, (6) rotating abrasive surface, (7) rotating iron disk, (8) husked rice and hull outlet, (9) disk housing, (10) frame, (11) base, (12) drive shaft, (13) upper bearing, (14) drive belt, (15) drive pulley, (16) lower bearing, (17) disk clearance adjustment handwheel, (18) shaft supporting arm.

rator, which consists of flat bins divided into three tiers of irregular compartments (Fig. 113). The cars are tilted in such a way that when they are rapidly shuttled, the lighter, bulkier, rough rice or paddy is concentrated at the raised side while the heavier brown rice migrates to the lower opposite side. The process is continuous, and streams of brown and rough rice are removed simultaneously. The unshelled paddy is then fed into another pair of shelling

Courtesy of Farmers Rice Growers Cooperative

FIG. 113. PADDY SEPARATORS SHOWING ARRANGEMENT OF COMPARTMENTS

stones set closer together than the first set, and the above process of shelling, aspiration, and separation is repeated.

Rough rice may also be shelled between rubber rolls or with a rubber belt operating against a ribbed steel roll (Fig. 114). This device is quite effective, but the abrasive hulls wear down the rubber, which must be replaced after a time. It is less bruising on the bran layers of the brown rice, however, than shelling stones are, and brown rice produced in rubber shelling devices is more resistant to fatty acid development in the oil fraction (Houston *et al.* 1952).

Whether produced by stone or rubber shellers, the brown rice is now ready for milling. Brown rice is first milled to remove the coarse outer layers of bran and germ. The most common device used in the United States for this purpose is erroneously called a huller because it was adapted from a machine used for hulling coffee. It would be more appropriate to call it a milling machine. Unlike the milling of wheat flour in which fine particle size is desired in the product, rice milling must be conducted so as to break as few kernels as possible because whole grains or head rice are of more commercial value than broken grains. The difference in price may be from 1.5 to 6 cents a pound.

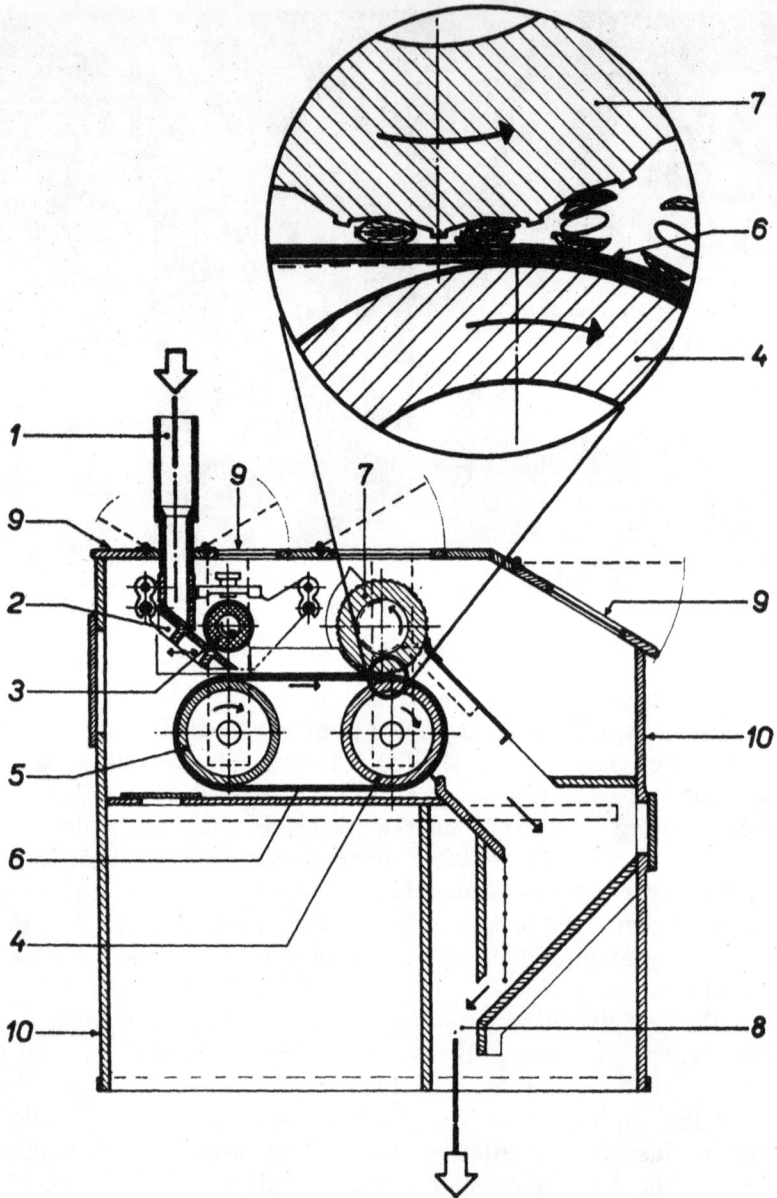

FIG. 114. RUBBER BELT HULLER

(1) Feed hopper, (2) feed plate, (3) feed roller, (4) stationary roll, (5) belt tension roll, (6) rubber hulling belt, (7) ribbed steel hulling cylinder, (8) discharge for hulls and hulled rice, (9) inspection doors, (10) housing and frame.

TABLE 80
COMPOSITION OF RICE AND RICE PRODUCTS[1]

	Brown Rice	Rice Bran	Rice Polish	Milled Rice	Parboiled Rice
Carbohydrates, %[2]	87.2	46.6	66.8	91.5	...
Protein, %[2]	8.3	14.6	13.2	7.6	...
Fat, %[2]	2.0	13.4	10.7	0.3	...
B-Vitamins, μg/gm					
Thiamine	4.2	27.9	23.9	0.80	2.57
Niacin	47.2	408.6	384.7	18.1	39.8
Pyridoxin	10.3	32.1	30.8	4.5	...
Pantothenic acid	17.0	71.3	92.5	6.4	...
Riboflavin	0.53	2.68, 1.77[3]	1.34	0.26	0.36

[1] Carbohydrate, protein, and fat figures are taken or calculated from data of Yampolsky (1944); carbohydrates for bran and polish are the percentage of nitrogen-free extract. All vitamin values under Brown rice, Rice bran, Rice polish, and Milled rice, except those for riboflavin, are averages for samples of three varieties (Blue Rose, Fortuna, and Early Prolific) (Williams et al. 1953). All values for riboflavin (except under Parboiled rice) are averages of 7 samples (2 Supreme Blue Rose, 2 Early Prolific, 1 Fortuna, 1 Lady Wright, and 1 Improved Blue Rose) (Kik and Van Landingham 1943). All vitamin values under Parboiled rice are averages of 6 samples of "Converted Rice" (Kik and Van Landingham 1943).

[2] On moisture-free basis.

[3] First and second break brans, respectively.

The conventional milling machine is illustrated in Fig. 115. The outer part consists of a hollow horizontal cylinder, the lower half of which is slotted. The upper half contains an adjustable longitudinal blade. Inside the outer shell is a ribbed steel rotor. The blade is adjusted so as to force the grains of rice against the rotating ribs. Brown rice is fed into the milling machine at the top until the annular space is filled, and the grains are rubbed by the ribbed rotor and by each other to remove bran and germ which are forced through the slots by internal pressure. There is always a certain amount of breakage in rice milling, and the finer bits of broken endosperm pass out of the mill along with the bran and germ. The milled rice is removed after it has been carried through the machine by the spiral action of the ribs.

Milling may be accomplished in one or two "breaks," that is, by a single pass through a mill or by consecutive passages through two mills, depending on plant practice. In some plants, as many as four breaks have been used.

After the rice is milled, it consists of almost white whole kernels mixed with broken kernels of different sizes. It is now ready for the brush—a device for removing the white inner bran layers and the proteinaceous aleurone layer. The brush is essentially a large vertical stationary cylindrical screen inside of which rotates a drum to which is attached overlapping leather flaps (Fig. 116).

The rice enters the annular space of this machine at the top and, as it progresses toward the bottom, is rubbed against the

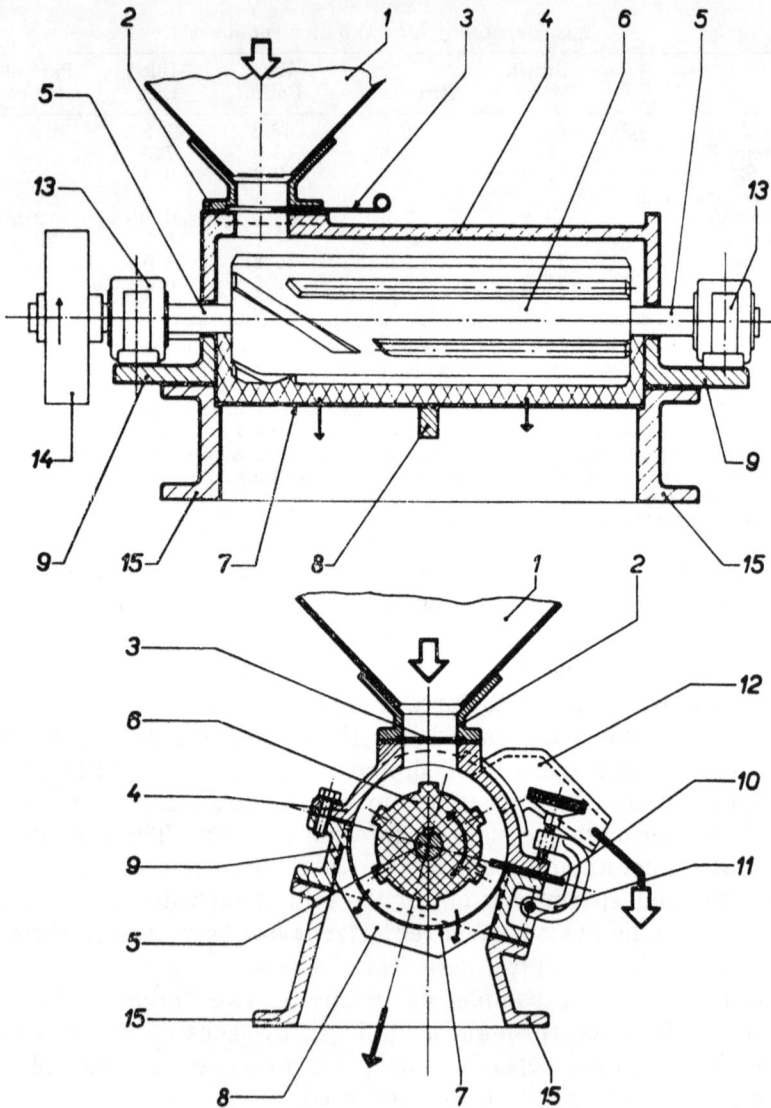

Courtesy of FAO

FIG. 115. RICE MILLING MACHINE

(1) Feed hopper, (2) hopper seat, (3) feed regulation gate, (4) cover, (5) shaft, (6) ribbed rotor, (7) slotted screen, (8) screen holder, (9) upper frame, (10) blade, (11) cover clamp, (12) discharge, (13) bearings, (14) drive pulley, (15) lower frame.

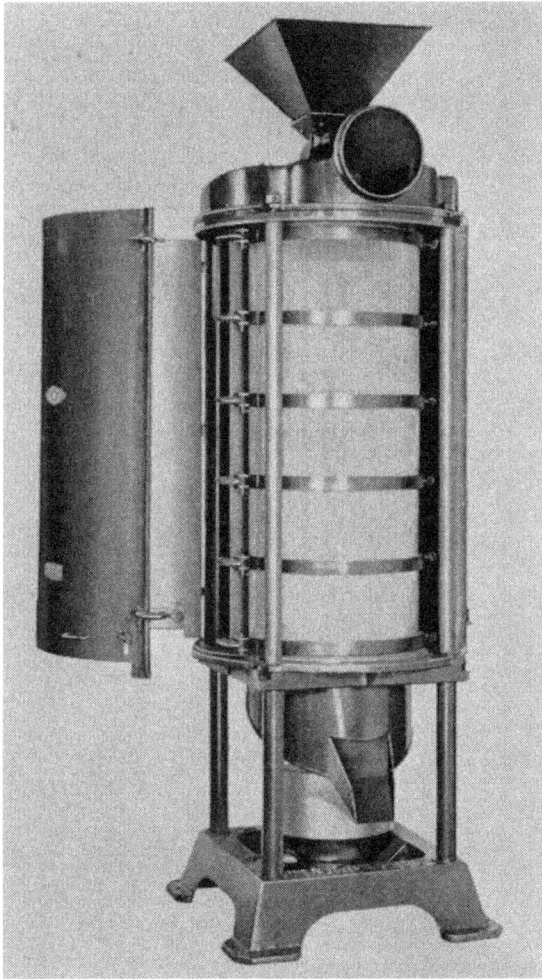

Courtesy of Engelberg, Inc.

FIG. 116. RICE BRUSH

screen by the leather flaps. The white floury mixture of fine bran and aleurone layer removed by abrasive action is forced through the screen and is collected and sacked.

At this point, the rice is fully milled. Some trade outlets, however, require that the rice have a high luster. This operation is performed in trumbols (Fig. 117), which are large, horizontal, rotating drums fitted with lengthwise baffles. The milled rice is charged into the trumbols and treated with talc and glucose solu-

TABLE 81

PERCENTAGES OF NITROGEN AND THE MORE IMPORTANT AMINO ACIDS IN RICE
AND RICE PRODUCTS[1]

	Brown Rice	Rice Bran	Rice Polish	Milled Rice	Parboiled Rice
Nitrogen	1.23	2.14	1.98	1.02	1.15
Cystine	0.090	0.137	0.141	0.073	0.13
Methionine	0.23	0.34	0.43	0.21	0.15
Tryptophane	0.074	0.096	0.107	0.086	0.08
Lysine	0.260	0.443	0.444	0.280	0.26
Arginine	0.254	0.344	0.273	0.251	0.41
Histidine	0.054	0.090	0.071	0.059	0.14
Threonine	0.27	0.37	0.39	0.30	0.28
Valine	0.50	0.61	0.63	0.49	0.41
Leucines	0.90	1.18	1.22	0.93	0.81
Phenylalanine	0.31	0.44	0.48	0.39	0.31

[1] Percentages of nitrogen, cystine, tryptophane, lysine, arginine, and histidine under Brown rice, Rice bran, and Rice polish were determined on Supreme Blue Rose variety (Kik 1942). Percentages of methionine, threonine, valine, leucines, and phenylalanine under Brown rice were determined on a commercial sample (Kik 1956), under Rice bran and Rice polish, on commercial samples (Kik 1956), under Milled rice on an unidentified sample (Kik 1954). All analyses under Parboiled rice were determined on a sample of "Converted Rice" (Edwards and Allen 1958).

Courtesy of Farmers Rice Growers Cooperative

FIG. 117. TRUMBOLS FOR COATING RICE

tion while the trumbols are rotated. After the coating is evenly distributed on the kernels and dried with warm air, the rice emerges from the equipment with a smooth glistening luster and is known as coated rice.

The mixture of whole and broken kernels obtained in the milling process is usually separated to meet certain grade requirements for milled rice. This operation is performed in two types of equipment. One is a rotating horizontal cylinder, the inside surface of which contains closely spaced depressions so shaped and of such a depth that when the cylinders are rotated, broken grains lodge in the depressions, are carried upward as the cylinder revolves, and fall out when it has rotated through a sufficient arc. They are collected in a discharge device and then carried out of the cylinder. Unbroken grains are too long to lodge in the depressions, so they move along the lowest part of the cylinder and discharge at the end opposite the point of loading.

In recent years, the indented cylinder has been replaced in many plants by the disk separator (Fig. 118). This is also a horizontal cylinder in which revolves a series of parallel indented disks. The indentations are designed with one deep side and one shallow side. As the disks rotate through a moving bed of rice in the lower part of the cylinder, the broken grains lodge against the deep side of the depressions, are carried upwards, and fall out at or slightly beyond the zenith of rotation. Like the indented shell-type of separator, the operation of the device is continuous. By proper choice of conditions, such as rate of feed and speed of rotation of the disks, an almost complete separation of broken grains from head rice can be achieved. These machines are provided with sets of disks having depressions of different size for use on different kinds of rice. Disks designed for the short grain Caloro rice, for example, would not be suitable for long grain types such as Bluebonnet.

ABRASIVE MILLING OF RICE

It has been shown that the composition of the rice kernel varies with the distance of the sample from the exterior surface of the kernel. When powder is collected and analyzed as the whole kernel is ground smaller and smaller, the fractions will be found to differ in percentages of protein, fat, starch, and other components. No commercial use has been made of this finding, so far as is known.

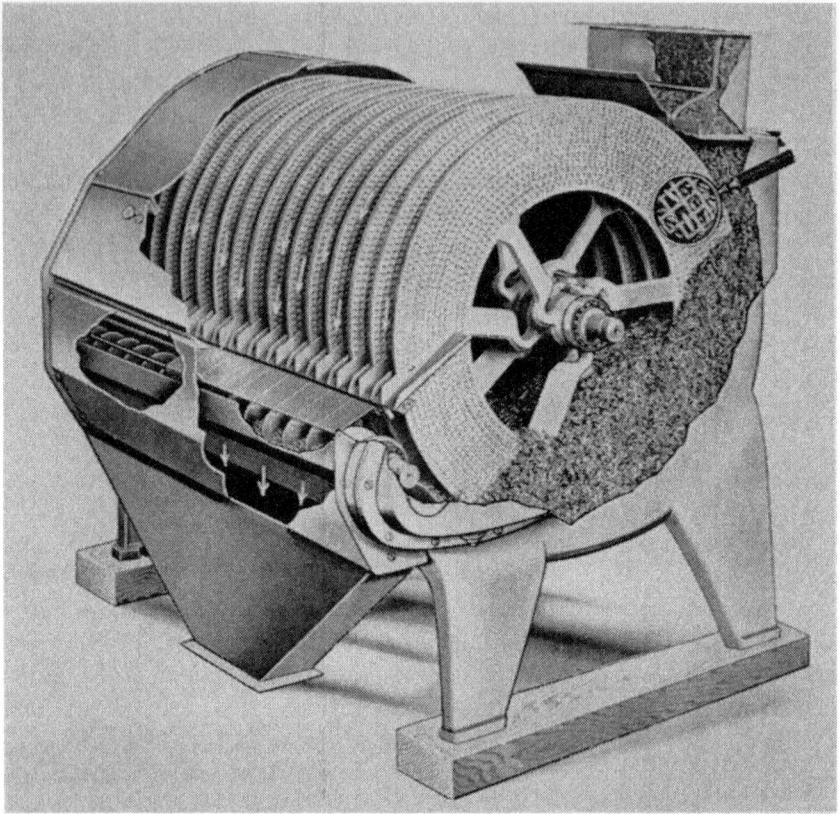

FIG. 118. DISK SEPARATOR FOR SEPARATING WHOLE AND BROKEN KERNELS OF
RICE

According to Deobald (1967A), the outside layers of milled
rice contain more protein, calcium, phosphorus, thiamin, niacin,
riboflavin, and lipids than the layers nearer the center. Starch is
lower near the outside. Flours containing as much as 20 to 21%
by weight of protein were obtained from kernels having an average protein content of not more than 8%. The amino acid composition of the proteins obtained from the various layers was substantially the same.

The differences in mineral and vitamin content were very striking. Calcium was 20 times higher in the outer layer than in the
whole kernel; phosphorus 10 times; the vitamins were 4 to 8
times higher. The lipid content was 5.8% compared with 0.2% in
the original milled kernel.

If the flour removal is held to about 8%, the remaining grain does not suffer in appearance and the cooking quality is close to that of the original rice. The flour contributes unusual viscosity effects to pastes and in this application behaves more like pre-gelatinized rice flour than it does unheated rice flour. These pastes show very little viscosity when tested in the Brabender viscometer.

A commercial Satake rice whitening machine has been used to prepare pilot plant quantities of the "peripheral flour."

LYE-PEELING

The so-called WURLD process for lye-peeling of cereal grains has been applied to bran removal from rice (Pence 1967). Hot sodium hydroxide solution is sprayed on the grain, allowed to stand a short time, and then washed away with the loosened bran layers. In the laboratory procedure, brown rice is first soaked in water that is gradually heated almost to boiling. The soaked rice is then transferred to a rotating drum, and hot 2 to 5% lye solution is sprayed on it. After the treated mass has been tumbled for about 3 min, it is dumped into water to remove the excess lye and to commence the washing process. The grain is next passed through a device in which it is subjected to an intense turbulence and whirling in additional water. This removes the loosened bran and further washes the peeled kernels. They are finally rinsed in a dilute acid to neutralize any remaining alkali, and dried. The yield of head rice by the lye-peeling process can be higher than from the usual dry milling methods if conditions are optimal, and it may approach 100% of the total rice yield. The aleurone layer and germ remain on the kernel which is duller or yellower in appearance than conventionally milled rice. This process would be more expensive to apply than dry milling and thus would have little to recommend it unless the yield of head rice was sufficiently increased to compensate for the extra cost.

TURBO-MILLING

Attempts were made by some US Dept. of Agr. laboratories to devise an air-classification method which would permit the separation of finely ground rice flour into protein-enriched and carbohydrate-enriched fractions, analogous to the turbomilling processes so successfully applied to wheat flours. Because rice starch granules are tenaciously embedded in the protein matrix and are also much smaller than the granules of other cereals, the separation of protein-rich from protein-poor material proved to be very difficult. The equipment used consisted of impact or turbulence grinders for reducing the rice flour to about single granule size and air-classifiers to separate the fine powders into fractions based on particle size. So far as can be determined, no commercial installations exist for air-classifying rice flour.

EXTRACTIVE MILLING

When clean hulled brown rice is soaked in certain oily or aqueous liquids under carefully controlled conditions, the bran coats are softened while the endosperm retains its initial properties. As a result, certain improvements can be made in the subsequent processing and the final product. This phenomenon forms the basis of a new milling technique now being used for large scale commercial production of milled rice (Anon. 1967).

A recent patent (Wayne 1967) describes the salient features of the procedure and lists many examples of bran-softening agents, including oils (both mineral and vegetable), soaps and esters of fatty acids, glycols, aqueous solutions of certain alkalies and salts, and acids. The essential steps are: (1) apply softening agent to hulled rice; (2) allow to stand until the bran layers are penetrated by the liquid; (3) pass through milling machines; (4) separate bran; (5) remove solvent; and (6) grade and size. The bran goes through a "desolventizer" which removes the oil mixed in with the solvent and then through two centrifuges. In the second centrifuge, the bran is washed with hexane to remove most of the remaining oil. Finally, the bran goes through another desolventizer and a cooler. The oil is passed through a solvent recovery system.

The softening agent used in the commercial process is a mixture of rice oil and hexane. Since both of these substances will be present in the system in any event, use of the combination avoids the introduction of a third component which would have to be removed later. Furthermore, rice oil-hexane rapidly and completely penetrates the bran layers while rice oil alone, in the proportion of 1.5% of the finished rice, may require 6 hrs to exert its maximum softening action.

The softening agents are used in amounts varying from 0.25 to 5% of the rice weight, and the time of contact is adjusted so that the bran is thoroughly penetrated but the endosperm is not much affected. According to the inventor, a sufficient degree of softening of the bran coat is indicated when it may be easily removed from the brown rice kernel by scraping with the fingernail. The softened bran coat is easily removed, exposing the white endosperm. The milling action in the process equipment need not be more severe than the fingernail scraping test.

For the most part, conventional equipment is being used for dehulling and the actual milling steps. Because of the softened condition of the bran, however, the removal or detachment of the pericarp is carried out under conditions far less severe in impact, mechanical, and abrasion effects and thermal stresses than when using conventional dry-milling hullers. The major economic advantage which results from these changes is an increase in yield of head rice of up to 10%.

Rice prepared by the process contains less fat than white rice milled conventionally but otherwise it is nutritionally similar to ordinary milled rice. The bran is creamy white and contains 17 to 20% protein but less than 1.5% fat. It is recommended for use in dietetic foods, snacks, etc. by the manufacturer. For each barrel of rice wet-milled, about 2.5 lb of oil is obtained, and this is more than the yield when bran obtained from the conventional milling process is extracted.

The patent claims that a rice product superior to brown rice in color, stability, and cooking response can be made if the processing is conducted in a manner calculated to retain most of the bran layers.

THERMAL METHODS OF RICE PROCESSING

About 15% of the national rice crop is processed into prepared foods of various kinds, such as parboiled, quick-cooking, enriched, and canned rice, canned soups, canned rice and vegetable mixtures, dry soup mixes, breakfast cereals, baby foods, and frozen dishes. When ground into flour, rice is sometimes used as a thickening agent in canned goods because of the smooth texture of its pastes, and its bland flavor. Rice flour is also used occasionally in commercial bakeries for dusting loaves of bread before they go into the oven to give a golden brown color to the crust. Because rice, unlike wheat, is lacking in gluten, it cannot be used by itself for making bread and, even when admixed with wheat flour in any appreciable quantity, reduces loaf volume in proportion to the amount of rice flour present. Broken grades of rice are frequently used as a source of carbohydrates in brewing. Because natural antioxidants present in rice are usually destroyed in processing rice, protective agents are often added to rice products to inhibit rancidity development. Butylated hydroxanisole (BHA) and butylated hydroxytoluene (BHT) have been found particularly effective for stabilizing rice products (Stuckey 1955). Even well-milled rice contains a small percentage of fat which is not removed by the usual processing treatments and may cause trouble when rice products are stored. The natural nutritional deficiencies of milled rice may be still further increased by certain types of processing but may be compensated for by adding to the products vitamins, iron salts, and protein supplements containing essential amino acids.

Parboiled Rice

The parboiling of rice has been practiced for hundreds of years in a primitive way in foreign countries, particularly Burma, India, and Pakistan. The method in its original form was to boil rough rice for an hour or so in open kettles; then it was spread on the ground to dry and hand pounded for use when needed. Manufacturing processes for parboiling rice have been introduced into the United States, Italy, and British Guiana since 1940. More recently, mechanized methods similar to those of the United States have been adopted in Burma.

The processes currently being used to manufacture most of the parboiled rice in the United States have been described in several technical articles (Anon. 1948; Court 1946; Havighorst 1947; Jones and Taylor 1935; Jones et al. 1946; Mecham 1961; and O'Donnell 1947) and are covered by numerous patents (Baumgartner 1917; Huzenlaub 1942, 1949; Huzenlaub and Rogers 1941, 1944, 1951, 1952; Landon et al. 1951; and Yonan-Malek 1943).

Rough rice is parboiled by soaking in water and then pressure-cooking to gelatinize the starch completely. The rice is then dried and milled to remove the outer layers. The conditions used in the soaking and cooking steps are critical with respect to the properties of the milled product, particularly for its appearance and the yield of head rice.

The soaking step is carried out in warm water. In one variation, rough rice is elevated from storage bins to an automatic scale hopper which weighs and dumps the rice into an accumulating hopper. When sufficient rice is assembled for a batch, it is dumped into a steeping tank (Fig. 119). This vessel is connected to a vacuum system, a water system, and a compressed air system. When the batch of rice is dropped, the tank is evacuated to remove air from the grain. Then sufficient water at a temperature of about 200°F is introduced to cover the rice. The tank is pressurized to about 100 psi and the rice is steeped about 190 min. The temperature and time may be varied

Courtesy of Converted Rice, Inc.

FIG. 119. PRESSURE STEEPING TANKS USED IN PARBOILING RICE

somewhat depending on the specific characteristics of the rice used, its moisture content, time in storage, etc. During the steeping operation, water soluble vitamin B components and minerals are infused into the endosperm from the bran, germ, and hull. At the end of this step, the water is drained off and the rice is discharged into a jacketed rotating vacuum drier equipped with steam tubes (Fig. 120).

Variations in properties of different lots of rough rice may affect their response to soaking conditions, although acceptable products may often be obtained over a wide range of conditions. Poor results are obtained at soaking temperatures above the gelatinization temperature of the starch. Incomplete soaking or tempering is reflected in excessive breakage of kernels when the

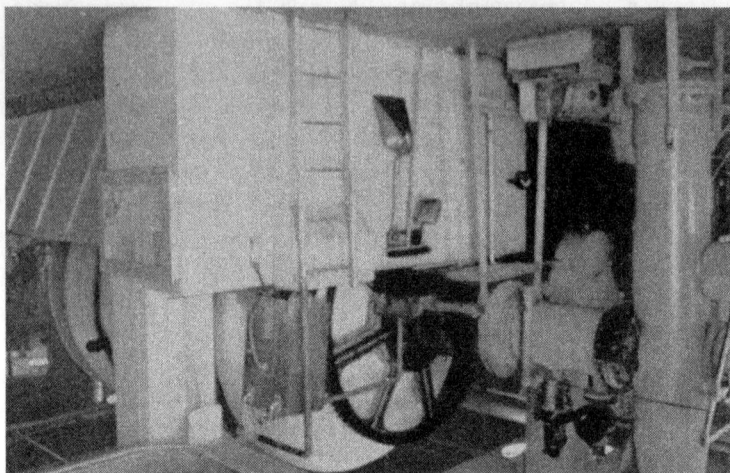

FIG. 120. ROTARY STEAMER

Shows trunnion end through which steam is admitted and through
which moisture vapors are removed.

rice is cooked, dried, and milled. For example, Calrose rice soaked
at 150°F for 2 hr followed by 1 hr of tempering before it was
cooked and dried yielded more than 30% broken kernels in the
standard milling test (Mecham 1961).

In the drier, the soaked rice is vacuumized and heated with
steam to remove excess moisture. Dry steam is then injected to
gelatinize the starch in the grains, after which the vessel is
vented and evacuated until the moisture of the rice is low enough
to permit it to be milled successfully. The dried rice is conveyed
to bins where it is cooled by drawing air through it, and tempered
to equalize the moisture content of the batch. Finally, the rice
is milled to remove hull, bran, and germ. Examples of conditions
which have proved to be successful for cooking are steam pressures of 20 psi for 5 to 8 min. Drying might be conducted at
120°F in a crossflow air drier.

In another process, cleaned rice is steeped in 2 parts of water
at 130° to 150°F in open steel tanks for 9 to 12 hr until it has
absorbed 30 to 35% moisture on the wet basis. The soaked rice
is transferred continuously to a vertical pressure vessel equipped
with rotary valves on the inlet and discharge openings, and
steamed at 230° to 245°F for 8 to 20 min, depending on the

degree of parboiling and the cooking quality and color desired in the end product. Shorter cooking times result in lighter colored rice. Little additional water is absorbed in the steaming process, and the rice is discharged with a moisture content of about 35%. It is dried in a steam tube drier and a series of hot air driers to 11 to 13% moisture, then milled in conventional equipment. Yields are from 66 to 71 lb of total milled rice and 58 to 67 lb of whole grains per 100 lb of starting rough rice. The product is said to contain 2.0 μg of thiamine, 0.40 μg of riboflavin, and 44.0 μg of niacin per gram of dry material. Its storage life is from 2 to 3 yr. Milling by-products are disposed of in the regular commercial channels for these materials. The waste water from the steeping process is not utilized.

A third process is similar to the second in general principles, except that both the soaking and steaming steps are performed in rotating cylinders (Court 1946; Havighorst 1947).

A recent variant of the parboiling process has been put into commercial production at the Sacramento plant of the California Rice Growers' Association (Anon. 1963). Rice is tempered in hot or cold water, depending on the variety, and then conveyed to soaking tanks each holding 15,000 lb of rice. Hot water at 100° to 200°F is added, and the rice soaked for 1 to 10 hr, depending on the variety. After soaking, the rice is transfered through a rotary valve to a screw conveyor passing through a pressure vessel. Here the grain is cooked at 15 to 100 psi steam pressure for 10 sec to 3 min. The cooked kernels exit through another valve, and are cooled and dried before milling.

Parboiled rice has a somewhat rubbery texture, and for that reason resists breakage when it is milled. The better head rice yields obtained in the milling of parboiled rice, as compared to raw rice, defrays to considerable extent the cost of parboiling so that the parboiled product sells for little more than white rice.

Although parboiled rice is not quick cooking, it has certain advantages over raw rice—it is more nutritious, it is more resistant to insect infestations, and it can be used in canned formulations such as soups and puddings without disintegration. It can be cooked with less danger of becoming mushy than white rice. Parboiled rice is darker than raw milled rice and has a slightly different flavor, but it is widely accepted and is often preferred to white rice. In some rice-eating areas of the world, however, attempts to introduce it have not been successful. Its color, which is usually a light tan, may be an adverse factor

in instances where extreme whiteness of rice is a criterion of
quality.

When parboiled rice is stored, rancidity occurs following an
induction period. This effect is accompanied by a rise in fat acidity and in monocarbonyl and peroxide values. Both monocar-
bonyls and peroxide rise to a maximum, then decrease, and
simultaneously the rancid odors begin to subside and ultimately
disappear. Fatty acids continue to rise during storage, but the
rate of increase diminishes as storage advances.

The behavior of the fat fraction in Pearl and Patna parboiled
rice, with respect to development of oxidation products and free
fatty acids, has been experimentally studied under a variety of
conditions including dark and light storage, closed and open
containers, and at several temperatures (Houston *et al.* 1954).
In open containers, dark storage of both rices showed maximum
carbonyl values at 77°, 100°, and 180°F in about a year, a
month, and a week, respectively. Figure 121 illustrates changes
in monocarbonyl, peroxide, and free fatty acid values at 77°F.
In closed storage at 77° and 100°F, peroxide and monocarbonyl
changes were similar to those for open storage, but induction
periods were longer. At 140°F no significant rise of either func-

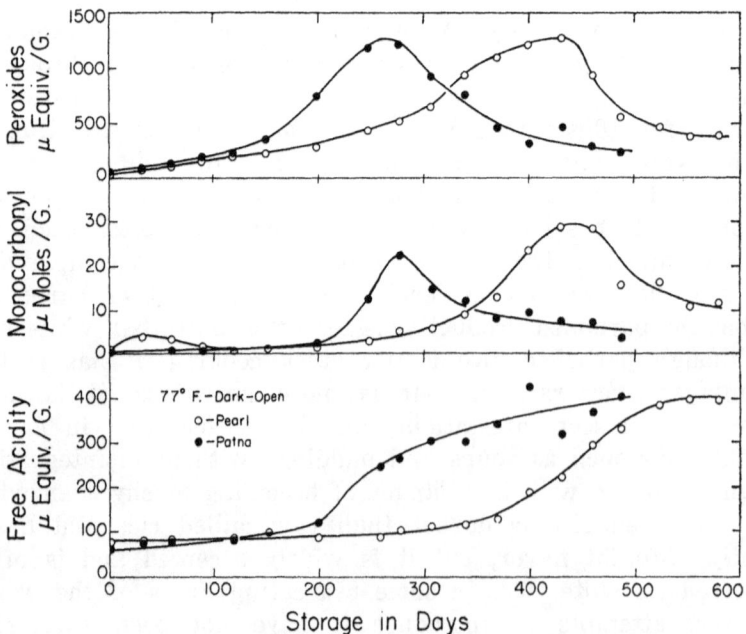

FIG. 121. CHANGES IN OIL DURING OPEN STORAGE IN THE DARK
AT 77°F

tion was observed. The accelerating effect of light on rancidification of fat-containing foods was confirmed for parboiled rice during storage at 77°F.

In a separate study (Houston et al. 1956), the effect of storage on color changes in parboiled rice at different temperatures was measured. For both Pearl and Patna rices, color changes were negligible for about a year at 77°F in either open or sealed containers. At 100° and 140°F color changes were observed after 3 or 4 months due to nonenzymatic browning, accompanied by losses of reducing sugars. Amino nitrogen and free amino acids were also reduced in amount (Hunter et al. 1956). Sulfur dioxide added during processing inhibited browning of parboiled rice but did not delay rancidification.

As parboiled rice is frequently sold in small permeable packages, and moves from factory to consumer rather rapidly, little difficulty is experienced with rancidity on retail store shelves. But when stored in bulk, it occasionally "goes off" in large quantities, and manufacturers of this product now stabilize it with high-potency antioxidants, which have proved quite effective. Close milling of the rice to reduce its fat content is also practiced, but always a few tenths percent fat remains and this is enough to cause trouble unless it is stabilized in some way.

Parboiled rice expands to a light porous product when heated in hot air or oil (Roberts 1952, Roberts et al. 1951). Roberts et al. (1954) made an intensive study of the influence of parboiling conditions on various properties of rice, including its color, the solubility of the amylose fraction, and the extent of expansion upon heating under controlled conditions. The steaming step in the parboiling of rice had a particularly great influence on these properties, the expansibility of the grains, depth of color, and starch solubility in the final product increasing decidedly with the severity of the treatment.

Expanded parboiled rice is usable as a dry breakfast cereal but for that purpose its comparatively bland flavor should be enhanced with other ingredients such as honey, malt syrup, salt, etc. Precooked brown rice expanded by toasting and ground to farina size has been patented as an instant-cooking breakfast cereal (Kester and Ferrel 1957).

Quick-Cooking Rice

The modern trend in processed foods is toward convenience items. The day of long, drawn-out preparation of food is fast disappearing. Housewives are relying more and more on canned, packaged, or frozen foods to serve their families. Food processors and research agencies have been eminently successful in developing tasty products in these categories. Rice, because of its bland flavor, lends itself readily to various forms of preparations and formulations. Parboiled rice, described in preceding pages of this chapter, is a case in point. This form of rice, however, although fully precooked, is not a quick-cooking rice for the reason that the grains are still quite dense and therefore do not take up water readily. Other dry forms of precooked rice are now available with a more open texture and may be prepared for serving in from 2 to 15 min. Some are made by precooking in water and drying under special and closely controlled conditions; others are made with dry heat. Following are some of the methods used for preparing quick-cooking rice.

In one commercial method, the cooking time of both brown and white rice is shortened by heating the dry grains in a current of air at elevated temperatures. The rice kernels are so modified by fissuring of the surface layers, by dextrinization, and, in the case of brown rice, loosening of the bran layers, that the cooking time of the product in the home is reduced to about ½ or ⅓ that of the raw rice. This represents an extremely important saving of cooking time, especially for brown rice, which, when not so treated, requires 35 to 45 min of cooking to make it soft enough to eat. Heat-processed brown rice is improved nutritionally by coating it with thiamine, riboflavin, and iron phosphate, and is stabilized against rancidification with antioxidant. These substances are applied at the same time in a water-vegetable oil emulsion. Unlike fortified or enriched raw rice, enriched heat-processed rice is not coated with a protective film and it should not be rinsed before being cooked for serving.

Other forms of quick-cooking rice are made by precooking white rice and either rapidly drying the cooked product in a current of hot air or slow drying followed by expansion of the dried grains with heat. A large number of patents (Alderman et al. 1953; Campbell and Hollis 1954; Flynn and Hollis 1955; Hollis et al. 1958; Knoch 1955; Mickus and Brewer 1957; Muller 1952; Ozai-Durrani 1948, 1950, 1955, 1956, 1958; Roberts 1952, 1955; Shuman and Staley 1954) have been taken out to cover different variations of the wet process for making quick-cooking rice.

One widely sold form of quick-cooking rice can be prepared for serving in about 5 min. It is enriched with thiamine, riboflavin, niacin, and iron. Actual details of the commercial method for manufacturing this product are not available, but they probably follow the general methods outlined above and are continually being modified to obtain improvement in texture and nutritional value of the product and to reduce processing losses.

In the original form of this process, the rice was gelatinized in one step by completely cooking it in water (Anon. 1947; Anon. 1949). This procedure resulted in extensive losses due to over-cooking of the surface layers of the grains which dispersed into the cooking water. In one modification (Flynn and Hollis 1955), the soaked rice is steamed under pressure to partially gelatinize the grains, which are then slightly compressed. They are re-treated with hot water to enlarge the grains and complete the gelatinization process. After these treatments the rice is dried under conditions which produce a slight puffing action.

Quick-cooking rice may also be made by first heating the raw grains in hot air or by infrared radiation (Campbell and Hollis 1954; Shuman and Staley 1954) then gelatinizing them with mois-ture and heat, followed by quick drying to set the grains in their enlarged porous condition.

In another patented process (Carman and Allison 1953) cooked rice is expanded by heating it under pressure and suddenly re-laxing the pressure into a vacuum chamber.

Still another process for making quick-cooking rice is a novel variation of prior methods. It consists of soaking and cooking rice, then freezing, thawing, and dehydrating it (Keneaster and Newlin 1957). The product is said to recook in about 5 min. De-tails of a similar process have been discussed by Roseman (1958).

Presence of oxidizable iron in the cooking system results in the development of grayness in the final product. McGinley (1968) proposes to eliminate this problem by the addition of a chelating agent such as sodium acid pyrophosphate to the water. His patent also teaches the use of a starch-complexing agent (e.g., mono-glycerides of fatty acids) in the cooking water to prevent matting and adhesion of the rice during drying.

Pence (1967) described an improved process (developed by Huxsoll and Ferrel) based on the use of microwave energy to achieve the degree of starch gelatinization essential for a rapidly rehydrating convenience product. After raw rice has been soaked in water until it reaches a moisture content of about 35%, it is passed through a microwave oven to cook it rapidly and com-pletely. Some rices require a second soaking and cooking cycle for optimum results. After cooking has been completed, the grain is again immersed in excess water to allow it to come to 68% moisture content, about equivalent to normal cooked rice. Because of the method of heating, the outer surfaces of the kernels are not overcooked and not extremely sticky and loose as is often the case with steam cooked rice. Therefore, the rice can now be dried in such a fashion as to leave the desired degree of porosity and fissuring in the finished product. Additional advantages of the microwave process are that washing and leaching losses of solids can be kept low and the original flavor of the rice can be rather well maintained. Furthermore, the reduction in gumminess and stickiness minimizes mechanical losses due to clumping and breaking during processing.

The property possessed by dried gelatinized rice of expanding under heat treatment (Roberts 1951, 1952) is most advantageous in the preparation of quick-cooking rice products. The degree of expansion in hot air may be as high as four-fold when the proper conditions are used. The grains retain their normal shape after expansion but are crisp and porous. Addition of boiling water causes them to shrink to about the size of boiled white rice grains. The texture is not the same as for rice cooked in the usual manner.

Canned Rice

Canned cooked rice may also be considered a form of quick-cooking or convenience rice. Formerly it was made almost exclusively from parboiled rice because this form, unlike cooked white rice, does not disintegrate when the cans are retorted. Roberts (1954) and Roberts et al. (1953) of the Western Regional Research Laboratory surmounted the difficulty of canning white rice by limiting the moisture content of the product to about 55%. Briefly, this process consists of: (1) soaking the rice to its equilibrium moisture content (about 30 to 35% water); (2) cooking it 4 to 5 min in excess water; (3) draining and packing the rice into cans; and (4) vacuum sealing and retorting. This rice may be prepared for serving by adding a small amount of additional water and heating for about 2 min until the added water is absorbed.

The critical points of this process are: (1) limitation of the moisture content as described; (2) adjustment of the pH in the product before canning to about 5.5 to 6.0; (3) packing in C-enamel cans or glass jars with C-enameled lids; and (4) sealing under at least 26 in. of vacuum.

It is important that the moisture content of the white rice

cooked for canning be in the range of 50 to 55%. Less moisture than 50% makes the rice too dry and it doesn't absorb the additional water readily, while moisture higher than 55% causes the rice to become pasty during canning and retorting. Doubtless these limits would vary for different kinds of rice.

In recent years, there has been increasing interest in the aseptic canning of rice puddings and the like. The critical feature here is the separate sterilization of the rice kernels, in a minimum amount of liquid, and the "sauce" portion of the pack. These two components are then combined in the can. Separation of the rice and the sauce is necessary because of the different sterilization treatments required by the two products. The liquid can be rapidly sterilized by almost any convenient means—swept wall heat exchangers, Spiratherms, Uperizers, or triple-tube heat exchangers —while the requirements for the rice are much more stringent due to the necessity for allowing a time interval sufficient for heat to thoroughly penetrate the kernels. If the combined product is heat treated sufficiently to sterilize the rice, there is a tendency for the sauce to become overheated with excess browning and the development of off-flavors.

There are very significant variations between different varieties of rice in their resistance to breakdown or disintegration as the result of heat processing received during canning. These variations are related, at least in part, to the amylose content of the starch (Hagberg 1967). Rice with low amylose content will always have poor stability, but, on the other hand, not all rice with high amylose content will have good resistance to breakdown. Some varieties of rice such as Jojutla become extremely firm. The best varieties have been Nira, Rexoro, and Texas Patna, while Century Patna completely lacks canning stability.

Frozen Cooked Rice

Tressler *et al.* (1968) give a complete review of the literature relating to freezing and frozen storage of cooked rice. They quote suggestions for commercially processing frozen cooked rice:

"(1) Place rice in an excess of water at 130° to 140°F which has received enough citric acid to give it a pH of 4.0–5.5. Enough water should be used to cover the rice after it has soaked for 2 hr.

"(2) After 2 hr, drain off the soak water and rinse with more of the same pH adjusted water to remove fines.

"(3) Drain thoroughly, tapping the screen to shake loose the adhering water, or by blowing the rice layer with air.

"(4) Meanwhile, place a small volume of water in the bottom of a pressure cooker and heat to boiling with the cover on to heat up the apparatus. Place the soaked drained rice in layers 2 in. deep or less over screens which are supported above the water in the vessel. Close the vessel and heat with the vent open until steam is emitted, then close off the vent and raise the pressure to 12–15 psi and hold for 12 to 15 min. Then blow off steam gradually enough to prevent violent boiling and flashing of the hot water.

"(5) Remove the hot, steamed rice. It should have a rubbery texture. Place it in an excess of hot water at 200° to 210°F and without stirring let it imbibe water until the grains are large, tender, and quite free. Stirring will cause it to become sticky. The rice should be held in a perforated vessel so that water may circulate freely through it. Do not boil the rice.

"(6) Cooking should require only 10 to 15 min following the step described in (4). Drain off the hot water, rinse twice with cool or cold water having the pH adjustment described in (1) above.

"(7) Tap and shake to remove free water; or better still if you have the equipment suck off free water over a vacuum filter.

"(8) Convey the cooked rice on a stainless steel mesh belt through an air-blast cooler to reduce it to approximately room temperature, then package in cartons. Freeze in air-blast freezers or by other appropriate means."

Boiled and steamed white and brown rice which have been frozen and reheated are virtually indistinguishable from their unfrozen counterparts. Frozen storage for up to about 1 yr appears to have no deleterious effects on quality (Boggs *et al.* 1951, 1952). In fact, white rice actually seems to improve in different respects with frozen storage according to the verdict of a taste panel.

There are a number of excellent frozen rice products on the market today. Some of these are combination dishes which can be reheated by the boil-in-bag method. Evidently the sticking together which created problems in some earlier versions has been avoided by individual quick freezing of the grains together with the inclusion of oily type sauces.

ENRICHMENT

The use of highly polished rice has been a public health problem in many places where rice is a main article of diet.

Beriberi or severe B-complex deficiencies are often endemic in these areas. A test in the Bataan province of the Philippines from 1947 to 1950 (Salcedo *et al.* 1950) demonstrated conclusively the value of rice enriched with thiamine, niacin, and iron for reducing the incidence of and mortality from beriberi. The results of the test were so dramatic that the Philippine government continued enrichment of rice in Bataan and has gradually extended it to other provinces. In 1951, the Puerto Rican government enacted legislation requiring that all rice sold on the island be enriched. And in our country, South Carolina, which has a high per capita consumption of white rice, passed a law requiring that only enriched rice be sold within the state after July 1, 1956. Enriched rice has also been made available in Cuba, Venezuela, Colombia, Dominican Republic, Taiwan, Singapore, Australia, Hawaii, and Thailand.

Rice may be enriched in many ways. The parboiling of rice is actually equivalent to an enrichment process because bran vitamins diffuse into the endosperm during the soaking and cooking of rough rice. In areas where parboiled rice is consumed as the main form of the cereal, the population receives a satisfactory dosage of these nutritional substances.

Approximately 80% of the thiamine, 56% of the riboflavin, 65% of the niacin, 60% of the pantothenic acid, and 55% of the pyridoxine are removed from brown rice when it is milled. Not all of these constituents are retained in parboiled rice and it must be supplemented with small amounts of vitamins and iron if it is to conform to the Standards of Identity for Enriched Rice.

Some of the methods which have been used for adding dietary supplements to rice are (Hammes 1967): (1) Mixing with the kernels a powder containing the vitamins and minerals. It is necessary that the consumer be cautioned against washing the rice, for such an operation removes the enrichment. (2) Supplying a wafer which must be added to the rice at the time it is cooked. The problems which would be involved in securing the cooperation of distributors and consumers in this procedure are rather obvious, but the method has been used successfully under military conditions. (3) Adding a small amount of rice grains which have been coated or impregnated with the enriching ingredients. This is the method most commonly used.

The highly enriched grains, or premix, are made by two somewhat similar processes. In one of these, which is representative

(Furter and Lauter 1946; Mickus 1955), a 2150 lb lot of milled rice is charged into a trumbol fitted with baffles (Fig. 122). A vitamin solution made up of 39.5 lb niacin, 6 lb thiamine, 21 lb sulfuric acid, and 58 lb water is fed into the trumbol by a perforated stainless steel pipe that runs the full length of the unit. In about 10 min the rice grains are uniformly coated with the mixture, and air is then blown into the trumbol to dry the coating so that the grains are free flowing. Next, half of an enrobing solution consisting of 13.75 lb of abietic acid, 16.25 lb of a solid fatty acid, 20.25 lb of zein, 18 gal. of isopropyl alcohol, and 1 gal. of water, is sprayed on the rice and the kernels are again dried. Open scoops containing 270 lb of dry ferric pyrophosphate with 405 lb of talc are then run into the trumbol and inverted.

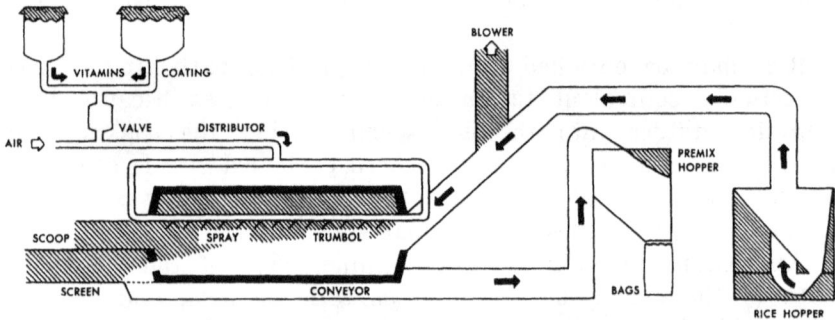

FIG. 122. DIAGRAM SHOWING PRODUCTION OF RICE PREMIX

When the solid material is well-dispersed over the rice, the second half of the enrobing solution is applied. With these proportions of ingredients, the final product has a content per pound of 440 mg thiamine, 3,200 mg niacin, and 2,600 mg iron. The premix is diluted incrementally with 60 lb quantities of untreated rice in the 1:199 ratio by proportional weighing.

The quantities of enriching agents specified in the above description apply to one procedure commonly used in the United States. In the Far East, where rice enrichment is practiced, different amounts of vitamins are used, and the composition of the coating solution is modified to meet specific local needs, both nutritional and environmental.

To give the grains of enriched rice a whiter appearance than they have when treated by previous methods, and to permit addition of riboflavin to the enriching mixture, the protective coating may contain white pigments such as calcium phosphate, talc, and titanium dioxide according to patents of Antoshkiw (1958) and

La Pierre (1955). When riboflavin is included in the coating, its yellow color diffuses into the water when the coated kernels are boiled, causing unsightly blotches in the cooked rice.

A simpler one-step process of enriching rice is also practiced. In this process, the powdered fortifying agents are merely tumbled with the dry rice until they are uniformly distributed over the surface of the grains, to which they cling with considerable tenacity. Such a coating would not be expected to pass a washing test as specified in Puerto Rico. In other localities where this form of enriched rice is sold, precautions against rinsing the rice before cooking appear on the package.

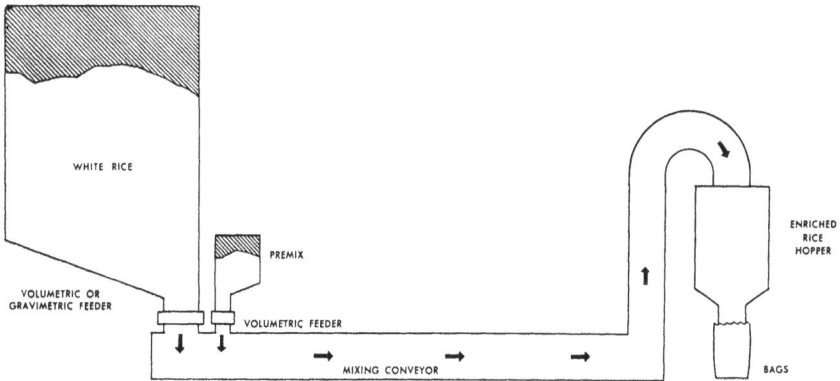

FIG. 123. DIAGRAM OF BLENDING OF PREMIX WITH WHITE RICE

USE OF WAXY RICE FLOUR IN FOODS

In tests of numerous thickening agents for white sauces, gravies, and puddings, waxy rice flour had superior properties in preventing liquid separation (syneresis) when these products were frozen, stored, and subsequently thawed (Hanson *et al.* 1951, 1953, 1957). Regardless of the thickening agent used, storage temperature proved to be an important factor. At 0° F, a white sauce thickened with waxy rice flour showed little or no separation when thawed at the end of a year; puddings made with this flour and egg yolk had a storage limit of about four months, and when made with the flour and whole egg, about two months. When these products were stored at 10°F, however, the storage life was only about $\frac{1}{5}$ that at 0°F.

The fact that waxy rice contributes this desirable property to frozen foods of the kind described above is doubtless because its starch is almost pure amylopectin. By way of partial explanation of its behavior, the following observation of Kerr (1950) is cited: "Solutions of starch which have aged at room or lower temperature undergo the phenomenon of retrogradation. A part of the starch aggregates progressively, and finally forms an insoluble microcrystalline precipitate . . . The retrogradation process may be hastened by freezing aqueous solution; in this way, ordinarily stable solutions may be forced to retrograde. Although some amylopectin preparations have a tendency to retrograde from solution, the property is greatly exaggerated in pure amylose solutions."

The inference drawn by Hanson *et al.* (1951) is that the common starches containing unbranched molecules would be expected to retrograde and eliminate water more rapidly than amylopectin starches under comparable conditions.

Bates *et al.* (1943) reported that amylopectins from different starches do not behave alike in an iodine titration. In their tests, the more highly branched the amylopectin, the less was its affinity for iodine. Of the polysaccharides investigated, glycogen had the least affinity for iodine, followed by waxy rice, waxy corn, waxy barley starch, and potato and corn amylopectin. Meyer and Fuld (1941) found that the degree of branching of waxy rice starch is between that of ordinary amylopectins and of glycogen and that the waxy rice starch molecules are of low and intermediate molecular weights. In the investigations at the Western Regional Research Laboratory, liquid separation in sauces containing waxy corn and waxy sorghum starches occurred in shorter storage times than in sauces containing waxy rice starch or flour. This is further evidence that the waxy cereal starches differ in molecular size and degree of branching.

In a separate investigation, it was found that waxy rice flour had desirable properties when used as a thickening agent in canned products (Davis *et al.* 1955). Pastes prepared from it showed essentially no increase in gelation or separation of liquid during storage. Other waxy cereals possess this property also, but waxy rice has the advantage of imparting a "short" or nonstringy character to the paste. Use of waxy rice flour with wheat flour for pastes is indicated when a certain amount of initial gelation is required without increase in gelation during storage.

UTILIZATION OF HULLS, BRAN, AND POLISH

The hull, the outermost layer of rough rice, differs greatly from the bran and polish in structure and composition (see Table 82).

Hulls are about 20% ash, and the ash is over 90% silica, which is a very high content for plant materials. Hulls have been more of a liability than an asset to millers, who often have to pay for their disposal. Many suggestions have been made for ways to economically utilize hulls. They have a peculiar gridlike structure which, together with the high content of silica, makes them rigid and abrasive. Consequently, they have been used for soft grit blasting of metals and for polishing of semiprecious gems. Their k value of about 0.25 puts them between asbestos and mineral wool in thermal insulating properties, and they burn slowly, so they might be of some value as an inexpensive insulating material. Their fuel properties are not particularly outstanding (they yield 6.27 Btu per lb, about $\frac{1}{2}$ as much as bituminous coal and $\frac{1}{3}$ as much as fuel oil), but many rice mills have used them as boiler fuel. The ash of the burned hulls can be used as a grease-absorbing sweeping compound, as a mild abrasive in certain types of hand soap, as a filtration medium, and as an ingredient of ceramic ware.

Attempts have been made to extract the cellulose of hulls, but, although it is of high quality, the fibers are short and do not felt well, limiting its application to paper-making, etc. Pentosans (about 12%) can be extracted by acid hydrolysis and used for furfural production, but hulls are less suitable for this purpose than oat hulls or corncobs (about 22% pentosans).

The Soil Conservation Service discovered that hulls could be

TABLE 82

COMPOSITION OF RICE HULLS AND RICE HULL ASH[1]

	%
Rice hulls	
Moisture	2.4–11.0
Ash	15.7–21.3
Protein	2.4–3.6
Ether extract	0.9–1.2
Crude fiber	39.0–45.7
Nitrogen-free extract	24.7–29.4
Rice hull ash	
Silica	94.0–94.5
CaO	0.3–2.3
MgO	0.2
K_2O	1.1–3.2
Na_2O	0.8
Fe_2O_3	Trace to 0.1
SO_3	1.1
P_2O_5	0.5
Al and Mn oxides	Trace

[1] Range of reported values.

used as a seed diluent for drill planting, and they are said to be superior in this respect to sand or sawdust in maintaining uniform distribution, particularly of mixed seeds and forage grasses.

Hulls have also been used or proposed as stock litter, soil conditioners, and raw materials for activated charcoal. They are low in food calories but can be used in limited amounts in feeds without injury to animals. "Rice mill feed" is the entire byproduct obtained in the milling of rice, consisting of hulls, bran, polish, and broken grains. Since sale of this material eliminates the separate handling of individual by-products in the plant, it is an attractive outlet for the miller (Beagle 1961).

In contrast to hulls, rice bran is a nutritious fraction of the grain and is used almost exclusively as a feedstuff. It contains about 14% of oil which undergoes rapid hydrolysis once it is separated from the grain. The free fatty acids resulting from this reaction are readily attacked by oxygen with the formation of fat peroxides and other oxidation products which are detrimental to the use of the bran as a stock feed because they may destroy fat-soluble vitamins or cause digestive disturbances. The rancid flavor also reduces feed consumption in most cases.

Rice bran can be stabilized by solvent extraction of the fatty materials. The defatted bran has a higher percentage of protein than the original material, but it is quite dusty and is often pelleted with molasses. The oil is relatively stable in its natural state and becomes extremely stable after hydrogenation. When refined, bleached, and deodorized, it can be used for general culinary purposes, deep fat frying, and in formulated foods. It has some acceptance as a salad oil.

Rice polish is also high in nutrients. Considerable quantities are sold for baby food formulations and in health stores, but most of the annual production goes into feeds.

FUTURE TRENDS

The desire of governmental agencies here and abroad to improve the nutritional status of persons who subsist almost entirely on rice will undoubtedly lead to demands for types which contain a balanced complement of vitamins, minerals, and amino acids so as to be a completely adequate food. Worthwhile advances in protein supplementation may occur as the result of the development of improved strains, and many agronomists are working with this goal in mind, but experience with other grains

TABLE 83

COMPOSITION AND OTHER CHARACTERISTICS OF RICE BRAN OIL[1]

Analysis of Crude Oil	
Fatty acids, %	
Saturated	17.6
Oleic	47.6
Linoleic	34.0
Linolenic	0.8
Iodine value	102.3
Unsaponifiable matter, %	2.7
Characteristics after Refining and Bleaching	°F
Smoke point	415
Flash point	615
Fire point	665
Cloud point	34
Solid point	18
Characteristics after Refining, Bleaching, and Winterizing	
Cold test, hr	9
Cloud point, °F	20
Pour point, °F	18

[1] Adapted from Mickus (1959).

as well as with rice has shown that such improvements are usually accompanied by undesirable changes, e.g., decreased yields or deterioration in cooking quality. A more direct approach would be to fortify the grains with proteins, modified proteins, or combinations of amino acids. The technology for supplementing the vitamin and mineral content is already available, but the protein deficiency is more difficult to resolve, primarily because of the greater quantity of material required.

If rice flour could be formed into particles similar to whole milled rice in appearance and cooking response, a valuable new outlet for broken rice would be opened up. Such a process would also permit more control over nutritional content and organoleptic quality through the incorporation of various additives. The success of quick cooking rice, with its altered appearance and texture, has shown that most consumers will accept products which differ considerably from milled rice in visual characteristics. Processes and equipment familiar to the macaroni processing industry could doubtless be adapted to shape the pieces, but methods for obtaining the desired cooking quality are not as obvious.

The present kinds of quick-cooking rice find a ready acceptance by many consumers, but improvements are certainly to be desired. Organoleptic properties more closely approaching steamed long grain rice, and a greater tolerance to changes in cooking conditions would be the goals. Methods by which these improvements can be obtained are not readily apparent, but might include freeze-drying of partially cooked rice with subsequent addition of wetting agents, or opening of channels into the kernel by selective enzyme attack.

Modification of the texture of short grain rice in the direction of the best long grain varieties would have some value and might be accomplished by various starch-complexing additives. Distributing these compounds throughout the unmodified kernel would be difficult, but could be accomplished if fissures were opened in the endosperm.

BIBLIOGRAPHY

ALDERMAN, M. W., MASSMAN, W. F., JR., and MICHAEL, E. W. 1953. Rice process. US Pat 2,643,951. June 30.

ANON. 1947. Instant rice—from an idea to the perfect product. Food Ind. *19*, 1056–1061.

ANON. 1948. Special process by Walton mill. Rice J. *51*, No. 11, 12–13.

ANON. 1949. The story of minute rice. Rice J. *52*, No. 3, 20–21.

ANON. 1963. Continuous pressure processing. Food Process. *24*, No. 9, 64–65.

ANON. 1967. Extractive milling yields 10% more whole grain rice. Food Process. Marketing *28*, No. 6, 34–35.

ANON. 1968. National Food Situation. US Dept. Agr. Circ. NFS-*124*.

ANTOSHKIW, T. 1958. Enrichment of cereal grains. US Pat. 2,831,770. April 22.

ATEN, A., and FAUNCE, A. D. 1953. Equipment for the processing of rice. Food Agr. Organ. U.N., Develop. Paper *27*.

AUTREY, H. S., GRIGORIEFF, W. W., ALTSCHUL, A. M., and HOGAN, J. T. 1955. Effects of milling conditions on breakage of rice grains. J. Agr. Food Chem. *3*, 593–599.

BATCHER, O. M., DEARY, P. A., and DAWSON, E. H. 1957. Cooking quality of 26 varieties of milled white rice. Cereal Chem. *34*, 277–285.

BATCHER, O. M., HELMINTOLLER, K. F., and DAWSON, E. H. 1956. Development and application of methods for evaluating cooking and eating quality of rice. Rice J. *59*, No. 13, 4–8.

BATES, F. L., FRENCH, D., and RUNDLE, R. E. 1943. Amylose and amylopectin content of starches determined by their iodine complex formation. J. Am. Chem. Soc. *63*, 142.

BAUMGARTNER, M. M. 1917. Process of treating rice and product thereof. US Pat. 1,239,555. Sept. 11.

BEAGLE, E. 1961. Uses for rice hulls. U.S. Dept. Agr. Circ. ARS *74-24*.

BEVENUE, A., and WILLIAMS, K. T. 1956. Hemicellulose components of rice. J. Agr. Food Chem. *4*, 1014–1017.

BOGGS, M. M., SINNOTT, C. E., VASAK, O. R., and KESTER, E. B. 1951. Frozen cooked rice. Food Technol. *5*, 230–232.

BOGGS, M. M., WARD, A. C., SINNOTT, C. N., and KESTER, E. B. 1952. Frozen cooked rice. II. Brown rice. Food Technol. *6*, 53–54.

BORASIO, L., and GARIBOLDI, F. 1957. Illustrated Glossary of Rice Processing Machines. Food Agr. Organ. U.N., Rome.

CAMPBELL, H. A., and HOLLIS, F., JR. 1954. Process of preparing quick-cooking rice. US Pat. 2,696,156. Dec. 7.

CARMAN, C. R., and ALLISON, J. E. 1953. Quick-cooking cereal and method for making same. US Pat. 2,653,099. Sept. 22.

COURT, A. B. 1946. Rice processing by Growers Association of California. Rice J. 49, No. 7, 1.

DAVIS, J. G., ANDERSON, J. H., and HANSON, H. L. 1955. Starchy cereal thickening agents for canned food products. Food Technol. 9, 13–17.

DEOBALD, H. J. 1967A. Manufacture and usage of rice flour. Eighth Ann. Symp. Central States Sect. Am. Assoc. Cereal Chem. 1967. St. Louis, Mo., Feb. 17–18.

DEOBALD, H. J. 1967B. Recent rice research development at the Southern Reg. Res. Lab. US Dept. Agr. Circ. ARS 72–53.

DESIKACHAR, H. S. R. 1956. Changes leading to improved culinary properties of rice on storage. Cereal Chem. 33, 324–328.

EDWARDS, C. H., and ALLEN, C. H. 1958. Cystine, tyrosine, and essential amino acid content of selected foods of plant and animal origin. J. Agr. Food Chem. 6, 219–223.

FLYNN, C. W., and HOLLIS, F., JR. 1955. Production of quick-cooking rice. US Pat. 2,720,460. Oct. 11.

FURTER, M. F., and LAUTER, W. M. 1949. Fortifying grain products. US Pat. 2,475,133. July 5.

FURTER, M. F., LAUTER, W. M., DE RITTER, E., and RUBIN, S. H. 1946. Enrichment of rice with synthetic vitamins and iron. Ind. Eng. Chem. 38, 486–493.

HAGBERG, E. C. 1967. Canned rice products. US Dept. Agr. Circ. ARS 72–53.

HALICK, J. V., and KELLY, V. J. 1959. Gelatinization and pasting characteristics of rice varieties as related to cooking behavior. Cereal Chem. 36, 91–98.

HALICK, J. V., and KENEASTER, K. K. 1956. The use of a starch-iodine blue test as a quality indicator of white milled rice. Cereal Chem. 33, 315–319.

HAMMES, P. A. 1967. Enrichment of rice. US Dept. Agr. Circ. ARS 72–53.

HANSON, H. L., CAMPBELL, A., and LINEWEAVER, H. 1951. Preparation of stable frozen sauces and gravies. Food Technol. 5, 432–440.

HANSON, H. L., FLETCHER, L. R., and CAMPBELL, A. 1957. The time-temperature tolerance of frozen foods. V. Texture stability of thickened precooked frozen foods as influenced by composition and storage conditions. Food Technol. 11, 339–343.

HANSON, H. L., NISHITA, K. D., and LINEWEAVER, H. 1953. Preparation of stable frozen puddings. Food Technol. 7, 462–465.

HAVIGHORST, C. R. 1947. Age-old process modernized. Food Ind. 19, 1192–1195.

HOLLIS, F., JR., MILLER, F. G., and MILLER, F. J. 1958. Process of preparing a quick-cooking rice. US Pat. 2,828,209. Mar. 25.

HOUSTON, D. F., HUNTER, I. R., and KESTER, E. B. 1956. Storage changes in parboiled rice. J. Agr. Food Chem. *4*, 964–968.

HOUSTON, D. F., HUNTER, I. R., McCOMB, E. A., and KESTER, E. B. 1954. Deteriorative changes in the oil fraction of stored parboiled rice. J. Agr. Food Chem. *2*, 1185–1190.

HOUSTON, D. F., IWASAKI, T., MOHAMMAD, A., and CHEN, L. 1968. Radial distribution of protein by solubility classes in the milled rice kernel. J. Agr. Food Chem. *16*, 720–724.

HOUSTON, D. F., and KESTER, E. B. 1958. Plant acids in cereal seeds. Presented Meeting Am. Chem. Soc., San Francisco, Apr. 13–18.

HOUSTON, D. F., McCOMB, E. A., and KESTER, E. B. 1952. Effect of bran damage on development of free fatty acids during storage of brown rice. Rice J. *55*, No. 2, 17–18.

HUNTER, I. R., FERREL, R. E., and HOUSTON, D. F. 1956. Free amino acids of fresh and aged parboiled rice. J. Agr. Food chem. *4*, 874–875.

HUNTER, I. R., HOUSTON, D. F., and KESTER, E. B. 1955. Adsorption-dialysis, an extraction technique and its use in recovery of amino acids. Anal. Chem. *27*, 965–968.

HUZENLAUB, E. G. 1942. Treatment of wheat and kindred cereals. US Pat. 2,287,737. June 23.

HUZENLAUB, E. G. 1949. Process of enriching the endosperm of cereal grains with natural vitamins. US Pat. 2,472,426. June 7.

HUZENLAUB, E. G., and ROGERS, J. H. 1941. Apparatus for the treatment of cereals, starch, or the like with fluids or by heating. US Pat. 2,239,608. Apr. 22.

HUZENLAUB, E. G., and ROGERS, J. H. 1944. Process for the treatment of rice and other cereals. US Pat. 2,358,251. Sept. 12.

HUZENLAUB, E. G., and ROGERS, F. H. 1951. Altering the flavor of cereals. US Pat. 2,539,999. Jan. 30.

HUZENLAUB, E. G., and ROGERS, F. H. 1952. Apparatus for the treatment of grains. US Pat. 2,598,915. June 3.

JOHNSON, R. M. 1963. Long-grain varieties of milled rice. Cereal Sci. Today *8*, 84, 90.

JONES, J. W., and TAYLOR, J. W. 1935. Effect of parboiling rough rice on milling quality. US Dept. Agr. Circ. *340*.

JONES, J. W., ZELENY, L., and TAYLOR, J. W. 1946. Effect of parboiling and related treatments on the milling, nutritional, and cooking quality of rice. US Dept. Agr. Circ. *752*.

KENEASTER, K. K., and NEWLIN, H. E. 1957. Process for producing a quick-cooking product of rice or other starchy vegetable. US Pat. 2,813,796. Nov. 19.

KERR, R. W. 1950. Chemistry and Industry of Starch. Academic Press, New York.

KESTER, E. B., and FERREL, R. E. 1957. Method of preparing precooked puffed brown rice cereal. US Pat. 2,785,070. Mar. 12.

KIK, M. C. 1942. Nutritive studies of rice and its by-products. Arkansas Univ. Expt. Sta. Bull. *416*.

KIK, M. C. 1954. Nutritive value of rice germ. J. Agr. Food Chem. *2*, 1179–1181.

KIK, M. C. 1956. Nutrients in rice bran and rice polish and improvement of protein quality with amino acids. J. Agr. Food Chem. *4*, 170–172.

KIK, M. C., and VAN LANDINGHAM, F. B. 1943. Riboflavin in products of commercial rice milling and thiamin and riboflavin in rice varieties. Cereal Chem. *20*, 563–569.

KNOLL, W. L. 1967. The handling of rice. US Dept. Agr. Circ. ARS *72–53*.

KNOCH, H. 1955. Quick-cooking cereals. Brit. Pat. 722,333. Jan. 26.

KOHLER, G. O. 1961. Air-classification milling. US Dept. Agr. Circ. ARS *74–24*.

LANDON, R. W., TALMEY, P., and GUTZEIT, G. 1951. Treating rice prior to milling. US Pat. 2,546,450. Mar. 27.

LAPIERRE, R. 1955. Coated cereal products and process for preparing the same. US Pat. 2,712,499. July 5.

LITTLE, R. R., HILDER, G. B., and DAWSON, E. H. 1958. Differential effect of dilute alkali on 25 varieties of milled white rice. Cereal Chem. *35*, 111–126.

MCCALL, E. R., HOFFPAUIR, C. L., and SKAU, D. B. 1951. The chemical composition of rice—A literature review. US Dept. Agr. AIC *312*.

MCCALL, E. R. *et al.* 1953. Composition of rice. J. Agr. Food Chem. *1*, 988–993.

MCGINLEY, F. A. 1968. Improvements in or relating to the processing of rice. Brit. Pat. 1,121,893. July 31.

MECHAM, D. K. 1961. Parboiling conditions for Calrose rice. Proc. Natl. Rice Util. Conf. 1961. US Dept. Agr. Circ. ARS *74–24*.

MEYER, K. F., and FULD, M. 1941. Starch studies XVII. The starch of glutinous rice. Helv. Chim. Acta *24*, 1404.

MICKUS, R. R. 1955. Seals enriching additives on white rice. Food Eng. *27*, 91–93, 160.

MICKUS, R. R. 1959. Rice. Cereal Sci. Today *4*, 138–141, 144–148.

MICKUS, R. R., and BREWER, G. W. 1957. Rice treating process. US Pat. 2,808,333. Oct. 1.

MONTGOMERY, C. J. 1967. Conventional milling of rice. Proc. Natl. Rice Util. Conf. 1966. US Dept. Agr. Circ. ARS *72–53*.

MULLER, F. P. 1952. Rice product. Australian Pat. 146,945. June 20.

NORMAND, F. L., HOGAN, J. T., and DEOBALD, H. J. 1964. Improvement of culinary quality of freshly harvested rice by heat treatment. Rice J. *67*, No. 13, 7–11.

O'DONNELL, W. W. 1947. Conversion process retains rice vitamins. Food Ind. *19*, 763–768, 892–896.

OZAI-DURRANI, A. K. 1948. Quick-cooking rice and process for making same. US Pat. 2,438,939. Apr. 6.

OZAI-DURRANI, A. K. 1950. Method of treating rice. US Pat. 2,498,573. Feb. 1.

OZAI-DURRANI, A. K. 1955. Quick-cooking rice. Brit. Pat. 737,372, 737,-446, and 737,450. Sept. 28.

OZAI-DURRANI, A. K. 1956. Quick-cooking rice and process therefor. US Pat. 2,733,147. Jan. 31.

OZAI-DURRANI, A. K. 1958. Method of treating rice. US Pat. 2,829,055. Apr. 1.

PECORA, L. J., and HUNDLEY, J. M. 1951. Nutritional improvement of white polished rice by the addition of lysine and threonine. J. Nutr. *44*, 101.

PENCE, J. W. 1967. Recent rice research developments at the Western Reg. Res. Lab., Proc. Natl. Rice Util. Conf. 1966. US Dept. Agr. Circ. ARS *72-53*.

RAO, B. S., MURTHY, A. R. V., and SUBRAHMANYA, R. S. 1952. The amylose and the amylopectin contents of rice and their influence on the cooking quality of the cereal. Proc. Indian Acad. Sci. *36B*, 70–80.

ROBERTS, R. L. 1951. Expanded rice product. A new use for parboiled rice. Food Technol. *5*, 361–363.

ROBERTS, R. L. 1952. Production of quick-cooking rice. US Pat. 2,610,124. Sept. 9.

ROBERTS, R. L. 1954. Process of canning rice. US Pat. 2,686,130. Aug. 10.

ROBERTS, R. L. 1955. Preparation of precooked rice. US Pat. 2,715,579. Aug. 16.

ROBERTS, R. L., HOUSTON, D. F., and KESTER, E. B. 1951. Expanded rice, a new use for parboiled rice. Food Technol. *5*, 361–363.

ROBERTS, R. L., HOUSTON, D. F., and KESTER, E. B. 1953. Process for, canning white rice. Food Technol. *7*, 72–80.

ROBERTS, R. L., POTTER, A. L., KESTER, E. B., and KENEASTER, K. K. 1954. Effect of processing conditions on the expanded volume, color, and soluble starch of parboiled rice. Cereal Chem. *31*, 121–129.

ROSEMAN, A. S. 1958. The effect of freezing on the dehydration characteristics of rice. Presented at Ann. Meeting of Inst. Food Technologists, Chicago, Ill. May 26–29.

ROSENBERG, H. R., and CULIK, R. 1957. The improvement of the protein quality of white rice by lysine supplementation. J. Nutr. *63*, 477–487.

SALCEDO, J., JR. *et al.* 1950. Artificial enrichment of white rice as a solution to endemic beriberi. J. Nutr. *42*, 501–523.

SASAOKA, K. 1957. Chromatographic determination of amino acids. III. Amino acid composition of glutinous rice glutelin. Mem. Res. Inst. Food Sci. Kyoto Univ. No. *13*, 26–31.

SHUMAN, A. C., and STALEY, C. H. 1954. Method of preparing quick-cooking rice. US Pat. 2,696,158. Dec. 7.

SORENSON, J. W., JR., and DAVIS, W. C. 1955. Drying and storing rough rice in farm storage bins, 1954–1955. Texas Agr. Expt. Sta. Progress Rept. *1819*.

SREENIVASAN, A., and GIRI, K. V. 1939. Quality in rice. IV. Storage changes in rice after harvest. Indian J. Agr. Sci. *9*, 208–222.

STUCKEY, B. N. 1955. Increasing shelf-life of cereals with antioxidants. Food Technol. *9*, 585–587.

SURE, B. 1953. Relationships between milled rice and milled flour and between milled rice and milled white corn meal. J. Agr. Food Chem. *1*, 1207–1208.

TRESSLER, D. K., VAN ARSDEL, W. B., and COPLEY, M. J. 1968. The Freezing Preservation of Foods, 4th Edition, Vol. 4, Avi Publishing Co., Westport, Conn.

WAYNE, T. B. 1967. Rice milling process. US Pat. 3,330,666.

WILLIAMS, K. T., and BEVENUE, A. 1953. A note on the sugars in rice. Cereal Chem. *30*, 267–269.

WILLIAMS, V. R., KNOX, W. C., and FIEGER, E. A. 1953. A study of some of the vitamin B-complex factors in rice and its milled products. Cereal Chem. *20*, 560–563.

WILLIAMS, V. R., WU, W., TSAI, H. Y., and BATES, G. 1958. Varietal differences in amylose content of rice starch. J. Agr. Food Chem. *6*, 47–48.

YAMPOLSKY, C. 1944. Rice. II. Rice grain and its products. Wallerstein Lab. Commun. *7*, No. 20, 7–26.

YONAN-MALEK, M. 1943. Method and control system for treating and canning rice. US Pat. 2,334,665. Nov. 16.

Index

A

Abrasive milling of rice, 353–355
Acetic acid, 49
Acidity, of corn products, 27
Active dry yeast, 48
Adlupulone, 178
Aeration, during malting, 142, 146–148
Aerobacter aerogenes, 199
Agglomerated flour, 16
Air classifiers, 12
Air conveying. *See* Pneumatic conveying
Air permeation techniques, 23
Air-slide cars, 16
Albumin rest, 187
Albumoses, 187
Ale, 174, 182
Aleurone, 29
Alfalfa, 101
Alum, 49
Amflow system, 66–67
Amino acid composition of rice, 352
Amylases, 50, 71, 185–186, 251
Amylograph, 25–26
Amylopectin, 312, 371
Amylose, 312
Anderson expeller, 321
Antibiotics, 126, 197
APV tower fermenter, 204
Artofex mixer, 53
Ash, 19, 21–22, 29
Attemperators, 196
Attenuation, 183, 196
Attrition grinding, 13, 305
Attrition mills, 102–103
Augers, for macaroni presses, 255, 258

B

Bacterial enzymes, 185
Bakeries, 16
Baking powder, 49–50
Baking tests, 27
Barley, 2, 93, 113, 129, 173, 176–177, 183–184
  varieties, 129–130, 140

Batters, 44, 53, 71, 87–88
Beall degerminator, 17
Beer analyses, 213–216
Bentonite, 106
Bins, 16
Biscuits, 2, 14, 32
  of macaroni products, 265
Bleaching agents, 32–33, 324
Bleaching of oil, 324–325
Blenders, for feeds, 106
  for macaroni products, 261
Blight damage, effect on semolina, 251
Blish-Sandstedt pressuremeters, 48
Bock beer, 175
Bologna style, 264–265
Borazzio cooking test, 296
Bottom yeast, 174, 182, 194–195
Brabender moisture tester, 21
Braibanti production system, 282–283
Braibanti short cut drying system, 287
Bran, 28–29, 46
Bread, 2, 14, 29–30, 43–44, 59, 69
Break milling system, 9–10
Break rolls, 9, 18–20
Breakfast cereals, 36
Bromelin, 177, 223
Bronze, use in dies, 268
Brookfield viscometer, 333
Bucket elevator, 2, 14
Buhler Bassano drying line, 281–282
Buhler macaroni line, 280–281
Buhler short cut drying system, 286–287
Buhr mills, 307
Burtonizing, 181
Butter cookies, 79

C

Cakes, 2, 14, 29, 31, 45, 69, 71
Calcium, 181
Calcium peroxide, 50
Calcium phosphate, 49
Canned rice, 366–367
Caramelization, 73
Caramelized malt, 175–176
Carbohydrates, 22, 48

Carbonation, 201
Carrageenan, 247
Carter disk separator, 134–135
Carter-Simons moisture tester, 21
Cattle feed, 115, 123
Cellulose, 173
Centrifuges, for starch recovery, 308–309
Cereal granules, 235
Chapatties, 1
Charcoal adsorbents, 182
Cheese crackers, 80, 82
Chemical leavening, 1, 49–50, 71–72
Chill haze, 201
Chillproofing, 174, 201–202, 211
Chlorides, 181
Chlorine, 26
Chlorophenols, 182
Cleaning, barley, 132–137
    wheat, 4
Cleaning house, 5–6
Clear flour, 28, 45–46
Clermont continuous production line, 284
Clermont sheeter, 264
Coagulable protein test, 214
Coating reel, 243
Cocoa, 223, 239
Cold break, 194–195
Cold test, 334
Color, 19, 22
    durum products, 39, 251–252, 259
    oils, 333–334
Colupulone, 178
Compressed yeast, 47–48
Conditioning, corn, 17–18
    feedstuffs, 101
    wheat, 6–8, 29
Cones, corn, 34
Continuous bread-making processes, 66
Conversion, 188
Cookies, 2, 30, 45, 77–80, 84–87
Coolers, 18, 194–195
Coolships, 194
Copper, 181, 209
Corn, 16–18, 92–93, 98, 111–112, 173, 240
    meal, 225
    oil, 33, 317, 318–319, 321–327, 329
    syrup, 52, 77
        components, 314
        uses, 328–329
Corn Industries' viscometer, 333
Cornflakes, 221, 226–229
Couching, 148–149

Cracked wheat, 224
Crackers, 31, 77, 80–84
Creaming, 52
Cross-grain molder, 65
Crumbles, 106–107
Crystal malt, 176
Cutting machines, 78

D

Decoction mashing, 187
DeFrancisci macaroni line, 278–280
DeFrancisci short cut production line, 285–286
Degerming, of corn, 18, 305–307
Deodorization, 325–327
Deposit cookies, 77, 86
Dextrinizing rest, 188
Dextrins, 186, 251
Dextrose, 316
Diastatic power, malt, 176
    sprouted durum, 252
Die inserts, 266–267
Dies, cleaning and storage of, 269–270
    for macaroni, 256, 266–269
Disodium phosphate, 223, 246
Distilled beverages, 216–217
Distillers' malt, 176
Dividers, dough, 54
Domalt system, 160–162
Donuts, 69
Dormancy, 131–132
Dortmund beer, 174
Dough breaks, 263
Dough improvers, 50
Doughs, 44, 52–53
    for pretzels, 76–77
Drying, corn, 18
    macaroni products, 270–276
    rice, 343–344
    wheat, 4
Drying theory, 271–272
Durum, 20–21, 38–40, 248
    granular, 250

E

Egg whites, 246, 253
Eggs, 44, 53, 247, 253–254
Electronic color graders, 21
Electrostatic cleaning, 17
Energy requirements for milling, 6
English draft beer, 199
Entoleters, 17
Enzymes, *See also* specific enzymes, 25, 47, 50, 72, 87, 129–130, 132, 150–151, 171, 251
Ergot, 36–37

Essonica press, 284
Ethanol, 48, 72, 173, 183
Eureka cleaner, 134
Extractive milling, 356–357
Extruders, 76, 78, 108, 240–241, 252, 254–257, 266
  heat generation in, 258–259
Extrusion, puffing, 240–242
  theory, 255

F

Farina, 223
Farinograph, 24, 47
Fat, determination, 23
Feed constituents, 111–113
Feed formulation, 114–115
Feed industry, regulation, 125–126
Feed mills, 91–92, 95–96
Feedstuffs, composition, 124
  nutrients, 116–121
Fermentable carbohydrates, test, 27–28, 50
Fermentation, beer, 195–199
Fermentation rooms, 59–61
Fiber, determination, 22
Ficin, 177
Fillings, for sugar wafers, 89
Flaked grain, 107
Flaked wheat, 224
Flaking, breakfast cereal preparation by, 226–231
Flaking rolls, 228
Flavobacterium proteus, 183, 199
Flaxseed, 93
Flocculence, 183
Flours, 2, 28–33, 44–47, 79
  durum, 39
  for pretzels, 77
  rye, 37
Fluoride, 181
Foam sigma, 207
Foam stability, 187, 207, 212–213
Foots, 321, 323
Formol proteins test, 214
Formula feeds, 91
Frozen rice, 367–368
Fructose, 48, 186
Fungal enzymes, 185

G

Garbuio drier, 288
Gelatinization, 26
Germ rolls, 18
Germinating chambers, 145–147
Germination, of barley, 145–153

Gibberellic acid, 166
Glucose, 48
Glutelin, 201
Gluten, 4, 7, 249
  development, 52
  quality test, 23–24
Gluten feeds, 91, 93, 112
Glyceryl monostearate, 246
Grading, barley, 132–137
  grain, 3
Graff system, 162–163
Graham crackers, 83–84
Grant, 174, 191, 202
Granula, 221
Gravimetric feeders, 260
Grinding, feedstuffs, 101–102
  wheat, 8–13, 45–46
Grist grinding, 183–184
Grits, corn, 35, 166, 177
Gruels, 221
Gum gluten, 246
Gun-puffing, 237–240
Gushing, 181

H

Hammer mills, 101–102
Hard wheat, 6, 44, 248, 249
Hart Uni-flow separator, 134, 136–137, 354
Hominy feed, 33, 36
Honey, 83
Hop strainer, 174
Hopper design, 259–260
Hops, 174, 178–180
  extracts, 179–180
Hordein, 201
Hordeum distichum, 176
Hordeum hexastichum, 176
Hot break, 192
Huller, 347–348
Humidity control, 59, 146–147, 158
Humulone, 178
Humulus lupulus, 178
Hydroclones, 309
Hydrogen ion concentration, 26
  of pretzels, 77

I

Ice cream cones, 87
Impact mills, 13
Indicator time test, 211–212
Infection, of wort, 195, 197, 199
Infusion mashing, 187
Insects, 3, 14
Intermediate proofers, 59, 61

Iron, 209
Isomaltose, 186

K

Kafir, 93
Kilning, barley, 138, 153–160, 168–170, 176, 186
Kjeldahl test, 21, 24, 46
Kraesen, 195–196
Krausening, 201
Kropf carbon dioxide system, 152

L

Labatts continuous fermentation system, 204
Lactic acid, 49, 81, 184–185, 201
Lactic organisms, 181, 303
Lactobacillus pastorianus, 199
Lager beer, 174, 195–196
Lasagne, 269
Lauter tub, 174, 189, 191, 202
Leavening agents, 44, 47–50
reactions, 71–72
Lignin, 106
Linear programming, of feed formulation, 111
Long goods continuous production lines, 277–284
Lupulin glands, 178
Lupulone, 178
Lye-peeling, rice, 355

M

Macaroni, 246
composition, 291–293
drying, 270–276
enriching, 247
production, 254–260
quality control tests, 293–298
Macaroni press. *See* Extruders
Macaroons, 85, 87
MacMichael test, 25
Mafalda, 246, 269
Magnesium, 181
Magnetic separation, 5, 103, 302
Malt, 173, 223
extract, 27
syrup, 77
Maltase, 186
Maltose, 48, 185–186, 216
test, 3, 12, 25
Maltotetraose, 186
Maltotriose, 186
Martin process, 310
Mash filter, 191

Mash-type feeds, 103–104
Mashing, 173, 181, 183–188, 201, 216
schedule, 189
Maturation, barley, 132
Maturing agents, 32–33
Meal, corn, 35
Mechanical benches, 65–66
Melanoidins, 209
Membrane filtration, 203
Metering devices, 51
Microscopy, 23
Milk, 16, 44
crackers, 82–83
Millet, 93
Millfeed, 10, 28, 32, 94
durum, 39–40
Mineral content of rice, 354
Mixers, 52–54
for feedstuffs, 103–105
for pretzels, 77
Mixing, doughs and batters, 51
reactions, 70–71
Modification, barley, 129, 150
Moisture, 21
Molasses, 105
Molders, 61–65
Monoglycerides, 242
Morton whisk, 53–54
Munich beer, 174
Myrcene, 179

N

Niacin, 227, 247
Nidis, 265–266
Nitrates, 250
Nonenzymatic browning, 73
Nonprotein nitrogen test, 216
Noodles, 20, 247, 252, 261–262
production, 262–266, 288–290
Nucleoproteins, 194, 210
Nutritional content of cereals, 245

O

Oakes mixers, 53
Oat flakes, 231
Oatmeal, 221
Oats, 2, 93, 98, 112, 224–225, 231, 241
Oil, corn, 18
Oilseed meal, 91
Oven-puffing, 237
Ovens, 66–68, 74–76
Overgrinding, 12
Oversteeping, 139–140
Oxidation, 71
Oxidation haze, 211
Oxidizing agents, 25, 50

P

Pablo continuous lautering system, 203–204
Packaging, feed, 107–108
  flour, 16
  macaroni products, 280
  shredded wheat biscuits, 234–235
Panose, 186
Papain, 177
Parboiled rice, 341, 359–363
Particle size, determination, 22–23, 47
  effect on flour quality, 29, 47
Pasteurization, beer, 203
Pastry, 2, 14, 30–31, 69
Patent flour, 28, 45–46
Pavan drying system, 283–284, 287–288
Pearl hominy, 35
Pearling, 2
Pectins, 184
Pediococcus cerevisiae, 199
Peel ovens, 67
Pekar color test, 22, 47
Pelleting, 105–107
Penicillin, 197
Pentosans, 301
Peptones, 187
Peroxide value, 334
Phlobaphene, 179
Phosphates, 184
Phosphatides, 323
Phosphomolybdate precipitable proteins test, 215
Pie doughs, 45, 53, 71
Pigments, 50
Pilsener beer, 174, 180–181
Plansifters, 9, 11, 21
Pneumatic conveying, 14–15, 51
Polymyxin, 197
Polypeptides, 187
Popp malting system, 160, 163–164
Porridges, 221
Porter, 174, 182
Potassium acid tartrate, 49
Potato flour, 241
Potatoes, 300
Poultry feed, 93, 97, 107
Precision grader, 136
Preservation techniques, 1
Pretzels, 73–77
Product segregation, 16
Proofing cabinets, 59
Protein, content, 12, 19, 131, 250
  determination, 21, 46
  permanently soluble test, 214

Proteolytic enzymes, 25, 50, 187, 211, 223
Puffed cereals, 235–243
Puffing guns, 238–239
Pumpernickel, 37
Purifiers, 9, 11–12, 21

Q

Quern, 1
Quick-cooking rice, 363–366

R

Racking, 202
Radley cooking test, 296–297
Rancidity, 177
Ravioli, 288–289
Reactions, baking, 68–73
Reaumur scale, 177
Reciprocating cutter, 81
Reducing sugars, 25
Reduction rolls, 10–11, 18–20
Reel ovens, 67–68
Reels, 17, 308
Research, feedstuffs, 109
Resteep malting system, 165–168
Reverse-sheeting molder, 64–65
Riboflavin, 247
Rice, 2, 113, 173, 177, 225, 240
  composition, 349
  enrichment, 368–371
  oil, 318–320
Rice bran, 374
  oil, 375
Rice brush, 349, 351
Rice hulls, 372–373
Rice polish, 374
Rigatoni, 246
Rodents, 17
Rolinox process, 281
Rolled oats, 224
Roller mills, 2, 8–11
Rolls, for sheeting noodle dough, 263
Ro-tap, 23
Rotary hearth ovens, 67
Rotary molded cookies, 77–80
Roto-therm, 290
Rough rice, 341
Rounders, 56–58
Rye, 19–20, 112

S

Saccharometer, 196
Saccharomyces carlsbergensis, 182
Saccharomyces cerevisiae, 47–48
Saccharomyces diastaticus, 199
Saddle stones, 1

Salt, 44, 75
Saltines, 80–81
Sandwich base cakes, 79
Scalperator, 134, 137
Scalping, 9–10, 103
   machine, 133–134
Scouring, 5–6
Scratch system, 12
Screw conveyor, 2
Sedimentation techniques, 23
Semolina, 38–39, 246, 248–249
Separation, by particle size, 3
Shortbreads, 85
Shortening, 44, 79
   bulk storage, 51
Shredded wheat biscuits, 231–235
Sieve analysis, 22–23, 47
Sieves, 3
Silica, 181, 190
Smoke point, 334
Soap stock, 329
Sodium acid pyrophosphate, 49, 72
Sodium bicarbonate, 49–50, 71
Soft wheat, 6, 44
Sorghum, 113, 310
Soundness, barley, 131
Soup crackers, 80
Soy macaroni, 247
Soybeans, 36, 241
Spaghetti, 246, 265
   canning, 290
Sparging, 188, 190
Specialty rations, 126–127
Specifications, corn products, 33–36
   durum products, 38–40
   rye products, 36–38
   wheat products, 28–33, 45–47
Spoilage, grain, 3
Spring wheat, 44
Sprout damage, effect on macaroni
   products, 251
Stamping machines, 78, 264–265
Starch, 12, 241
   conversion, 311–316
   damage, 12, 26, 249
   hydrolysis, conditions, 315
   quality, 25–26
   rice, 311
   tables, 308
   wheat, 310–311
Steel-cut groats, 224
Steep tanks, 143–144, 303
Steeping, barley, 137–144, 150
   corn, 303–305
Steepwater, 305, 327, 330
Stem rust, 248, 252

Stick pretzels, 76
Stone-ground meal, 16–17
Stone sheller, 345–346
Storage, beer, 199–201
   flour, 15–16
   rice, 342–345
   wheat, 4
Stout, 174
Straight flour, 28, 45–46
Sucrose, 48
Sugar, 44
   bulk storage, 51–52
   liquid, 52
Sugar-coating, cereals, 243, 245
Sugar rest, 188
Sugar wafers, 78, 87–89
Sulfur dioxide, 170–171, 209, 332
Sulfuring, 154, 170–171
Sweating, barley, 132
Sweet dough products, 65–66
Swine feed, 115, 122
Syrups, sugar, 52

T

Tannin, 179, 190, 193, 201
Tannin precipitatable proteins test,
   215
Tapioca, 241
Taraxanthin, 250
Teflon, 59, 270
Tempering, 6–8, 20
Terpenes, of hops, 179
Theby test, 24
Thermal processing of rice, 357–358
Thiamine, 247
Tip caps, 18
Top yeast, 182, 194
Total soluble protein test, 213
Tower fermentation, 204
Transport, of feedstuffs, 100–101
Tray ovens, 67–68
Troughs, dough, 59
Trumbols, 351
Tufoli, 246
Tunnel ovens, 68
Turbo grinders, 13, 355–356
Twisted goods, 265
Tyrothricin, 197

U

Understeeping, 138–139
U.S. Standards, rice, 339–341

V

Vacuum presses for macaroni products, 259, 275–276

Variety breads, 69
Vegetable macaroni, 247
Vermicelli, 247, 265
Vertical malt houses, 164–165
Vienna beer, 174
Viscometers, 333
Vital gluten, 247
Vitamin supplements, 100
Volumetric feeders, 260
Votator, 51

W

Washing, wheat, 6
Waste effluent, from malting, 143
Water, 44, 143
  in brewing, 180–182
Waxy rice, 371–372
Wheat, 2–16, 28–29, 44–45, 93, 112
  winter, 44
Wheat flakes, 229–231

Wheat germ oil, 320
Whiskey, 217
Whizzers, 18
Whole wheat cereal, 224
Whole wheat macaroni, 247
Winterizing, 325
Wire-cut cookies, 84–87
Wort, 138, 151, 155, 188–195
  boiling, 191
  cooler, 174
WURLD process, 355

X

Xanthophyll, 250

Y

Yeast, 25, 47–48, 193
Yeast foods, 50, 77
Yeast leavening, 1, 47–48, 59, 71–72
Yolks, color, 253